ROADSIDE GEOLOGY of COLORADO

SECOND EDITION

Halka Chronic and Felicie Williams

Mountain Press Publishing Company
Missoula, Montana
2002

ROADSIDE
GEOLOGY

Roadside Geology is a registered trademark
of Mountain Press Publishing Company.

© 2002 by Halka Chronic and Felicie Williams

Fifth Printing, May 2009

Maps © 2002 by Halka Chronic and
Felicie Williams unless otherwise credited

Cover art © 2002 by Felicie Williams

Library of Congress Cataloging-in-Publication Data

Chronic, Halka.
    Roadside geology of Colorado.—2nd ed. /Halka Chronic and Felicie Williams.
        p. cm. — (Roadside geology)
    Includes bibliographical references and index.
    ISBN 978-0-87842-447-4 (pbk. : alk. paper)
        1. Geology—Colorado. 2. Roads—Colorado. I. Williams, Felicie, 1953- II. Title.
III. Roadside geology series.
QE91.C49 2002
557.88—dc21

                                                                        2002070952

Printed in the United States of America

MP Mountain Press
PUBLISHING COMPANY
P.O. Box 2399 · Missoula, MT 59806 · 406-728-1900

"*I have taken the liberty . . . of attacking the reader
through his imagination, and while trying to amuse his fancy with
pictures of travel, have thought to thrust upon him unawares
certain facts which I regard of importance.*"

—Clarence Dutton, Geologist
*United States Geological Survey
Monograph 2,* 1882

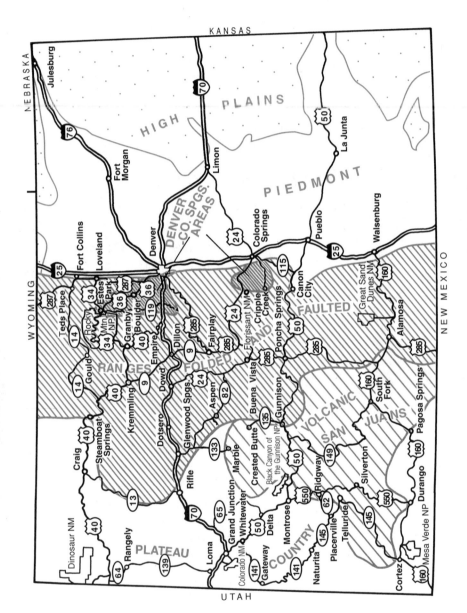

Roads and sections of Roadside Geology of Colorado.

Contents

vi

The 2,300-foot Painted Wall more than rivals the Rock of Gibraltar (1,330 feet) and the Empire State Building (1,250 feet). As the ancient rock, nearly 2 billion years old, cooled and contracted, leftover fluids penetrated cracks and fissures, forming light-colored veins. Several lengthy periods of erosion leveled the horizontal surface at the top of the photograph; more recent cutting laid bare these rocks in Black Canyon of the Gunnison National Park. —Halka Chronic photo

Preface

Colorado's colorful scenery springs from colorful geology. The eastern plains, rugged central mountain ranges, volcanic regions, and western deserts all have starkly different scenery because rocks underlying each, and geologic processes that have acted upon them, are different in each region.

Colorado's prehistory—an exciting story of stability and upheaval, of rest and unrest—began billions of years ago and involves all the forces that have shaped our planet Earth. It involves the building of mountains by unimaginable heat and pressure and their destruction by rain, snow, wind, and creeping tongues of ice. It includes the creation of new rocks from fragments of old ones. It embraces the birth and evolution of living things, first in the sea, later on land, and still later in the air. It encompasses earthquakes and floods, landslides and volcanoes, framed by long, long periods of quiet.

We tell Colorado's story in simplified form, with easy-to-read maps and diagrams and a glossary to help with unfamiliar words. In Colorado, rocks are center stage in the scenery, standing bare and brazen as they rarely do in more humid parts of the world. They beckon you to look at them, invite you to stop often and examine them closely. The best geology lies beyond the cities—in the mountains, plateaus, and canyons that ornament this state.

With few exceptions, road guides in this book read from east to west or north to south, and refer to features like towns, rivers, passes, and mileposts as reference points. For finding your way around, use any good highway map in conjunction with maps in this book. Most Colorado maps identify and give elevations for towns, peaks, and passes.

As you travel, stop often for a better appreciation of both geology and scenery. Along interstates, we point out features near rest areas—the only places where nonemergency parking is permitted. Along other highways, stop where you can safely do so. And when possible, vary your trip—seek out a path or a trail and look more closely at the rocks. Watch for fossils and minerals. With this book as your guide, fit what you see into Colorado's entire geologic picture.

We've used several types of illustrations in this book: photographs, geologic maps, sections, and stratigraphic diagrams. Geologic maps give formal names and ages for rocks present at the surface or just below soil layers. In sections, vertical dimensions are exaggerated, so don't be startled if mountains look too high or valleys too deep!

The stratigraphic diagrams illustrate rock-layering history. You probably won't see all the layers in any one spot, but as you drive a stretch of road, you may pass sequentially through the layers. The diagrams show vertical cliff faces for rocks that *tend* to but don't always form cliffs and sloping faces for rocks that *tend* to but don't always form slopes. Symbols on the diagrams represent rock types such as limestone, shale, or sandstone, and layers shown in red are more colorful in outcrop than layers shown in black.

Chapter I is a minicourse in geology—just the rudiments to help you understand the rest of the book. It includes a geologic timescale and a time chart for Colorado, listing easily recognized rock units, called formations. Use the geologic time chart as a reference when you want to know how a local rock fits into the big picture. Rock and rock symbol charts are on the front and back inside covers. Before starting each trip, read the introduction to the chapter covering the area you'll be traveling through to get a geologic overview of the region.

Material in this volume comes largely from published geologic literature. To the many colleagues who were kind enough to discuss their work with us directly, or to provide us with photographs, our very special thanks. National forest, national park, and national monument personnel, and librarians at the University of Colorado, the U.S. Geological Survey, the Colorado Geological Survey, the Colorado Historical Society, Mesa County Library, Mesa State College, and the Rust Geotechnical Library are on our thank-you list as well. Our special gratitude goes to Ted Walker, Peter Lipman, Jim Cappa, and Rex Cole, who kindly read and commented on portions of the manuscript.

Maps in this book are derived from the *Geologic Map of Colorado,* published by the U.S. Geological Survey and the Colorado Geological Survey in 1979. The map summarizes years of geologic studies by thousands of geologists. Wall-sized and in full color, it is available from the U.S. Geological Survey at the Federal Center in Denver, Colorado.

As you travel, please respect private property, and keep in mind that any land that is not private belongs to you and me, as well as to posterity. Enter but do not destroy, deface, or desecrate with litter. National parks and monuments deserve special care so that they will remain as beautiful and interesting for our children and our children's children as they are for us.

On Colorado's increasingly crowded highways, common sense, courtesy, and safety come first—way ahead of learning about geology! Read about your route *before* you drive. If you're behind the wheel, let a passenger do the reading, or find safe spots to pull off the road to "study up" on what is ahead. In using this book you assume responsibility for your own safety and that of your passengers and other vehicles.

Have a good trip!

GEOLOGIC TIME CHART

ERA	PERIOD	AGE (millions of years ago)	EVENTS IN COLORADO
CENOZOIC	Q: Quaternary — Holocene / Pleistocene (Ice Age)	0.01	Glaciation and concurrent erosion of high country, erosion of Colorado Piedmont. Periodic faulting and volcanism in San Luis Valley and southwestern Colorado.
	T: Tertiary — Pliocene	1.6	Erosion of new mountains with deposition of sand and gravel in intermontane valleys and on plains.
	Miocene	5	Miocene-Pliocene regional uplift (mid-Tertiary): about 5,000 feet of uplift to present elevations.
		23	Clockwise rotation of Colorado Plateau, causing rifting of San Luis Valley and thrust faulting in northern Rockies.
	Oligocene		Lots of volcanic activity in San Juans and other volcanic centers. Numerous small intrusions, many of them volcanic conduits.
	Eocene	35	Deposition of shale and oil shale in a huge western lake between Rocky Mountains and Uncompahgre Plateau.
	Paleocene	57	Laramide Orogeny: 72 to 40 million years ago. Intense mountain building created most of the present ranges. Intrusion of mineral-rich magmas in a northeast-trending belt across the mountains. Huge volumes of sediment are washed off the mountain flanks.
		65	Meteorite impact in Gulf of Mexico causes massive extinctions.
MESOZOIC	K: Cretaceous		Sea covers the land rapidly then recedes, leaving sandstone beaches in its wake. The sea covers the land again, staying for most of the period, and depositing thick layers of shale with fossils of marine life including ammonites and plesiosaurs. As the sea recedes to the east it leaves thick layers of sand and beds of coal, sometimes with dinosaur tracks.
	J: Jurassic	146	A moister climate washes sands into streams and rivers, gradually replacing dune fields with floodplain, river, and dune deposits. The humid lowland climate supports lush vegetation and many kinds of animals, including dinosaurs. Plates colliding at the western edge of the continent cause mountain building there, and volcanic ash from the west occasionally blankets Colorado. Pangea begins to divide.
	Ŧ: Triassic	208	Colorado is above sea level. Shales are deposited in lakes and rivers, then dry weather and wind build huge sand dune fields. Dinosaur tracks, but few fossils, are left. Earth's continents are united into Pangea.
		245	

This four-page time chart summarizes geologic events in Colorado and shows what rocks remain from each event. It provides a key to the maps in the rest of the book, explaining where in the state each formation is and how it relates to other formations. To follow events sequentially through time, turn to the next page and start at the bottom of the Precambrian, as geologists do, reading upward from oldest to youngest.

GEOLOGIC TIME CHART

WESTERN COLORADO Plateaus and San Juans	CENTRAL COLORADO Rocky Mountains	EASTERN COLORADO High Plains and Piedmont	ERA

Qb: Dotsero volcanic rocks

uncemented sediments found across the state:

Qa: alluvial fan | Ql: landslides | Qg: glacial sediment | Qlk: lakebeds

Ogallala sand and gravel

CENOZOIC

Tbp: Browns Park sandstone and siltstone

volcanic rocks

Tvl: Late Phase basalt and rhyolite related to rifting

Tui: intrusive igneous rocks

Ts: sediments in intermontane basins

Twr: White River sandstone and siltstone

Tvm: Middle Phase ash from explosive volcanism

Tve: Early Phase flow and ash from huge volcanoes

Tc: Castle Rock conglomerate

Tt: Telluride conglomerate

Tu: Uinta sandstone and siltstone

Tg: Green River shale

Tw: Wasatch shale and sandstone

Ts: sediments in intermontane basins

TKi: Laramide age intrusive rocks

Td: Denver and Dawson conglomerate, sandstone, and siltstone

Tb: Table Mtn. basalt

TKa: Animas shale and conglomerate

Kk: Kirtland sandstone

Kf: Fruitland shale

Kpc: Pictured Cliffs sandstone

Kmv: Mesaverde sandstone, shale, and coal

Km: Mancos shale

Ks: sandstone

Kp: Pierre shale

Km & Kp

Kn: Niobrara shale and limestone

Kg: Greenhorn limestone

Kb: Benton shale

Kd: Dakota sandstone and shale

Kbc: Burro Canyon sandstone

MESOZOIC

Jm: Morrison shale and sandstone

Jw: Wanakah shale

Je: Entrada sandstone

Jn: Navajo sandstone

Jk: Kayenta sandstone

J℞w: Wingate sandstone

Js: shale and sandstone

℞d: Dolores shale and sandstone

℞c: Chinle shale and sandstone

℞m: Moenkopi shale and sandstone

℞Pl: Lykins shale

GEOLOGIC TIME CHART

ERA	PERIOD	AGE (millions of years ago)	EVENTS IN COLORADO
PALEOZOIC	P: Permian	245	Continued erosion of Ancestral Rockies and Uncompahgria down to a smooth plain.
		290	
	ℙ: Pennsylvanian		Ancestral Rockies Orogeny: 320 to 290 million years ago. Uplift of two great island ranges, Frontrangia and Uncompahgria, in response to the collision of Africa with North America. Erosion of new mountains, with saltwater basins between and west of them. Away from the ranges, deposition of marine shales with the remains of many animals, especially shellfish.
	M: Mississippian	323	Widespread open sea deposits thick gray limestone then uplift causes weathering of limestone.
	D: Devonian	363	Kimberlite pipes preserve a few Silurian rocks. The rest are eroded. Marine limestone and shale deposited.
	S: Silurian	409	Marine deposition.
	O: Ordovician	439	Deposition of limestone layers bearing some of the oldest fish remains; some mudflat deposits. Colorado was near the equator.
	Є: Cambrian	510	Deposition of marine sandstone and limestone as the sea crept east across denuded land.
		570	
PRECAMBRIAN	Late Proterozoic		A very long period of stability and erosion with mountains beveled to their roots, possibly partly by global ice ages.
		929	Rifting along east-west weak zone in northwestern corner of state. Rift filled with thick sediments.
			Pikes Peak granite intruded 1.1 billion years ago.
	Y: Middle Proterozoic	1,200	Rifting along northwest-trending weak zone in southeast; rift valley filled with thick sediments.
		1,400	Granitic rocks of the Berthoud plutonic suite intrude 1.48 to 1.35 billion years ago as more crust is added to continent in faraway Texas and Oklahoma.
		1,440	Rift in southwestern corner of Colorado fills with sediment.
		1,650	Early Proterozoic Orogeny: 1.73 to 1.65 billion years ago. Metamorphism of older rocks and intrusion of granite of Routt plutonic suite caused by collision with second island arc in New Mexico.
	X: Early Proterozoic	1,700	
		1,800	300-mile-wide island arc is added onto the southeastern margin of the Wyoming Province.
		2,500	
	W: Archean		Deposition of beach and delta sands at the margin of the Wyoming Province, mostly weathering to a smooth surface.
			Formation of early continents, locally the Wyoming Province, from igneous rocks.
		4,600?	

GEOLOGIC TIME CHART

WESTERN COLORADO Plateaus and San Juans	CENTRAL COLORADO Rocky Mountains	EASTERN COLORADO High Plains and Piedmont	ERA

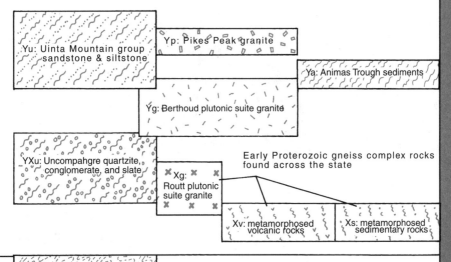

Pl: Lyons sandstone

Pc: Cutler Fm. Pw: Weber sandstone

PPm: Maroon sandstone and shale and P Minturn shale and limestone

PPf: Fountain conglomerate and sandstone (north)

Pp: Paradox evaporites

Pe: Eagle Valley evaporites

PPs: Sangre de Cristo conglomerate and sandstone (south)

Ph: Hermosa shale

Pm: Molas formation

Pb: Belden shale

Ml: Leadville limestone

limestone and sandstone

Do: Ouray limestone

MDc: Chaffee shale, limestone, and dolomite

De: Elbert sandst. & siltst.

Di: Ignacio conglomerate

kimberlite pipes

Of: Fremont dolomite

Oh: Harding sandstone

Om: Manitou limestone and dolomite

Cl/Ci: Lodore and Ignacio quartzites

Cs: Sawatch quartzite

PALEOZOIC

Yu: Uinta Mountain group sandstone & siltstone

Yp: Pikes Peak granite

Ya: Animas Trough sediments

Yg: Berthoud plutonic suite granite

YXu: Uncompahgre quartzite, conglomerate, and slate

Xg: Routt plutonic suite granite

Early Proterozoic gneiss complex rocks found across the state

Xv: metamorphosed volcanic rocks

Xs: metamorphosed sedimentary rocks

Wr: Red Creek quartzite

PRECAMBRIAN

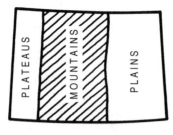

Colorado's geography is simple. The state is an almost perfect rectangle divided into plains, mountains, and plateaus.

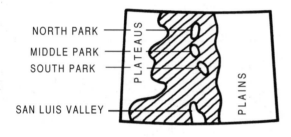

NORTH PARK

MIDDLE PARK

SOUTH PARK

SAN LUIS VALLEY

Now wiggle the lines and insert four large valleys in the mountain region.

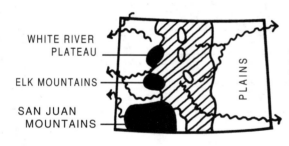

WHITE RIVER PLATEAU

ELK MOUNTAINS

SAN JUAN MOUNTAINS

Add some volcanic mountains and a few rivers. Just about all of Colorado's rivers flow outward into neighboring states.

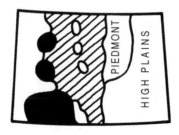

Erosion cut a valley east of the mountains, dividing the plains neatly into High Plains and Piedmont.

Colorado's geography

I
A Foundation to Rest On

How old is Colorado? A thousand years? A million? Some rocks in Colorado, dated by measuring daughter elements produced by radioactive decay, tip the calendar at close to 2.5 billion years.

Two and a half billion is a staggering figure: more than half the age of Earth, time enough for 100 million human generations. If each page of this book were to represent a single year, 2.5 billion years would build a pile of pages twenty-five times as high as Mt. Everest, fifty-one times as high as Colorado's loftiest peak, Mt. Elbert.

But Colorado's topography as we see it today—with plains in the east, mountains through the center, basins and plateaus in the west, most of them a mile or more above sea level—goes back a paltry 20,000 years, a short life in geologic terms.

Colorado's lowest point, where the Arkansas River flows into Kansas, is 3,300 feet above sea level. The plains rise from an average elevation of 4,000 feet along Colorado's eastern border to 5,500 feet partway to the mountains; they then drop off to about 5,000 feet closer to the mountains. One of the steps of the State Capitol in Denver is exactly a mile high.

Fifty-three peaks in Colorado top 14,000 feet and are known affectionately as "Fourteeners." Many of them lie near or on the Continental Divide, an almost mystical dividing line between east and west, between streams flowing toward the Pacific and streams bound for the Atlantic.

To understand the events that shaped this land, we have to step back in time as far as we can and look at Earth as a whole.

Continents Adrift

Earth formed as a ball of matter as the solar system took shape. Maybe just a loose mass of space material pulled together by its own gravity, it eventually became molten, allowing molecules in it to move around. The heavier molecules gradually sank toward the center of Earth while the lighter ones rose toward the surface. Gases, the lightest molecules, formed the atmosphere.

As Earth cooled, the lighter liquids started to solidify into a rocky crust. Because rock shrinks as it cools and because it was floating on moving liquid, cracks formed in the crust. Currents in the underlying, still molten rock caused crust fragments to jostle each other, sometimes pulling pieces of crust downward into the molten layer and melting them again. Through eons of repeated melting, elements in the crust sorted into continental crust—an accumulation of lighter-weight rock-forming material—and into oceanic crust beneath newly formed seas. Made of heavier, darker rock, oceanic crust tends to be pulled downward when very large fragments of crust, called tectonic plates, collide. Most large plates include both continental crust and oceanic crust.

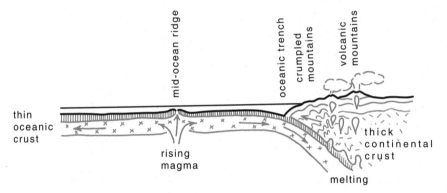

Deep within Earth's mantle, powerful, slow-moving convection currents cause upwelling at mid-ocean ridges, where submarine volcanoes generate new crust. Where plates meet, heavier oceanic crust is pulled downward and remelted.

Earth still has a layer of partly molten material, called the mantle, below its crust. It behaves like thick, superheated slush. The mantle's outermost part is coupled with the crust; together they are called the lithosphere, the "rocky sphere." Numerous convection cells in the mantle roll ever so slowly, rafting the stiffer lithosphere at speeds of up to 7 inches per year. Probably the rolling movement varies with time. Fueled, as far as we can tell, by the heat of Earth and atomic reactions deep within it, the whole process is like a slow boil. Colliding continents and occasional impacts of extremely large meteorites may vary the direction and speed of convective flow in the mantle.

Below the mantle is the core, which is mainly composed of the heavy element iron. Its outer layer is white-hot liquid and its center is solid. Gravity may have compressed all the atoms of the inner core until they fit together in the simplest, tightest possible way, forming a single sphere-shaped crystal.

Today, the surface of Earth consists of sixteen large rigid plates that resemble the mosaic segments of a turtle shell. The plates are a little over 3

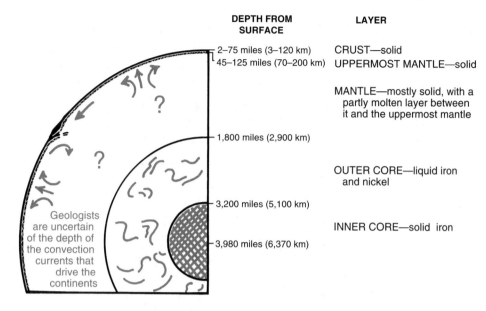

DEPTH FROM SURFACE	LAYER
2–75 miles (3–120 km)	CRUST—solid
45–125 miles (70–200 km)	UPPERMOST MANTLE—solid
	MANTLE—mostly solid, with a partly molten layer between it and the uppermost mantle
1,800 miles (2,900 km)	
	OUTER CORE—liquid iron and nickel
3,200 miles (5,100 km)	
	INNER CORE—solid iron
3,980 miles (6,370 km)	

Geologists are uncertain of the depth of the convection currents that drive the continents

Geologists learn about Earth's interior by studying the way earthquake waves travel through its different layers and the speed with which they trigger seismographs at widely scattered locations.

miles thick where they consist of oceanic crust, 20 miles thick where they are made of continental crust, and up to 40 miles thick where continental crust thickens into high mountains. Winding mid-ocean ridges and deep arc-shaped oceanic trenches border most plates. At mid-ocean ridges, plates slowly spread apart or rift as hot lava wells up from Earth's interior to form new oceanic crust. At the trenches, plate margins are drawn down or subducted under adjacent plates, creating zones where earthquakes are common. Subducted plates eventually melt; their rising magma may form strings of volcanoes or island arcs on Earth's surface.

Since continental crust weighs less than oceanic crust, continental crust overrides oceanic crust. Any continental crust on the downward-moving plate may jam up the process by refusing to sink beneath other rock. Instead, it adds itself onto the side of the other plate in a gargantuan fender bender, causing mountains to pile up and forcing the site of subduction to move. Plates can also slide by one another horizontally, as the western edge of California is sliding northward along the San Andreas fault. Tectonic plates have been on the move for most of Earth's existence, so it makes sense that their shapes and locations have changed over time.

Early in the study of geology, many geologists believed that our planet's features could best be explained by a series of catastrophes, major devastating changes that shaped Earth and the nature of life upon it. Other researchers were sure Earth had changed very gradually, by the accumulated effects of the same processes we see around us today. More and more geologists now recognize that both ideas are correct: long periods of little change have from time to time been interrupted, or punctuated, by profound catastrophes—like large asteroids crashing into our planet—that cause many major changes all at once. As if we divided a smoothly flowing. sentence. with periods, and commas, and perhaps! most of all! with exclamation points! This idea is called the theory of punctuated equilibrium.

History—The Underlying Theme

The central theme of geology is history. Colorado's geologic history began more than 2.5 billion years ago. We are not certain what Colorado or North America or even the planet as a whole looked like then. We do know that large bodies of water existed; some ancient rocks, now altered and recrystallized, still wear the ripple marks of water-deposited sediment. We know that no plants or animals lived on the land, though primitive life forms may have existed in the sea. We know that immense mountain chains formed because we can examine the distorted, twisted metamorphic rocks that were their roots. We know that volcanoes erupted because we find unmistakable rocks that were once lava flows. Did an atmosphere exist? Almost certainly because volcanoes produce gas as well as lava and ash. With equal certainty, we know the early atmosphere was not life-supporting or oxygen-rich like today's atmosphere, a result of plant metabolism.

Coming to Terms

Here at this end of geologic history, we need handles to keep track of the vast amount of geologic time. Geologists have divided Earth's life span into periods, which were named after places where rocks representing those time intervals were first described. Devonian rocks, for instance, were first described in Devon in southwestern England.

And just as we lump weeks into months, geologists lump periods into eras. Names given to the eras record the levels of development of life. Fossils, petrified remains or impressions of ancient animals or plants, enable geologists to recognize the different ages of rocks, as primitive plants and animals gradually evolved into new species.

We live in the most recent era, the Cenozoic, which means "recent life." It is also called the Age of Mammals. Mesozoic means "middle life," known as the Age of Dinosaurs. Paleozoic means "ancient life." The oldest era, with the sparsest record of primitive life, we call Precambrian, usually dividing it into two eras: the older Archean era, followed by the Proterozoic era. Life may have begun in Archean seas, perhaps in warm, sulfurous waters near mid-ocean ridges.

Fossils of early plants and animals help geologists understand Earth's evolving ecosystems and enable them to arrange rocks from all over the world in their correct order from oldest to youngest. But fossils don't tell us the age of rocks in years. For that we rely on radioactive elements and their gradual but remarkably steady decay. Reasonably exact dates for periods and eras are obtained by comparing the relative amounts of radioactive substances and their decay products. A radiometric date tells when an igneous rock last cooled, so it gives the youngest possible age for the rock. Current methods are accurate to within about 2 percent, so raw numbers are rounded off to millions or tens of thousands of years.

We must remember that geologic time is a long, slow process. Except for occasional catastrophic events, Earth has sailed through space with little perceptible change from year to year, century to century, millennium to millennium. Only when we look at millions of years at a time do we gain an idea of its geologic evolution.

The geologic time chart outlines the great events in the history of Colorado. And it lists the most recognizable rock units—formations or groups— used in this guide, also named after places where they are well developed and well exposed. The relative position of rock layers indicates their relative age: younger rocks usually rest on top of older rocks. According to the law of superposition, one of geology's basic tenets, geologic history begins at the bottom, with rock layers successively younger upward. (There are exceptions to this law, and we will point out a few in Colorado.)

A few other geologic terms are so common, you really can't get along without them. You may already be familiar with words used to describe breaks and folds:

- A fault is a break along which movement has taken place. A zone of more or less parallel faults is often called a fault zone.
- A joint is a rock fracture, usually much smaller than a fault, along which no perceptible movement has occurred.
- An anticline is an upward bend or fold in rocks, most easily visible in layered (stratified) rocks.
- A syncline is, conversely, a downward bend or fold. Anticlines and synclines are not always symmetrical—they may be lopsided or asymmetrical.
- A monocline is a simple fold where near-horizontal rock bends and then levels out again.

The word *orogeny* denotes mountain-building episodes that are caused by plates crushing together. Orogenies are named after geographic features, usually the mountains they created. The Laramide Orogeny created the Laramie Range of Colorado and Wyoming, for instance, as well as many other southern Rocky Mountain ranges.

The names of minerals and rocks are important geologic terms as well. Rocks are composed of minerals, natural substances that have definite chemical makeups and often crystallize in recognizable ways. In this book,

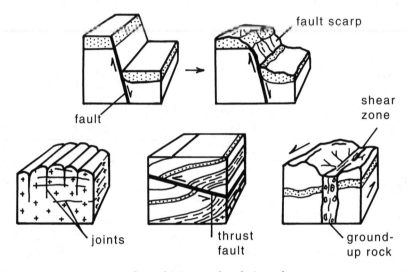

Faults and joints are breaks in rocks.

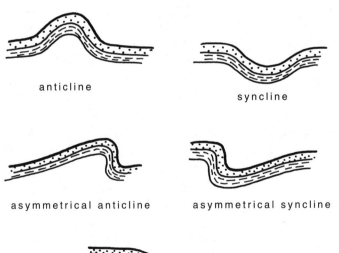

Several types of folds form in layered rocks.

COMMON MINERALS

name	description
quartz	hard (can't be scratched with a knife), shiny, glassy mineral; silicon dioxide; the main mineral in most sandstone and one of the main minerals in granite; quartz in rocks is commonly clear and colorless, but it may be pink **(rose quartz),** lavender **(amethyst),** snow white **(milky quartz),** or gray and clear **(smoky quartz)**
feldspar	a translucent pinkish, grayish, or whitish hard mineral; very common in granite; recognized by its tendency to break along flat faces that flash in sunlight
mica	soft (can be scratched with a fingernail) black to silvery minerals that separate easily into shiny paper-thin sheets or flakes; very common in granite and schist; black mica is **biotite**; white mica is **muscovite**
calcite	a white to light gray mineral (often colored by impurities) making up most limestone, including dripstone in caves; can't be scratched with a fingernail but can be with a knife; geologists identify this mineral with dilute acid—it fizzes
hematite	an iron oxide mineral easily recognized by its dark red color; occurring as tiny grains that tint sandstone and shale redbeds; when concentrated, hematite is a valuable iron ore
limonite	a brown to yellow oxide of iron common as tiny grains that impart an overall yellow or tan to sandstone and shale; in Colorado's mining areas, limonite colors old mine dumps; if concentrated it is an iron ore
kaolinite	a soft white clay mineral abundant in some shale and sandstone, formed by decomposition of feldspar, and easily recognized by its clayey, earthy odor (dampen the rock with your breath before sniffing)
hornblende	a black to dark green hard mineral occurring as needlelike crystals in igneous and metamorphic rocks
gypsum	a soft (scratch it with a fingernail), transparent or translucent, colorless evaporite mineral formed when seawater or salty ponds dry up; blade-shaped gypsum crystals are **selenite**
pyrite	"fools gold"; a brittle, metallic, brass-colored mineral (iron sulfide) often forming small cubes in igneous rock; sometimes mistaken for gold, pyrite is far more common, and when abundant it is an ore of iron and sulfur

COMMON ROCKS

class	rock	description
sedimentary	**evaporite**	rock formed by evaporation of mineral-laden water; includes **gypsum** and **halite** from seawater, and **dripstone**, **travertine**, and **siliceous sinter** from water flowing through limestone caves and hot springs
	limestone	rock made mostly of calcite (calcium carbonate), deposited as limy mud; usually white or gray, often containing fossils
	dolomite	rock similar to limestone but containing magnesium carbonate as well as calcium carbonate
	coal	Black flammable rock made of decomposed, slightly metamorphosed plant material
	shale	very fine-grained rock made of silt and clay cemented together, usually breaking into flat slabs
	sandstone	rock composed of sand grains cemented together
	conglomerate	rock made of sand and pebbles deposited as gravel and then cemented together; **agglomerate** is made up of rounded pieces of volcanic rock, often cemented with volcanic ash
	breccia	broken, unrounded, angular rock fragments cemented together
igneous extrusive (volcanic)	**latite**	fine-grained volcanic flows and ash particularly high in feldspar minerals
	rhyolite	light-colored, very fine-grained volcanic rock, either lava flows or volcanic ash, containing both quartz and feldspar
	dacite	medium-colored, very fine-grained volcanic rock
	andesite	medium- to dark-colored, very fine-grained volcanic rock
	basalt	very dark fine-grained volcanic rock, often with visible gas bubble holes called **vesicles**
igneous intrusive	**monzonite**	medium-grained rock (often with scattered larger grains) made mostly of feldspar, with some hornblende or quartz
	granite	common light-colored, coarse-grained rock with visible crystals of quartz and feldspar, usually peppered with black mica or hornblende crystals
	pegmatite	very coarse-grained rock made of the same minerals as granite, plus often a few other unusual ones; found in veins cutting across granite and other rocks
metamorphic	**marble**	recrystallized limestone, often with visible calcite crystals
	quartzite	sandstone so tightly cemented that it breaks through the individual sand grains
	schist	medium-grained rock with abundant parallel mica flakes and/or hornblende needles that give the rock a streaky appearance and a tendency to split along parallel planes; a dark grayish green rock made almost entirely of hornblende grains is called **amphibolite**
	gneiss	coarse-grained rock with alternating bands of flaky and granular minerals, more intensely metamorphosed than schist
	migmatite	rock with irregular bands of schist and granite; if granite magma is squeezed between schist layers, it is called **injection gneiss**

we use the accepted geologic names for rocks and minerals; take note that rock hounds and gem collectors may use less formal names based on geographic places or some aspect of the rock, like color. Geologists recognize hundreds of kinds of rocks, but twenty-one common types suffice for everyday use in Colorado.

Rocks fall into three main classes depending on their origin:

- Igneous rocks originate from molten rock material, or magma, that rises from deep within the crust or from the mantle. Magma may erupt onto Earth's surface as a volcanic or extrusive igneous rock or it may cool slowly below the surface as an intrusive igneous rock. Intrusive rocks are usually coarser grained than extrusive ones because they cool more slowly, giving mineral grains more time to grow.
- Sedimentary rocks form from broken or dissolved bits and pieces of other rock, deposited by water, wind, or ice, or as chemical precipitates. Most sedimentary rocks are layered, or stratified. They tend to harden with age.
- Metamorphic rocks are formed from older rocks that have been altered by heat, pressure, or chemical action. They may be only slightly altered or they may be changed so severely that it is difficult or impossible to figure out what they originally were. The words *schist, gneiss,* and *migmatite* describe little more than the shapes and orientations of mineral grains in the rock. Where possible in this book, we use terms that describe the pre-metamorphic nature of the rock: *metasediment* refers to metamorphosed sedimentary rocks and *metavolcanic* refers to metamorphosed volcanic rock.

We define other geologic terms where they first appear in this book, as well as in the glossary.

Colorado through the Ages

Our area of North America is now on a tectonic plate that reaches from the Arctic to the Caribbean and from the San Andreas fault and East Pacific rise to the mid-Atlantic ridge. When Earth was younger, the shapes and sizes of the continents and of the area we call Colorado were a lot different. Consult the geologic time chart as you read about our prehistory.

Precambrian Time

In the geologically tiny length of a human lifetime, very little happens to change the landscape. If we speed up the motion, starting our time machine when Earth's crust was just forming, the events may seem violent and mind-boggling. Remember how slowly and steadily each epoch really passed by, features changing as calmly and imperceptibly as they do now.

Colorado's geologic story began around 2.5 billion years ago. Several areas of continental crust, called *cratons* after the Greek word for "shield," had already formed. The oldest craton near Colorado, often called the Wyoming Province, underlies Wyoming and parts of Utah, Idaho, and Montana. Some geologists speculate that it extends south beneath part of Colorado's Front Range. A tiny sliver of 2.5-billion-year-old quartzite appears in the Uinta Mountains in northwest Colorado.

The Wyoming Province was the first area of continental crust to develop in this part of North America. A strip of island arcs, added to the province's southeastern edge around 1.8 billion years ago, formed the oldest rock known in most of Colorado.

During Proterozoic time, three sets of island arcs were added onto the Wyoming Province as a plate from the southeast plunged westward beneath it. The first is represented by a jumbled mass of metamorphosed volcanic and sedimentary rocks about 1.8 to 1.7 billion years old. The second collision, about 1.7 to 1.6 billion years ago in New Mexico, resulted in large bodies of igneous rock known as the Routt plutonic suite. Though the state has since remained far from any plate boundaries, roughly 1.4 billion years ago some distant landmass—possibly South America—collided with what is now Texas and Oklahoma. In Colorado that collision caused crustal melting and produced a scattered group of intrusions known to geologists as the Berthoud plutonic suite. These three collisions produced most of the Precambrian rocks visible in Colorado. And around a billion years ago, another mass of molten rock rose in south-central Colorado, the Pikes Peak granite.

Geologists often lump Precambrian rocks together as "basement" because they are poorly understood and lie beneath younger, less deformed rocks. We consider them "foundation" because younger rocks rest on them and are influenced by their strengths and weaknesses.

Movement along huge linear weak zones in the Precambrian foundation helped relieve stresses caused by our plate's travels through geologic time.

Between collisions, the crust of the growing continent pulled apart along several weak zones, allowing thick layers of sediment to pile up in low areas. This happened in the southwestern corner of the state 1.7 to 1.4 billion years ago, the southeastern part of the state 1.4 to 1.2 billion years ago, and the northwestern corner of the state 1.1 billion to 925 million years ago.

These tempestuous events were followed by 400 million years of relative peace and tranquility, during which Colorado and much of the rest of the world saw little tectonic activity—no dramatic plate collisions or even slow-motion displacements. We can't yet explain this change—this long period of inactivity—but it resulted in worldwide erosion as wind, waves, running water, and ice wore down the land to a fairly smooth, flat surface near sea level. You can still see this surface, called the Great Unconformity, in parts of Colorado.

Paleozoic Time

The rock record in Colorado picks up again 570 million years ago with Paleozoic, Mesozoic, and Cenozoic rocks, in which we see renewed evidence of drifting and jostling continents. Geologists are able to reconstruct the changing positions of continents and the growth and demise of various seas by measuring the alignment of iron minerals, which lined up with Earth's magnetic field as they crystallized in igneous rocks.

In Colorado the earliest Paleozoic rocks are Cambrian Ignacio and Sawatch quartzites, well-washed beach sands laid down by a receding sea as our continental plate moved gradually northward toward the equator.

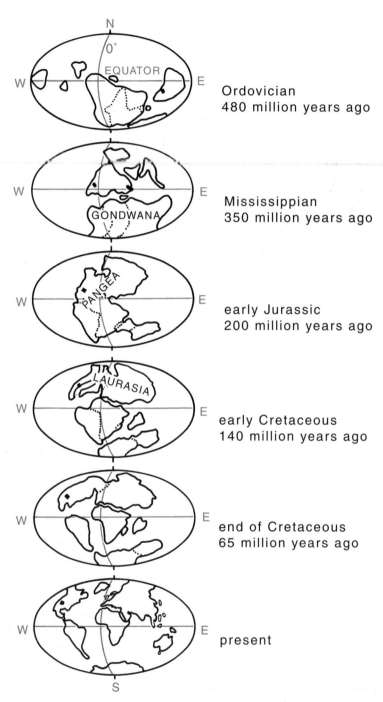

N
0°
EQUATOR
W   E

Ordovician
480 million years ago

W   E

GONDWANA

Mississippian
350 million years ago

W   E

PANGEA

early Jurassic
200 million years ago

W   E

LAURASIA

early Cretaceous
140 million years ago

W   E

end of Cretaceous
65 million years ago

W   E

present

S

During the past 480 million years, Colorado (small rectangle) has wandered the globe with our changing continent as it spun northwestward, crossed the equator, crunched into other continents to form the supercontinent Pangea, and then split off again and drifted northwest to its present position.

Only an area of the Gore Range in north-central Colorado remained an island. Keep in mind that Colorado was not so high then. This part of the continent was just above or just below sea level during the first half of the Paleozoic era.

In Ordovician time vast shallow mudflats covered North America, with only a few islands scattered from central Colorado northeastward to Minnesota. The continent drifted westward, rotating counterclockwise close to the equator. The mudflats solidified into the Manitou and Fremont dolomites; between them the Harding sandstone represents a brief time when enough land was above sea level to allow waves to rework older rocks along the shore. Ordovician fossils indicate abundant and varied marine life.

Just a few fragments of Silurian limestone have been found in Colorado, blocks that fell down volcanic conduits near the Wyoming state line just east of the Rockies. They tell us a shallow sea covered Colorado in Silurian time. All other traces of Silurian rocks were washed away in late Silurian or early Devonian time.

In Devonian time shallow seas covered most of the state. A long island, roughly where the Front Range and Wet Mountains are now, provided a source for silt and windblown sand that are interlayered with limestone and dolomite to form the Chaffee formation and the Ouray limestone. Though more than 140 degrees west of where it had started out 200 million years before, Colorado was still within 10 degrees of the equator.

Mississippian Leadville limestone was deposited as limy mud in a shallow tropical sea that covered most of Colorado, leaving only a low north-south ridge above sea level. Then the whole region rose slightly, and the top of the limestone weathered into caves, sinkholes, and red soil of the Molas formation, with features typical of tropically weathered limestone.

During Pennsylvanian time two huge mountain ranges were faulted upward in central and western Colorado. These ancient ranges were roughly in the same place as the present-day Rocky Mountains, so geologists call them the Ancestral Rockies, or more specifically, Frontrangia and Uncompahgria. Sediment that piled up along their flanks became the Fountain formation to the east, the Maroon formation between the two ancient ranges, the Sangre de Cristo formation in the southeast, and the Hermosa group and Cutler formation west of Uncompahgria. Close to the steep ranges, coarse, poorly sorted gravels were deposited; sand and silt accumulated farther away. The mountains were still surrounded by seas, in which limestone was deposited. In landlocked basins west of each range, salt water evaporated, forming the Eagle Valley and Paradox evaporites.

The rise of Frontrangia and Uncompahgria coincided in time with the collision of Africa and eastern North America as the continents merged into the supercontinent Pangea. Many geologists, though, believe the Great

Plains would have shielded Colorado from such a faraway event, and that creation of the ancestral ranges has not yet been adequately explained.

Permian rocks in Colorado document continued erosion of the Pennsylvanian mountains and reworking of their debris into finer and finer sediment. By the end of Permian time, much of Colorado was a flat, smooth plain again, just above sea level. Mountains that had once been 10,000 feet above the sea had eroded into low hills.

Permian time, and the Paleozoic era, ended abruptly and catastrophically, perhaps as a meteorite impact or a major volcanic event brought about worldwide changes and destroyed nearly all living things.

Mesozoic Time

Above the Permian rocks, dark red Triassic shales are overlain by red windblown sandstones: coastal plains gave way to great fields of desert sand dunes. Dinosaurs roamed Pangea, which remained the world's supercontinent for most of Triassic time, drifting northward through the latitudes where deserts form, 15 to 30 degrees from the equator.

As rifts in Pangea opened and widened in Jurassic time, the sea repeatedly lapped at Colorado's western shores, then gradually receded as land rose west and south of the state. Volcanic ash layers attest to tectonic events farther west. Early Jurassic dune sands were slowly covered by river-borne sand, then by varicolored shales and sands of the Morrison formation, laid down in streams, marshy areas, and lakes. Dinosaurs flourished in this moister climate, and their fossils and footprints are numerous in rocks of late Jurassic age.

At the beginning of Cretaceous time the sea rose rapidly, coming in from the northeast to cover most of the state. It then fell rapidly as well, and as it receded its waves, working in tandem with rivers draining new mountains in Utah, covered a vast area from Canada south into New Mexico with sand and pebble layers of the Dakota formation. The land then subsided and the sea returned, depositing thousands of feet of fine gray shale, the Mancos and Pierre formations. Limestone, such as the Niobrara formation, formed between shale layers where the sea deepened. Thick sand layers document brief times when the sea receded. The shales contain the remains of many marine animals, including ammonites, shellfish, and plesiosaurs.

The sea was receding as the land began to rise in the first pulses of a major mountain-building episode, the Laramide Orogeny. Waves washed the sandy shoreline, and nearby wetlands supported flourishing plant life that would later turn to coal. Numerous sandstone and coal layers near the top of the Cretaceous strata thicken in the western part of the state, where they are called the Mesaverde group.

In the 225 million years since the last major mountain-building episode, North America had moved northwestward, overriding more than 1,200 miles of the Pacific plate, scraping islands off the descending slab of oceanic crust, possibly even overriding part of the Pacific mid-ocean ridge. Mountain building that had begun in California in Jurassic time had moved eastward through most of Cretaceous time. Upon reaching Colorado, it created the Uncompahgre uplift and the Rocky Mountains. Tectonic activity became intense enough to cause melting deep within or below Colorado's crust. Magma rose along old faults, particularly along some of the Precambrian north-northeast trending ones that define the Colorado Lineament.

East of the two great ranges, as though compensating for the lifting of the mountains, two basins sank: the Denver Basin east of the Rockies and the Piceance Basin east of the Uncompahgre uplift. As the basins began to fill, the Cretaceous period ended abruptly with extinction of more than half of the animal and plant species then alive, including the dinosaurs. For many years the extinctions were a mystery, but there is now evidence that a large asteroid or comet struck Earth at that time, perhaps darkening the skies with dust and causing a global firestorm.

A large subsurface meteor crater at the edge of Mexico's Yucatán Peninsula may document the asteroid collision that ended Cretaceous time. North America is shown as it appeared 65 million years ago. Around the world, a thin layer of ash containing the element iridium, rare on Earth but abundant in meteorites, marks the Cretaceous-Tertiary boundary.

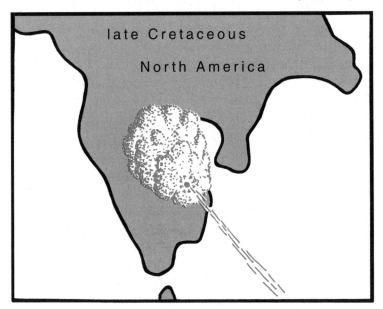

late Cretaceous

North America

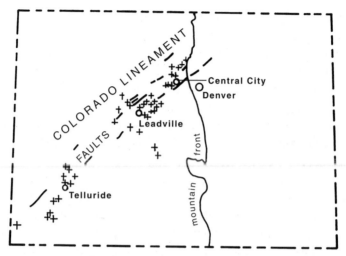

Intrusive rocks related to the Laramide Orogeny (+) fall in a belt that angles northeast across Colorado, roughly following the faults that form the Colorado Lineament.

Cenozoic Time

During the 32 million years of the Laramide Orogeny, which began in Cretaceous time and continued into Eocene time, pressures that caused mountains to build up slowly changed direction. Originally from the west, they swung counterclockwise until they came from the south. Thus the oldest Laramide mountain ranges generally north-south, the youngest east-west, and those of intermediate age trend northwest-southeast. The twisting motion tore loose and lifted the Colorado Plateau, rotating it clockwise to open up the Rio Grande Rift in southern Colorado and New Mexico. Farther north, the great plateau crunched into the northern Colorado Rockies and folded up more mountain ranges.

Sand and gravel that washed eastward from the Rockies into the Denver Basin now form the Denver and Dawson formations, Paleocene in age. Farther west, extensive lowlands between ranges filled with river-borne sediment, now the Wasatch formation. Erosion almost kept pace with uplift, so the mountains remained more like hills, their slopes grading gently toward the plains.

Mountain building petered out during Eocene time, and in 20 million years of relative peace and quiet, erosion cut gently inclined surfaces, called pediments, into bedrock at the bases of the mountains. In northwestern Colorado, movement along Proterozoic faults created a new basin that gradually filled with thick lake deposits—layer upon layer of silt, now the Green River formation, topped with river-borne sand and silt of the Uinta formation.

Dramatic changes were in store for Oligocene time. Fields of huge volcanoes grew, the largest forming the San Juan Mountains in the southwestern corner of the state. Next, explosive volcanism covered as much as a third of Colorado with ashflows.

Soon after, in Miocene time, regional uplift raised all of Colorado and adjacent parts of Utah, Arizona, and New Mexico into a broad dome 5,000 feet or more above sea level. The same area was stretched to the point of cracking, and in many places lava escaped through fissures to cover wide areas with basalt and rhyolite flows.

Oligocene and Miocene events are clearly related to each other, but geologists are still trying to figure out what caused them. Was the subducted Pacific plate stuck beneath the roots of the Rockies? Or was it pushing against the extremely stable area that underlies the Great Plains? We know only the sequence of events: Laramide mountain building, caused by compressive forces, had moved from west to east since Jurassic time. As it ended with uplift of the Front Range, an episode of stretching, faulting, and volcanism seemed to creep gradually westward again. And somehow, regional uplift raised Colorado to its present elevation.

Through Pliocene time, weathering and erosion worked with renewed vigor to shape the Colorado we see today. Streams were given new life by the regional uplift and cut faster and deeper than before. Heading in high mountains on the Continental Divide, streams that became the Colorado River and its tributaries—the Yampa, Gunnison, Green, and Dolores—trenched through layered sedimentary rocks into the ancient crystalline rocks of the continent's Precambrian foundation, carving spectacular gorges and scenic canyons. The South Platte and Arkansas Rivers and the Rio Grande made similar incisions into the mountains on their way to eastern and southern plains.

No doubt the stream cutting was further enhanced by the Pleistocene Ice Age, 1.2 million to 10,000 years ago, when glaciers enveloped the high peaks of Colorado and plowed down surrounding stream valleys. The ice deepened and straightened mountain canyons, while streams draining the ice were flooded with meltwater torrents white with glacial "flour" and loaded with glacially scoured rock fragments—the tools of erosion.

The Final Touch

Colorado presents an exciting variety of geologic scenery, ranging from large-scale features of plain and plateau to intricate erosional landforms. Of particular interest to geomorphologists (geologists who study surface landforms) is an old erosion surface still visible in the mountains. Called the Tertiary pediment, this surface is now represented by broad rolling uplands around 8,000 to 9,000 feet in elevation. It is quite evident in the Front Range as you approach Denver and Colorado Springs from the east or northeast. Once continuous with the High Plains, the surface is a remnant of the gently inclined pediment carved into the base of the mountains in Eocene time. When regional uplift in mid-Tertiary time increased stream flow near the mountains, it caused renewed erosion along the edge of the Tertiary pediment and separated it from the surface of the plains by a wide but uneven valley known as the Colorado Piedmont.

Below high peaks of the Front Range, the nearly horizontal surface of the Tertiary pediment was once smoothly continuous eastward with the High Plains.

Many other weathering and erosional features vary Colorado's roadsides: joint-controlled weathering of youthful lava flows and ancient crystalline rocks, different degrees of erosion of hard and soft sedimentary rock layers, relatively rapid and often intricate carving of soft volcanic ash, slow solution of limestone caves, and landslides that scar the mountainsides. Glacial erosion shaped many spectacular mountain peaks; several small glaciers still exist today. Wind-built sand dunes developed in intermontane valleys, vying with mountain streams for possession of the land.

Gold Is Where You Find It

Colorado's rocks and the treasures they contain drew the first great surge of settlers to the state. The rush to the Rockies began in 1859 after gold was discovered in the bed of the South Platte River near what is now Denver. Gold here comes in two forms—relatively pure native gold (sometimes in ornate crystals and sometimes as shiny yellow flakes and rounded nuggets) and gold combined with other elements in ore minerals.

The first gold produced in Colorado was native gold found in stream gravels, where, far heavier than other sand grains, it tends to settle. Miners separated the gold from these placer deposits by washing the gravels in gold pans, homemade sluices, or big sluices in floating dredges. Prospectors following the placer deposits upstream sometimes struck it rich, finding the lode from which the placer gold had come, usually in hard metamorphic and igneous rock. Gold production from placer and hard rock mines centered around Central City, Idaho Springs, Cripple Creek, Leadville, Lake City, Creede, and Summitville.

Silver bonanzas in Aspen, Central City, Creede, Gilman, Leadville, Silverton, and Lake City drew fortune-seekers to Colorado as well. As these bonanzas were exhausted, ores mined for lead and zinc became the principle sources of silver, especially near Leadville, Gilman, and Silverton.

Hot, mineral-rich solutions saturated old sedimentary rocks to form the deposits at Leadville, Gilman, and Aspen. Similar solutions deposited ore minerals in Precambrian rocks at Central City and Idaho Springs. In the San Juan Mountains and at Cripple Creek, the deposits accumulated in veins and fissures related to ancient collapsed volcanoes.

Colorado boasts two of the world's largest known concentrations of molybdenum ore, the Climax and Henderson Mines. From 1925 to the 1970s, the Climax Mine produced more than half of the world's molybdenum. The state has also produced uranium, vanadium, tungsten, copper, tin, and even diamonds!

Numerous Colorado towns have museums with displays of mining history and minerals. The Denver Museum of Nature and Science, the Colorado School of Mines Geology Museum in Golden, and the National Mining Hall of Fame and Museum in Leadville have outstanding displays of Colorado's minerals.

Fossil Fuels

Oil probably brought as many people to Colorado as did mining. Shortly after early settlers arrived, oil and gas seeps were found along the mountain front. On Oil Creek near Canon City, an 1862 oil well only 50 feet deep launched the second oil field in the United States. Initial production was one barrel a day! Later, several thousand gallons of petroleum were produced from the small field, and kerosene and lubricating oil were shipped by ox-drawn wagons to Denver and Santa Fe. Producing oil fields can be recognized by the grasshopper-like pumps that lift oil from deep wells. Where natural underground pressure forces oil to the surface, wells are topped by "Christmas trees" of pipes and valves.

One of the world's greatest known potential sources of petroleum lies in oil-saturated shales that cover a large area north of Rifle and the Colorado River, extending well into Wyoming and Utah. The oil shales are in the Eocene Green River formation, deposited in a big lake about 50 million

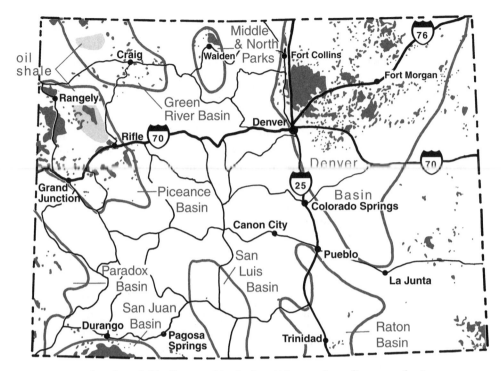

Oil and gas fields discovered in the last 130 years dot sedimentary basins of Colorado. Total statewide production has been more than 1.3 billion barrels of oil and well over 4 trillion cubic feet of natural gas.

years ago. Oily material called kerogen, waxy and too solid to flow through the fine pore spaces of shale, can be freed by a process in which oil-rich rocks are fractured and heated in place. But nearly a century of research has not lowered costs enough for kerogen to be competitive on the world market.

Coal in late Cretaceous rocks is mined from open pits and underground mines in many places in Colorado. Since coal is found in sedimentary rocks, which are much softer than igneous and metamorphic rocks, coal mines are known as soft rock mines. Most of Colorado's coal is of very high quality, with a high heat content and a very low sulfur content; it is used for generating power and for metallurgy.

Much of what we know about Colorado's geology we can credit to the human quest for wealth from the rocks. Conversely, knowledge of geology led to discovery of many of the state's metal mines and most of its oil fields and coal beds. Luckily, such knowledge also led to a greater understanding of the way the entire state formed—its magnificent journey as part of Earth.

II

Denver and Colorado Springs

Denver and Colorado Springs are well placed for an introduction to the geology of the state. They lie close to the Front Range of the Rockies, the largest of Colorado's faulted-anticline ranges, which rise sharply from the mile-high Colorado Piedmont to the Tertiary pediment at 8,000 to 10,000 feet, and then to summits 12,000 to 14,000 feet above sea level. Well east of the cities, the High Plains swing in an embracing arc from Wyoming to New Mexico.

The Denver Museum of Nature and Science, the Geology Museum at Colorado School of Mines in Golden, the South Platte Visitors Center at Chatfield Lake, Dinosaur Ridge north of Morrison, Henderson Museum at the University of Colorado in Boulder, and the Western Museum of Mining and Industry in Colorado Springs all present geologic exhibits well worth visiting. Many smaller museums and shops, especially in former mining towns, display museum-quality mineral specimens and exhibits about Colorado's mining history.

Geologic road guides in this chapter touch on a varied portion of Colorado's geologic story, from glacial features of the mountains to Colorado's dynamic mining history. For a general background to Front Range geology, read the introduction to Chapter IV.

Denver Area

More than 1,200 feet of sedimentary rocks—Paleozoic to Cenozoic shale, limestone, and sandstone—underlie Denver. Close to the mountains, strata of Paleozoic and Mesozoic age come to the surface, pushed up steeply as if to arch across the Precambrian core of the Front Range. Eroded edges of these tilted rocks, rows of steep hogbacks alternating with low, smooth swales, outline the mountain uplift.

Away from the mountains, sedimentary layers bow downward in a hidden sag, known from well cores and occasional outcrops, that geologists refer to as the Denver Basin. The center of the Denver Basin is filled with Tertiary sand and gravel, most of it also bowed down by continued sagging in the basin. The uppermost layer, the Denver formation, filled the basin's center and overflowed its edges as downward sagging ended. Its upper

In 1864 a major flood swept through the young city of Denver, destroying many buildings, including City Hall, and killing nineteen people.
—Courtesy Colorado Historical Society No. F26642

surface is now cut by the South Platte River and its tributaries, several of which flow through Denver.

In 1858 local Indians warned early settlers that great floods sometimes swept down these drainages. Denver grew anyway. Floods devastated low parts of the city in 1864, 1876, 1885, 1894, 1912, and 1921. Despite enlarged river channels, a 15-foot rise of Cherry Creek brought yet another flood in 1933.

After that disaster, Cherry Creek dam was built, designed to remain empty and ready to contain upstream deluges. It did its job in June 1965 when heavy rains caused upstream flow greater than the 1933 flood. But on that day the South Platte River flooded also, causing $508 million in damage and drowning six people. H. F. Matthai of the U.S. Geological Survey described what happened:

> The quiet was shattered by the terrible roar of the wind, rain, and rushing water. Then the thudding of huge boulders, the snapping and tearing of trees, and the grinding of cobbles and gravel increased the tumult. The small natural channels on the steep slopes could not carry the runoff, so the water took shortcuts, following the line of least resistance. Creeks overflowed, roads became rivers, and fields became lakes—all in a matter of minutes.

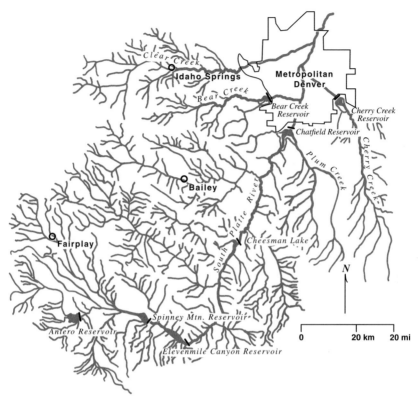

Rivers and streams from an area reaching 120 miles south and 120 miles west converge on Denver. Disastrous flash floods have come from the south, usually from Cherry Creek, Plum Creek, and South Platte drainages. Three flood-control reservoirs—Bear Creek, Chatfield, and Cherry Creek—now protect the city.

The flow from glutted ravines and from fields and hillsides soon reached East and West Plum Creeks. The combined flow in these creeks has been described as awesome, fantastic, and unbelievable; yet none of these superlatives seems adequate to describe what actually occurred. Large waves, high velocities, crosscurrents, and eddies swept away trees, houses, bridges, automobiles, heavy construction equipment, and livestock. All sorts of debris and large volumes of sand and gravel were torn from the banks and beds of the streams and were dumped, caught, plastered, or buried along the channel and flood plains downstream. A local resident stated, "The banks of the creek disappeared as if the land was made of sugar."

The flood reached the South Platte River and the urban areas of Littleton, Englewood, and Denver about 8 P.M. Here, the rampaging waters picked up house trailers, large butane storage tanks, lumber, and other flotsam and smashed them against bridges and structures near the river. Many of the partly plugged bridges could not withstand the added pressure and washed out. Other bridges held, but they forced water over approach fills, causing extensive erosion. The flood plains carried and stored much of the flood water, which inundated many homes, businesses, industries, railroad yards, highways, and streets.

After the 1965 disaster, Chatfield and Bear Creek dams were added to Denver's flood protection system.

Red Rocks Park—Dinosaur Ridge Area
16 miles (26 km) from State Capitol, Denver

In Red Rocks Park many layers of coarse, brick red sandstone and conglomerate separated by thinner layers of dark red siltstone are eroded into spectacular red monoliths, some of which dramatically frame Red Rocks Amphitheater. Part of an apron of debris washed off the eastern flank of the Ancestral Rocky Mountains nearly 300 million years ago, these Pennsylvanian and Permian rocks are part of the Fountain formation. Deposited by torrential streams, much of the sandstone is crossbedded, showing diagonal layers of sand and silt.

After the rock layers were tilted by Laramide mountain uplift, erosion shaped the great monoliths. Perhaps they stand so tall because the rock is cemented just a little better here than elsewhere. Similar monoliths occur in Roxborough State Park and elsewhere along the east side of the Front Range.

Deposited along the flanks of the Ancestral Rocky Mountains, sandstone and conglomerate of the Fountain formation were tilted by the rise of the present Rockies 72 to 45 million years ago. —T. S. Lovering photo, courtesy of U.S. Geological Survey

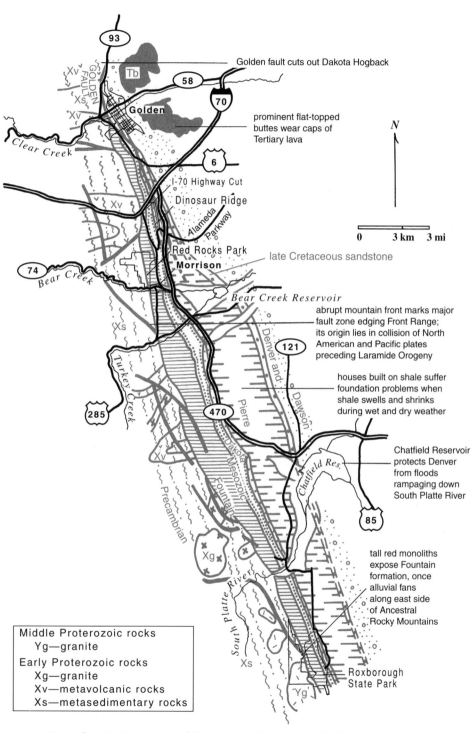

Golden fault cuts out Dakota Hogback

prominent flat-topped buttes wear caps of Tertiary lava

N

0   3 km   3 mi

I-70 Highway Cut

Dinosaur Ridge

Red Rocks Park

Morrison

late Cretaceous sandstone

abrupt mountain front marks major fault zone edging Front Range; its origin lies in collision of North American and Pacific plates preceding Laramide Orogeny

houses built on shale suffer foundation problems when shale swells and shrinks during wet and dry weather

Chatfield Reservoir protects Denver from floods rampaging down South Platte River

tall red monoliths expose Fountain formation, once alluvial fans along east side of Ancestral Rocky Mountains

Golden

Clear Creek

Bear Creek

Turkey Creek

Bear Creek Reservoir

Denver and Dawson

Pierre

Dakota

Mesozoic

Fountain

Precambrian

Chatfield Res.

South Platte River

Roxborough State Park

GOLDEN FAULT

Middle Proterozoic rocks
    Yg—granite
Early Proterozoic rocks
    Xg—granite
    Xv—metavolcanic rocks
    Xs—metasedimentary rocks

Pennsylvanian, Permian, and Cretaceous sedimentary rocks, bent up along the base of the mountains, are exposed at Red Rocks Park, Dinosaur Ridge, and the I-70 highway cut, all north of Morrison.

Red and white, sometimes concentric bands on these rocks cut across both bedding and crossbedding planes, especially on the northern wall of the amphitheater. The banding developed while the Fountain formation was buried under younger sediments, as groundwater seeped through the rock, dissolving and redepositing the dark red mineral hematite, which gives the rock its color. Notice the rough, corrugated surfaces of the monoliths. Sandstone layers are harder than siltstone layers and more resistant to erosion by wind and water, so they jut out more than the siltstone.

At the western edge of the park, with one hand you can span the contact between Pennsylvanian sedimentary rock and steeply rising Precambrian metamorphic rocks. Your fingers will span a time interval of nearly 1.4 billion years! Such a contact, where rocks from that interval of time were erased by erosion, is called an unconformity. This one represents one of the longest periods of erosion Earth has known.

Above and east of the Fountain formation is the younger Lyons sandstone, deposited as wind reworked the older sediments into sand dunes. It is not as red as the Fountain formation because the reworking process removed some of the fine iron minerals.

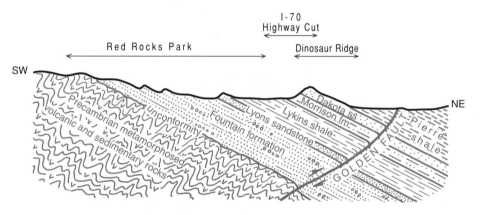

East-west section through the Red Rocks Park—Dinosaur Ridge area.

East of the Lyons sandstone is a valley underlain by brick red Lykins shale, which accumulated in salt water in Permian to Triassic time, and by the Morrison formation, deposited in winding streams of Jurassic time. For a close look at the Morrison formation, visit the large roadcut where I-70 slices through the Dakota Hogback.

Bordered on east and west by softer shale, this prominent hogback outlines the base of the mountains from Wyoming to New Mexico. Soft, easily eroded Morrison shale is well exposed in the roadcut. Its purple and grayish green layers accumulated when western North America was a flat,

Photographed when Morrison was a small town, the Dakota Hogback curves along the eastern side of the Front Range, partly enclosing Red Rocks Park. Hills to the left are Precambrian gneiss and schist. A small hogback of Cretaceous limestone catches sunlight in the right foreground. —T. S. Lovering photo, courtesy of U.S. Geological Survey

mountainless expanse with dinosaurs roaming among and feasting on conifers and palmlike cycads. Thin limestone layers formed in shallow lakes, and volcanic ash drifted in from far to the west.

Between 1876 and 1878 the Morrison formation brought international fame to the nearby town for which it was named. Several entire dinosaur skeletons were unearthed here and taken to eastern museums for study and exhibit. They included remains of the largest "thunder lizard" then known, *Apatosaurus*. On Dinosaur Ridge south of I-70, a quarry open to the public exposes some remaining dinosaur bones. Morrison Natural History Museum on Colorado 8 south of Morrison features more information on this area.

After the Morrison formation was deposited, an eastern sea crept across this part of the continent; a thin layer of river and beach sand, the Cretaceous Dakota sandstone, marks the sea's departure. On the east side of Dinosaur Ridge, walk along beds of the Dakota sandstone and see dinosaur footprints, ripple marks, and root patterns left by a mangrove swamp. Fine black marine mud deposited above the Dakota sandstone forms the valley east of the ridge.

Like most sedimentary rocks, rock layers above the Fountain formation were deposited in horizontal sheets; here they were bent up as the Rockies rose. Given enough time, as well as plenty of pressure from earth movements and the weight of thick overlying layers, rocks are more flexible than you might think. Myriad joints and tiny, almost microscopic faults usually accompany folding.

Faulting and bending along the mountain front dates from the Laramide Orogeny and possibly also from later regional uplift. The largest fault in this area, the Golden fault, runs east of the Dakota Hogback, though it is well concealed by overlying shale and soil. By measuring and matching rocks on opposite sides of the fault, geologists ascertained that rocks of the Front Range in this area have risen 11,000 feet and shifted eastward relative to rocks of the Denver Basin. The fault swings west near Golden, locally cutting out the Dakota sandstone and many of the older sedimentary rock layers.

At Dinosaur Ridge wide footprints of a large plant-eating dinosaur, perhaps an **Iguanodon,** *mark the Dakota sandstone. The biggest prints are from the animal's hind feet; smaller prints between them are from its front feet. Meat-eating dinosaurs made the thinner, distinctly three-toed prints.*
—Mike Williams photo

At Roxborough State Park, Fountain formation ridges flank the edge of the Front Range; walking trails weave in and out among them. —Felicie Williams photo

Mount Evans
26 miles (42 km) from Idaho Springs to summit

The Mt. Evans highway climbs from 7,500 feet elevation at Idaho Springs to about 14,200 feet. A short trail leads to the summit, 14,264 feet above sea level.

From Idaho Springs the road leads southwest for the first 5 miles, up a long valley that follows weak faults that formed in Precambrian schist as surrounding continental crust consolidated between 1.8 and 1.7 billion years ago. Each time major stresses built up in this area of the crust, the faults were remobilized.

Schist on either side of the fault zone probably originated as sediments deposited on the flanks of undersea volcanoes. Buried under as much as 7 vertical miles of sedimentary and volcanic deposits, the older sedimentary rocks were so altered by heat and pressure of the Early Proterozoic Orogeny that they recrystallized as metamorphic rock.

As the road turns south and begins to climb, it passes into pinkish granite of the huge intrusion, or batholith, that makes up most of Mt. Evans. During the Early Proterozoic Orogeny, after pushing and melting its way into the surrounding metamorphic rocks, the batholith cooled and solidified very slowly—its large mineral grains indicate the extremely slow growth of individual crystals. Here, you can easily see individual

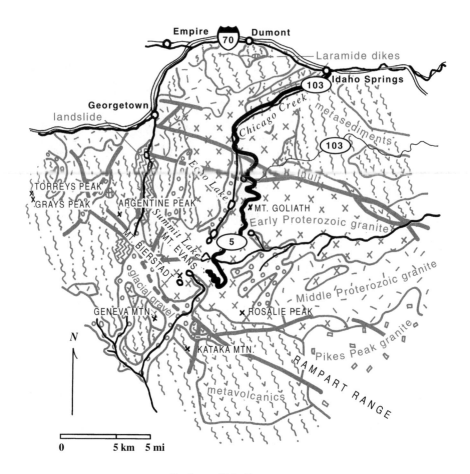

Geology of Mt. Evans area.

grains and identify the minerals of which the rock is made: glassy quartz, white and pink feldspar, and black mica or biotite.

Along with surrounding metamorphic rocks, the Proterozoic granite was probably faulted and pushed upward 1.43 million years ago. The resulting mountains were then beveled by a lengthy period of erosion at the end of Precambrian time. The granite rose again as part of the Ancestral Rockies 300 million years ago, was eroded once more, and then was faulted upward 72 million years ago as part of today's Front Range.

With stresses and strains of the repeated episodes of mountain building, several sets of joints and faults developed in the granite. Minerals, brought in by mineral-laden water, filled these openings, so that now the rock is crisscrossed with narrow veins of milky quartz and other minerals. During and after the Laramide Orogeny, ancient faults served as channels for rising mineral-rich fluids, thereby determining the location of future mining districts.

Weathering at or just below the surface has widened many joints, in many cases rounding granite into separate boulders in a process known as spheroidal weathering. Characteristic of granite and other even-grained rocks, this type of weathering is due to the way mica grains change gradually into clay when exposed to water. The grains enlarge during this change, and pop off adjacent quartz grains or even whole surface layers. Once cracks form, water seeps into them, aiding the weathering process as it freezes and thaws repeatedly. Weathering is most rapid at corners and sharp edges where the rock is exposed on more than one side, so gradually the boulders become more and more spherical. Coarse quartz and feldspar sand, known as grus, accumulates around the base of rounded boulders—the first step in making soil.

Above timberline, geologic features are no longer hidden by trees. Many geologic details such as joints and faults can be clearly seen. Exfoliation peels off the outermost curved layers of rock. Wind blasts small smooth pits on the western (windward) side of some exposed rocks, especially in the Alpine Garden area and along the trail to the summit. Watch for veins of milky quartz and for pegmatite dikes, bands of granitelike rock with very large crystal grains and often very light color.

Quartz veins form intricate patterns in Precambrian gneiss and schist of Grays Peak, near Mt. Evans. —T. S. Lovering photo, courtesy of U. S. Geological Survey

Summit Lake lies in a small basin below a horseshoe-shaped cirque carved by a glacier in the hard gray granite of Mt. Evans. Walk the short trail to the low divide beyond the lake to look down 1,000 feet into the magnificent cirque and valley of Chicago Creek. Valleys like this one, U-shaped in cross section, are hallmarks of mountain glaciation. Streams usually carve distinctly V-shaped valleys. Paternoster lakes, so-named because they are strung down the valley like beads on a rosary, decorate the valley floor, which is sculpted in steplike levels also typical of glaciated mountain valleys. A terminal moraine, an arc-shaped ridge of rock debris left by the last glacier, dams the lowest lake.

Straight across the valley, fallen rock forms long talus cones below steep cliffs. When water trickles into crevices and freezes, it expands and, with repeated freezing and thawing, wedges the rock apart. Though each step in this process is tiny, the effect over time is staggering.

The summit of Mt. Evans offers a striking panorama of the Front Range from Longs Peak 50 miles due north to Pikes Peak 58 miles south-southeast. These two peaks, as well as Mt. Evans, Mt. Bierstadt, Torreys Peak, and Grays Peak, are Fourteeners, more than 14,000 feet high. They stand 9,000 feet above the Colorado Piedmont and about 5,000 feet above the gently sloping Tertiary pediment, which shows up clearly from here, forming the surface of the Rampart Range between Mt. Evans and Pikes Peak.

Far to the south lie the peaks of the Sangre de Cristo Range, a younger fault-block range that stretches south to Santa Fe, New Mexico. To the southwest, U.S. 285 winds across Kenosha Pass into South Park, one of Colorado's four large intermontane basins.

Boulder—Estes Park—Central City Loop
U.S. 36/Colorado 7/Colorado 72/Colorado 119/U.S. 6
150 miles (241 km)

U.S. 36 angles northwest from Denver across Tertiary and Cretaceous sediments that fill the Denver Basin. Stop briefly at the scenic overlook a few miles before reaching Boulder, on top of a fault block of late Cretaceous sandstone. Boulder Valley, like much of the Colorado Piedmont, is floored with Pierre shale, also of Cretaceous age. Under the shale are older Mesozoic and Paleozoic sedimentary layers—a total of about 10,000 feet of sedimentary rock. Where not removed by faulting these layers rise to the surface along the mountain front, but only those that resist erosion are clearly exposed. The gently sloping, pine-covered surface halfway up the Front Range

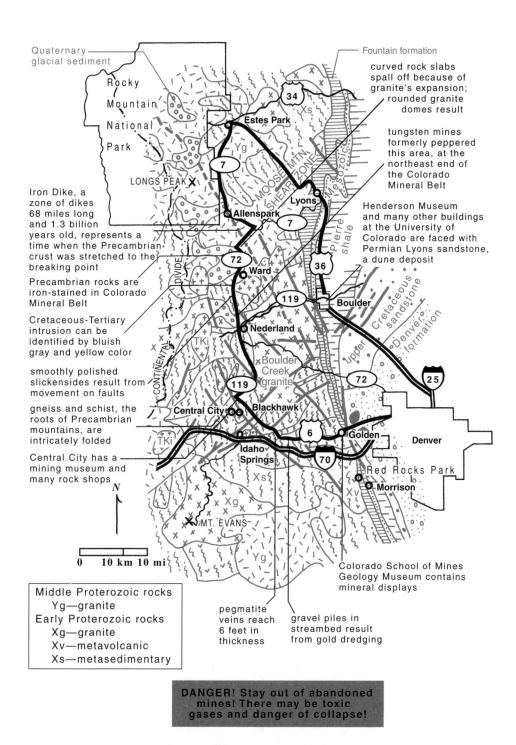

Geology along the Boulder–Estes Park–Central City loop.

is the Tertiary pediment. After regional uplift, rivers draining the mountains cut many canyons into it, including the canyon of Boulder Creek.

Boulder's Flatirons lean against the mountains west of town. The Fountain formation of Pennsylvanian red sandstone and conglomerate lies directly against Precambrian granite; older Paleozoic rocks were eroded off the Ancestral Rockies before the Fountain formation was deposited as alluvial fans. Along with younger rock layers, the Fountain formation was dragged and faulted upward by rebirth of the Rockies during the Laramide Orogeny, 72 to 40 million years ago. Erosion of small mountain canyons and removal of softer layers created the unusual shape of the Flatirons.

Mesas south and north of Boulder are parts of mountain pediments eroded in upturned Cretaceous sedimentary rocks, later veneered with 10 to 20 feet of gravel. The gravel slows erosion because water sinks into it quickly rather than running off on its surface. Beyond the pediments the Dakota Hogback borders the mountains; near Boulder Creek it is eliminated by faults.

North of Boulder, U.S. 36 parallels the Dakota Hogback, running on the edges of upturned Mesozoic sedimentary rocks that form smaller, less continuous limestone and sandstone hogbacks. Steeply dipping dark gray Pierre shale shows up on mesa slopes east of the road. The limestones and the shale contain marine fossils: clams, oysters, occasional bones of extinct

Boulder's Flatirons are composed of sandstone and conglomerate deposited as alluvial fans along the edge of the Ancestral Rockies. They are now bent up sharply at the edge of the Front Range. —Jack Rathbone photo

swimming reptiles, and tiny one-celled foraminifera. South of Lyons the limestones are quarried for cement.

Close to Lyons, U.S. 36 turns abruptly west through the Dakota Hogback and into valley-forming Jurassic Morrison shale, then into more resistant Permian-Triassic Lyons sandstone. Lyons sandstone is quarried north of town for flagstone and building stone. Strong, with fine, even grain and an attractive color, the stone breaks smoothly along crossbedding surfaces inherited from the sand dunes in which it originated. Footprints of small four-footed creatures have been found in quarries here.

Northwest of Lyons, U.S. 36 soon encounters granite, part of a batholith formed as a large mass of molten rock or magma slowly cooling. The granite is part of the Berthoud plutonic suite, intruded into older rocks about 1.4 billion years ago. Berthoud granite contains small, chunky crystals of glassy quartz, pink and light gray feldspar, and biotite. In places it passes gradually into gneiss, banded strongly in shades of black, gray, and pink. Schist also occurs northwest of Lyons, with flat-faced mica crystals aligned parallel to one another, a result of pressure during metamorphism. Watch for white quartz and feldspar veins and for pegmatite bands containing coarse, large crystals. The gneiss and schist are 1.7 billion years old, though the rocks from which they formed, probably a mix of volcanic and sedimentary rocks of an ancient island arc, are even older.

At mile 11.6, where a gravel road heads west to Big Elk Meadows, U.S. 36 crosses the Moose Mountain shear zone, a 3,000- to 4,000-foot-wide zone that trends northeast, parallel to the edge of the Precambrian continent. Motion along the shear zone 1.2 billion years ago ground the rock into

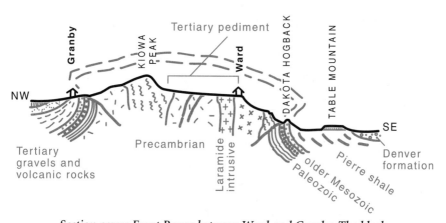

Section across Front Range between Ward and Granby. The block of Precambrian rock that forms the core of the range slid up and over sedimentary rocks on either side. Dashed lines show the hypothetical faulted anticline of Paleozoic and Mesozoic strata.

microscopic grains, forming fine, dense mylonite. This fault, and others roughly paralleling it, cut diagonally across Colorado's ranges; movement on them was renewed during mountain-building events in Paleozoic, Mesozoic, and Tertiary time.

The valley of Big Thompson Creek, below the overlook between mileposts 3 and 4, was carved during glacial times when rainfall, snowfall, and stream runoff were far greater than now. The present valley, formerly a deep, stream-carved canyon, is filled with debris washed from Ice Age glaciers. Big Thompson Canyon east of Estes Park was gutted by a flash flood in 1976, with 139 lives lost and practically all buildings—summer cottages, stores, and motels—demolished.

At Estes Park our route turns south on Colorado 7 to climb through more Precambrian rocks. In about 8 miles the mountains level off onto rolling, fairly open country 9,000 to 10,000 feet in elevation—the Tertiary pediment. The highway remains on this surface almost to Blackhawk, dipping down only to cross major stream valleys.

Longs Peak, west of Colorado 7, wears a 2,000-foot "Diamond" on its east face, high above Chasm Lake. Trails from Longs Peak Campground lead to the cirque and to the summit, 14,255 feet above sea level. —W. T. Lee photo, courtesy of U.S. Geological Survey

Magnificent views of Longs Peak to the west show the 2,000-foot granite cliff of the peak's east face, the headwall of a deep glacial cirque. The flat top of Longs Peak may be a 600-million-year-old relic of a time when much of the world had been eroded to a smooth, near-sea-level surface. The top of Precambrian rocks on Longs Peak is 22,000 feet higher than the corresponding surface in the deepest part of the Denver Basin. The combination of mountain uplift and basin subsidence is, altogether, at least 4 miles!

North and south of Allenspark the road crosses and recrosses a group of parallel, dark dikes that cut vertically northnorthwest across the Front Range. Collectively called Iron Dike, they are composed of igneous rock that filled cracks formed in the Precambrian crust during an episode of crustal stretching about 1.3 billion years ago.

Ward and Nederland mark the northeastern end of the Colorado Mineral Belt, a 50-mile-wide zone that extends from here to southwestern Colorado and that contains most of the state's metal mines. It follows the Colorado Lineament, a band of northeast-trending faults and shear zones that formed in the Proterozoic crust about 1.8 billion years ago. During and after Laramide mountain building, beginning 72 million years ago, this weak zone provided channels for numerous igneous intrusions and mineral-rich fluids. Gold, silver, lead, copper, zinc, molybdenum, uranium, and tungsten have been mined in the Colorado Mineral Belt.

Where did the minerals come from? We don't know for sure. Some may have originated along Precambrian mid-ocean ridges, perhaps as black smokers, sulfide-rich hot springs discovered recently on modern mid-ocean ridges. Deposited in Precambrian rocks, the minerals could have been carried up and redeposited by Tertiary igneous fluids. Along Colorado 72 just south of Ward is a fine-grained, gray Tertiary intrusion that may be related to mineralization there.

Leaving Nederland on Colorado 119, our route passes a terminal moraine left by a receding glacier. The side road to Eldora skirts the downstream edge of this moraine, then cuts westward through it.

The Central City–Idaho Springs Mining District has produced almost $200 million worth of gold, silver, lead, zinc, and copper, almost all of it when gold was worth $20 an ounce. In 1859 John Gregory, seeking the upstream source of Golden's placer gold deposits, struck it rich between Blackhawk and Central City. Mines dot slopes above these towns in "the richest square mile on earth." Most of the mines have long since shut down.

The gold occurred in vertical or nearly vertical quartz veins in shattered Precambrian rocks. Some of the veins around Central City and Blackhawk extend through the mountains toward Idaho Springs, following the trend of the Colorado Mineral Belt and its ancient faults. Formed by hot, mineral-rich waters associated with small Tertiary intrusions, the veins filled

In 1917 a tungsten mill operated just below the present Barker Dam in Boulder Canyon, a few miles east of Nederland. The nearby town of Ferberite, no longer in existence, was named after a tungsten ore mineral.
—F. L. Hess photo, courtesy of U.S. Geological Survey

hundreds of cracks and fissures. Locate them by looking for collapsed tunnels, where ore-bearing rock has been mined out. Do not enter old tunnels! They cave in easily, and some contain toxic gases. Many have been blocked off, and most mine drainage is now treated before it is released into nearby creeks.

Mine tours and jeep tours are offered here, and Idaho Springs boasts several small museums.

South of Blackhawk, Colorado 119 and Clear Creek (from which gold seekers still pan a bit of "color") follow the weakened rock along the Blackhawk fault zone. Gneiss and schist are exposed in Clear Creek Canyon. Rocks that were originally volcanic contain amphibolite, dark gray metamorphosed basalt in which the primary mineral is hornblende. Lacking mica, it breaks into chunky pieces rather than plates or flakes.

Our route soon joins U.S. 6 and, at the mouth of Clear Creek Canyon near Golden, leaves the Precambrian core of the Front Range. Here, most Paleozoic and Mesozoic sedimentary rocks, including the Fountain formation and the Dakota sandstone, have been eliminated by the fault zone that edges the range.

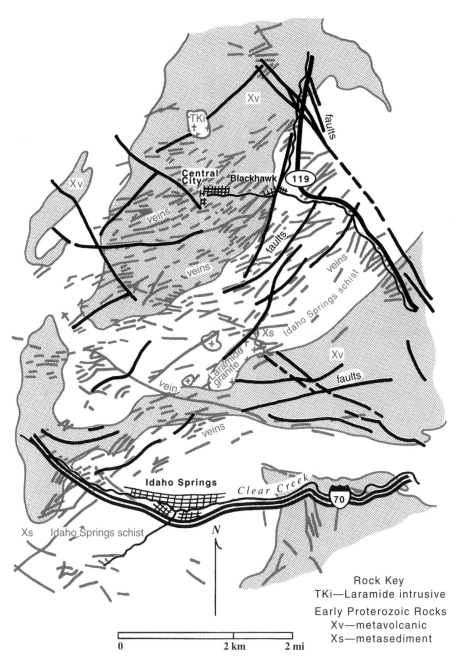

This much-simplified map of the Central City–Idaho Springs Mining District shows the three main rock types and most of the five hundred known ore-bearing veins. Notice the northeasterly trend of the veins; though far younger, they follow the grain, or fabric, of the Precambrian metamorphic rocks.

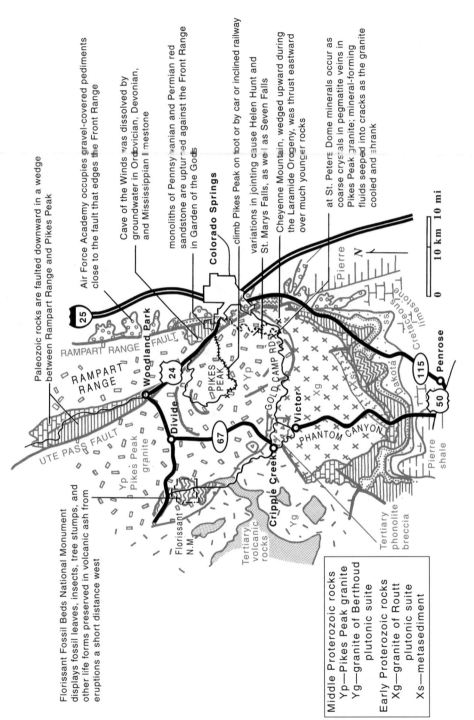

Paleozoic rocks are faulted downward in a wedge between Rampart Range and Pikes Peak

Air Force Academy occupies gravel-covered pediments close to the fault that edges the Front Range

Cave of the Winds was dissolved by groundwater in Ordovician, Devonian, and Mississippian limestone

monoliths of Pennsylvanian and Permian red sandstone are upturned against the Front Range in Garden of the Gods

climb Pikes Peak on foot or by car or inclined railway

variations in jointing cause Helen Hunt and St. Marys Falls, as well as Seven Falls

Cheyenne Mountain, wedged upward during the Laramide Orogeny, was thrust eastward over much younger rocks

at St. Peters Dome minerals occur as coarse crystals in pegmatite veins in Pikes Peak granite; mineral-forming fluids seeped into cracks as the granite cooled and shrank

Florissant Fossil Beds National Monument displays fossil leaves, insects, tree stumps, and other life forms preserved in volcanic ash from eruptions a short distance west

Middle Proterozoic rocks
Yp—Pikes Peak granite
Yg—granite of Berthoud plutonic suite
Early Proterozoic rocks
Xg—granite of Routt plutonic suite
Xs—metasediment

Geology of the Colorado Springs area. Minor faults are not shown.

Two flat-topped mesas, North and South Table Mountains, frame Golden. Their tilted Cretaceous and Tertiary sedimentary rocks are protected from erosion by flat caps of Tertiary basalt. Even from a distance you can see the vertical columnar jointing of the lava flows. North Table Mountain is known for delicate zeolite crystals found in little gas-bubble holes, or vesicles, in its basalt cap. Plant fossils occur in soil zones between the volcanic flows, and dinosaur bones and tracks have been found in the Cretaceous rocks below.

Colorado Springs Area

Originally the town of Colorado Springs nestled near the confluence of Monument and Fountain Creeks, between the southern end of the Front Range and a cluster of sheltering white sandstone buttes and pinnacles. The city has now overflowed eastward across the sandstone hills and crept up the pediment terraces that project like paws of giant sphinxes from the edge of the mountains.

The old part of town lies on dark gray Pierre shale, a soft, fossil-bearing rock that was deposited as mud 75 million years ago on the bottom of the shallow Cretaceous sea—a sea stretching from what is now the Arctic to the Gulf of Mexico. Here, the drag of the rising mountain mass tilted up the shale westward toward the mountains.

A network of particularly large faults separates the mountains from the city, chief among them the Ute Pass and Rampart Range faults. The Rampart Range fault trends north-south, edging the mountain uplift north of

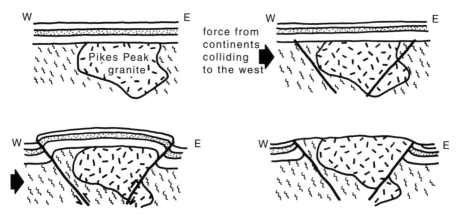

As North America was shoved against a plate to the west, extreme pressure formed faults along both sides of the Pikes Peak batholith, wedging up the mountain core, arching and faulting overlying sedimentary rocks. On both sides, Precambrian granite slid out over younger sedimentary layers. Erosion has removed all the sedimentary rocks that once lay above the granite batholith.

Fountain Creek. The Ute Pass fault fronts Cheyenne Mountain southwest of the city, then bends at Manitou Springs and runs northwest into the mountains, west of the Rampart Range. U.S. 24 follows the broken, weakened rock of the fault zone up Fountain Creek to Woodland Park.

At Manitou Springs, mineral water bubbles to the surface through a number of channels in shattered rock along Ute Pass fault. The naturally carbonated water has dissolved carbon dioxide from Paleozoic limestone at a depth where pressure is similar to that in a capped pop bottle. As the water rises rapidly along the fault zone, pressure on it decreases—just as it does when you uncap a bottle of pop—and the carbon dioxide comes out of solution to form the bubbles that give Manitou Springs water its refreshing fizz.

Garden of the Gods
West of Colorado Springs on U.S. 24

In the narrow point of land between two major faults that edge the mountains near Colorado Springs, a wedge of folded and faulted Paleozoic and Mesozoic rocks has been tilted steeply. The road entering Garden of the Gods from the east crosses a broad swale underlain by soft Cretaceous shale, passing a small wall-like hogback of white rock, a resistant sandstone layer in the Morrison formation. West of another swale eroded in soft red Triassic shale, the road slips through a narrow gateway into Garden of the Gods. The gigantic gateposts, as well as other tall monoliths, are eroded in salmon-colored Lyons sandstone, a coastal dune deposit of probable Permian age. Its sand grains are cemented together tightly by silica and tiny grains of hematite, so although the sandstone layers are tilted 90 degrees or even slightly overturned, they are strong enough to stand as tall, thin, bladelike towers.

The towers continue in regimental rows south of Garden of the Gods, but south of Fountain Creek they are cut off abruptly by the Ute Pass fault and steep slopes of Cheyenne Mountain.

West of the monoliths of Lyons sandstone, dark red sandstone alternates with deep red mudstone or shale and a few bands of pebbly conglomerate, all belonging to the Fountain formation. Here, the strata tilt less steeply and are shaped into mushroomlike pedestals and balanced rocks by differential erosion of harder and softer layers. The Rampart Range fault, concealed by the soil of the valley floor, separates them from the tall gateway slabs farther east.

All these pink and red rocks were deposited in alluvial aprons by streams draining Frontrangia, part of the Ancestral Rockies. Look closely for sedimentary details in them: ripple marks, thin bands of stream-rounded pebbles, impressions of mudcracks. In places you can see sweeping lines of crossbedding cutting diagonally across sandstone layers.

The lack of fossils (except for a few animal footprints) or volcanic material in the Fountain formation makes it impossible to date it precisely, so we have to deduce its age. We know it was deposited above Pennsylvanian shales and below Triassic ones. We know its older part is Pennsylvanian

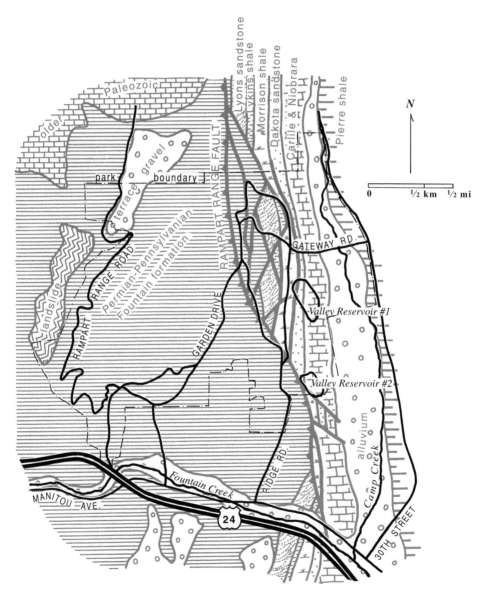

Geology of Garden of the Gods.

Where faults along the mountains' edge have tilted Paleozoic and Mesozoic strata 90 degrees or more, they have weathered into the spectacular towers, pinnacles, and mushroom rocks of Garden of the Gods. Pikes Peak looms in the distance. —Bob Rebello photo

because the lowest layers alternate with thin gray shales that contain Pennsylvanian fossils. But the upper part, the pink sandstone of the tallest pinnacles, resembles the Permian Lyons sandstone north of Boulder, suggesting that Fountain formation deposition continued into Permian time.

Cave of the Winds
West of Colorado Springs on U.S. 24

Cave of the Winds, near Manitou, formed where groundwater, made slightly acidic by passing through air, forest soils, and the granite of the mountains, slift. In layers of Paleozoic limestone, the water gradually dissolved and carried away some of the rock, forming many rooms and passageways.

The cave was probably shaped in Pleistocene time, when surface runoff and groundwater were much more abundant than they are at present. As groundwater supplies decreased at the end of the Ice Age, solution decreased too, until water in the cave was reduced to the mere trickles seen today. As solution ended, a new process began. Where calcite-laden water evaporated, each evaporating droplet left behind a tiny bit of calcite. Accumulating over

Stalactites and stalagmites developed at the end of the Ice Age, within the last 10,000 years, when solution ended and deposition began.

stalac**tite**s (with a **c**) hang "**tite**ley" to the **c**eiling

stalag**mite**s (with a **g**) "**mite**" reach the ceiling from the **g**round

time, the calcite built stalagmites, stalactites, and curtainlike and ribbonlike ornaments. Cave ornaments formed by this process may grow less than an inch in a human lifetime. Note how curtains and stalactites line up below joints in the ceiling rock.

The exit road passes through Williams Canyon, past the rocks that contain the cave. At the edge of the canyon floor, Cambrian strata are visible— Sawatch sandstone lying almost horizontally on coarse pink Precambrian granite and overlain by multicolored dolomitic sandstone. The top of the granite was smoothed during the long interval of erosion at the end of

In Williams Canyon northwest of Manitou, Sawatch sandstone and Manitou limestone lie directly on Precambrian granite, which is barely visible in the streambed. The lowest passages of Cave of the Winds are behind the cliffs in the left part of the photograph.
—N. H. Darton photo, courtesy of U.S. Geological Survey

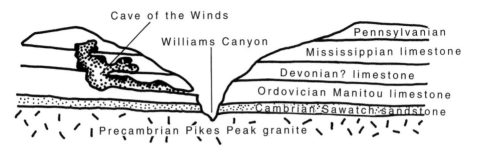

The rooms and passageways of Cave of the Winds were dissolved along joints and small faults in three limestone units, relics of Paleozoic seas.

Precambrian time. Above these rocks and forming the canyon narrows is the massive gray Manitou limestone, the Ordovician rock in which the deepest level of Cave of the Winds is carved. This rock contains many fossils, including trilobites, brachiopods, and straight-shelled or horn-like cephalopods, ancestors of today's chambered nautilus.

The Manitou limestone is overlain by limestone of probable Devonian age. It contains few fossils—only tiny toothlike conodonts. This rock walls middle-level rooms and passageways in Cave of the Winds. Limestone of the cave's uppermost chambers lies higher up the slope; fossils show it is of Mississippian age. Pennsylvanian marine shales, some of them with many fossils, are still higher up.

Pikes Peak
Pikes Peak Toll Road
19 miles (31 km) from U.S. 24 to summit

Pikes Peak consists almost entirely of Pikes Peak granite, an attractive pink granite with stubby, interlocking crystals of glasslike quartz and flat-faced white and pink feldspar, liberally sprinkled with hornblende and biotite. This Precambrian rock, about a billion years old, developed as a batholith, an immense mass of magma that pushed and melted its way upward through overlying rocks. It never reached the surface but cooled slowly over millions of years. No known tectonic event is associated with this intrusion; geologists do not know its cause. The batholith extends to the southeast another 50 miles beneath the plains.

In many places within the batholith, many of the mineral grains lie parallel to each other. As the magma flowed upward like thick pudding, newly formed and newly forming crystals lined up in the direction of

Pikes Peak, at an elevation of 14,110 feet, is the southernmost summit of the faulted anticline that forms the Front Range. Red monoliths in the Garden of the Gods rise in the foreground. —Jack Rathbone photo

flow. Geologists mapping the orientation of crystals in exposed parts of the batholith now recognize three centers of upward movement.

As the granite solidified, cracks formed parallel to flow directions. Watery, superheated fluids penetrated the cracks, cooling very slowly to form the large crystals of pegmatite dikes. The Pikes Peak granite is known worldwide for these pegmatites, some of which contain smoky quartz crystals, topaz, and an unusual blue green feldspar called amazonite.

The granite has had an eventful history. Probably formed as the core of a Precambrian mountain range, it may have been laid bare and smoothed during the erosion interval at the end of Precambrian time. Later covered with Paleozoic sedimentary rocks, it was pushed up again—not molten this time—to form part of the Ancestral Rocky Mountains of Pennsylvanian time. As those highlands were attacked by erosion, it wore down further, then was covered with late Paleozoic and Mesozoic sedimentary layers. Uplifted a third time during the Laramide Orogeny, it was cleaned off again, and today is slowly yielding anew to erosion.

One of the first steps in the erosional process can be seen easily from the road up the mountain: weathering of granite and its gradual conversion to coarse, sandy soil. Mica and feldspar crystals in the granite gradually decompose, especially when exposed to moisture, and as they decompose they expand into clay minerals. The expansion loosens quartz crystals that form

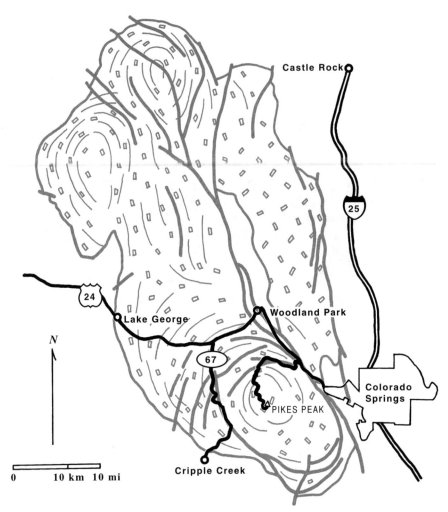

Sheets of fine-grained granite and veins of pegmatite (thin lines) *formed as the molten rock flowed upward and cooled. They define three main intrusive centers within the Pikes Peak batholith. Thick red lines are faults.*

much of the body of the rock. This swelling process is augmented by frost action when water that seeps into cracks freezes into ice. The two processes gradually weaken the granite, which falls apart into a deep layer of soft "rotten" granite, or grus.

Weathering takes place faster where joints and crevices cut the rock, and the Pikes Peak granite, thanks to its checkered billion-year history, has plenty of these. Weathering attacks sharp corners from three directions at once, eventually rounding jointed rock into separate boulders in a process known as spheroidal weathering.

Higher on the slopes of Pikes Peak are many features of glacial erosion created during the Pleistocene Ice Age. The spacious cirques of Bottomless Pit and the so-called Crater (a misnomer because the mountain is not a volcano) were carved by large valley glaciers. Steep-walled, U-shaped canyons extend as low as 8,000 feet, the level at which the downstream tongues of the glaciers melted.

The high slopes look down on the Tertiary pediment's roughly horizontal surface, sprawling some distance westward and all the way north to the slopes of Mt. Evans near Denver.

At overlooks near the top of the mountain, the granite is so intricately jointed that it looks as though someone has chopped it with a cleaver. Crisscrossing sets of joints represent different directions of stress to which the rock was at some time exposed.

In Pikes Peak granite, as well as other massive rocks, weathering along joints separates boulders and rounds their angular edges.
—T. S. Lovering photo, courtesy of U.S. Geological Survey

There is less weathering, and no spheroidal weathering, on top of Pikes Peak. Moisture here is often in the form of snow, which doesn't sink readily into the rocks to break down the mica crystals. But water from snowmelt seeps into joints, where it freezes and wedges the rocks apart. Years and years of such fracturing produces a barren, jagged surface with sharp-edged rock fragments tipped at crazy angles. Here, acids secreted by lichens that grow on some rock surfaces help to slowly disintegrate the granite. Violent mountain winds have whipped away all small rock particles.

In 1806 President Thomas Jefferson dispatched an expedition to explore the Pikes Peak region, part of newly purchased Louisiana Territory. The party was led by Zebulon Pike, who attempted to climb to the top of the mountain that he called "Grand Peak." He estimated its elevation as 18,581 feet. Unsuccessful, he declared it would never be climbed. Mapped first as James Peak, it was commonly called Pike's Peak by trappers and military men, and Pike's Peak it became—the U.S. Geological Survey now customarily leaves out the apostrophe in geographic names like Pikes Peak and Longs Peak.

The name was for a time used for the entire Front Range region, leading to the "Pikes Peak or bust" slogan of miners rushing toward goldfields near Denver. Gold was not discovered near Pikes Peak until 1891, thirty-two years after it was found in canyons west of Denver. The mountain was the inspiration for the anthem *America the Beautiful,* with its "purple mountain majesties above the fruited plain."

Cripple Creek and Gold Camp Road
73-mile (117-km) loop from Colorado Springs

Head west from Colorado Springs, following the road guide for **U.S. 24: Colorado Springs—Antero Junction** until you reach Divide. At Divide turn south on Colorado 67 to Cripple Creek, or continue 8 more miles on U.S. 24 and then turn south on the road through Florissant Fossil Beds National Monument, which also leads to Cripple Creek.

The Cripple Creek Mining District southwest of Pikes Peak was discovered in 1891, fairly late in Colorado's mining history. Veins exposed on the surface contained gold and silver tellurides—minerals not familiar to earlier miners. The district produced close to $450 million in gold and silver between 1891 and 1942, when gold was worth $20 per ounce. In recent years, mining has been revived in the Cripple Creek district with the discovery of the new Cresson gold deposit, not far from the historic Cresson mine, which produced over 2 million ounces of gold in the early part of this century.

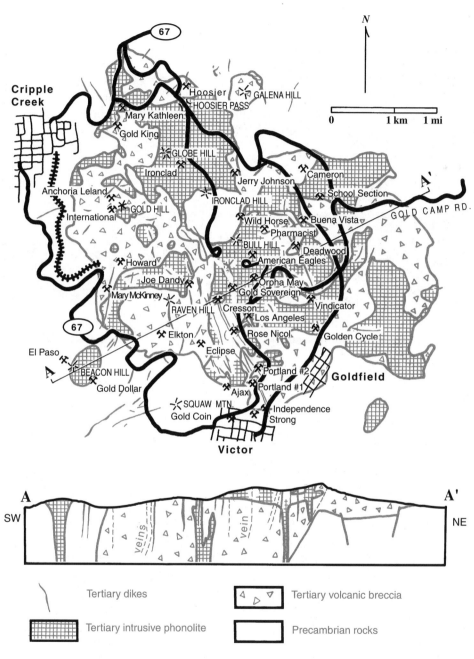

A–A' shows location of section

| | Tertiary dikes | | | Tertiary volcanic breccia |
| | Tertiary intrusive phonolite | | | Precambrian rocks |

Cripple Creek Mining District lies within the throat of a Tertiary volcano, an area cut by myriad dikes and veins. —CC&V Gold Mining Company map, courtesy of Jeff Pontius

The town of Independence, a few miles southeast of Cripple Creek, was dotted with prospect pits and mine dumps. By 1903, when this photograph was taken, most of the native forest had been cut for mine timbers and fuel. —F. L. Ransome photo, courtesy of U.S. Geological Survey

The ore deposits are in an area 4 miles long and 2 miles wide, between Cripple Creek and its early rival, Victor. Ore is present in Tertiary volcanic rocks surrounded by pink Precambrian Pikes Peak granite and older granite and metamorphic rocks. The Tertiary rocks are a mixed-up mass of volcanic and nonvolcanic rock fragments created within the throat of a Tertiary volcano. Known as breccia, the rock has been shattered, cemented together, and shattered again several times, suggesting that the volcano erupted repeatedly. The breccia is cut by many dikes and several irregular tubes of volcanic material, either black basalt or a fine-grained, feldspar-rich lighter rock called phonolite, a "Greekified" version of the picturesque miners' term *clinkstone,* referring to the sound the rock makes when struck with a hammer.

In periods between collapses, mineral-rich vapors and fluids from seething magma below seeped up along cracks and crevices in the breccia, crystallized on the rock walls, and gradually developed into ore-bearing veins. Most of the richest veins lie at the edges of the breccia mass; some extend into it or into surrounding granite. In addition to gold and silver ores, the veins contain crystals of fluorite, pyrite, galena, calcite, and other minerals, many of which can be found on mine dumps. The mines are private property; ask for permission before collecting. Keep away from old mine shafts and tunnels, which may harbor toxic gases and are often on the verge of collapse.

The new Cresson deposit, discovered in 1990, is thought to contain about 2 million ounces of gold at an average grade of 0.028 ounce per ton of rock, about one-thirtieth of the grade historically mined in the district. A ton of rock (roughly a cubic yard) yields a lump of gold about as big as a match head! Crushed ore from the low-grade deposit is piled on a large plastic- and clay-lined pad, where it is soaked with chemical solvents to dissolve the gold. The solution drains into a treatment plant, where the gold is precipitated. The solvents are repeatedly recycled through the pile to extract more of the gold.

In this 1995 aerial photograph the Cripple Creek Mining District is dotted with light-colored dumps from old underground mines. The leach pad for the new Cresson Mine is at lower right, across the main road from the crusher. As economically recoverable gold is processed, the land is restored to something like its original appearance. Pikes Peak rises in the background. —CC&V Gold Mining Company photo, courtesy of Jeff Pontius

Mine tours are offered in Cripple Creek, and there are mining museums in Cripple Creek, Victor, Florence, and Colorado Springs. Cripple Creek & Victor Gold Mining Company provides interpretive trails in the Vindicator Valley and an overlook and explanation of its operation at the American Eagles headframe about 2 miles north of Victor.

Three narrow-gauge railroads that once vied for Cripple Creek business were dismantled when the glory faded. The route of Colorado 67 between Divide and Cripple Creek follows one of the railroad grades. The Phantom Canyon railroad led to the south to carry ore to be processed in coal-burning smelters in Florence. The third route, now Gold Camp Road, skirts the southern flank of Pikes Peak, penetrating ridges and mountain spurs with cuts or tunnels just wide enough for narrow-gauge trains. Our route follows it to Colorado Springs.

Much of Gold Camp Road is on the Tertiary pediment. Along the route, examine weathering characteristics of the Pikes Peak granite. In places it is

Arcuate slabs peel off large masses of granite; the joints that separate them result from expansion of granite as it is relieved of confining pressure. —T. S. Lovering photo, courtesy of U.S. Geological Survey

cut by many joints, usually arranged in parallel sets—one set vertical, another set horizontal, still another vertical but running in a different direction. Spheroidal weathering turns such fractured granite into rounded boulders. At Cathedral Rocks, similar weathering has created huge turrets and pinnacles.

Eventually the granite decomposes completely into coarse sand or grus containing feldspar and quartz grains that were once individual crystals in the rock. With the addition of fallen leaves, grass roots, and other organic matter, the grus gradually becomes loose forest soil.

At Devils Slide and St. Peters Dome, huge curving sheets of granite are flaking off the batholith. When free from the weight of overlying rock and glaciers, the granite expands, causing a set of joints parallel to the exposed surface. Weathering of mica and feldspar crystals along the joints helps loosen the surface layer, and water seeping into the joints freezes by night and melts by day, further widening the once tiny cracks. Finally, the surface layer loosens and spalls off.

North of Old Stage Road, Gold Camp Road is closed to automobile traffic because the old tunnels are unsafe. It is accessible by foot, however, and passes through an area with many pegmatites where fluorite, zircon, smoky quartz, and amazonite can be found. Old Stage Road leads on to Colorado Springs.

Development of the Plains and Piedmont

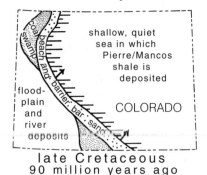

late Cretaceous
90 million years ago

Ninety million years ago, shoreline features slowly migrated east as land rose in the west.

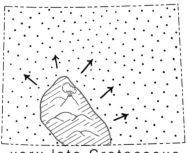

very late Cretaceous
72 million years ago

By 72 million years ago the Rockies had begun to form and the sea had departed from Colorado. Erosion spread an apron of debris outward from the Rockies, which continued to rise until 40 million years ago.

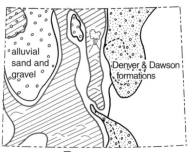

Eocene
50 million years ago

Between 65 and 50 million years ago, the Denver and Dawson formations were deposited in a shallow basin east of the Rockies. These formations later eroded back to the dashed line. By 28 million years ago the mountains were nearly covered with their own debris.

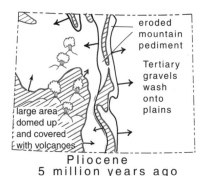

Pliocene
5 million years ago

Erosion, renewed by regional uplift between 28 and 5 million years ago, smoothed the Tertiary pediment, sweeping both old and new debris onto the Great Plains and into western basins. Volcanism left its mark in southwestern Colorado.

III
High Plains and Piedmont

The High Plains, appearing perfectly flat from a distance, are really a broad, gently rolling upland creased by streams and rivers. The plains rise westward from around 4,000 feet at the border with Kansas and Nebraska to over 5,000 feet at their western rim. Streambeds are dry most of the year because the plains are surfaced with porous, gravelly soil that readily absorbs rain and snowmelt.

The High Plains region of Colorado underwent little or no major faulting or folding through all of Paleozoic, Mesozoic, and Cenozoic time. The only changes came from distant tectonic events that warped the continent and elevated it or lowered it relative to sea level.

Beneath the sandy, pebbly soil of the High Plains is the Ogallala formation, composed of sand, gravel, and clay washed off the Rocky Mountains between 7 and 3 million years ago in late Tertiary time. Poorly compacted and only loosely cemented, this formation is an important aquifer that supplies hundreds of eastern Colorado water wells, most notably those that feed extensive sprinkler systems now prevalent here.

In places the Ogallala formation is as much as 700 feet thick. Below it are older sedimentary rocks, including several thousand feet of thick gray Pierre shale, deposited 80 to 70 million years ago on the muddy bottom of a shallow Cretaceous sea.

In Jurassic time the western edge of North America had collided with an arc of continental crust on the Pacific plate, initiating an eastward-moving wave of mountain building. The effects of the collision reached Colorado about 72 million years ago, fairly late in Cretaceous time, bringing about the event we now call the Laramide Orogeny. Land began to rise and the sea drained away from Colorado; the Rocky Mountains were born. And in the midst of the upheaval, the Mesozoic era came to its catastrophic end. Let's look at these events in order, starting with the Cretaceous sea.

Cretaceous Sea

Fine gray muds of the Pierre formation were deposited in quiet ocean waters that stretched across the continent from the Arctic to the Gulf of Mexico and far to the west of the present Rocky Mountains. Some of the sea's abundant marine life was preserved as fossils in the Pierre shale.

The sea's retreat—gradual and fluctuating—left the thick muds of the Pierre shale blanketed with layers of shoreline deposits that formed on a coastal plain quite like the broad, almost featureless slope of the Gulf Coast today. Beach and bar sand, lagoon mud, and plant-filled swamp deposits would eventually become interbedded sandstone, shale, and coal.

Rising Mountains and Sagging Basins

As the mountains continued to rise, land east of each range sagged a little as if to counterbalance the mountain uplift. As the mountains grew higher and the sags grew deeper, stream gradients steepened, so streams and rivers became increasingly powerful. Carving into the rising mountains, they dumped heavy loads of sand, silt, and gravel into the sagging basins.

The Denver Basin, a particularly deep sag east of the Front Range, does not show up well on the surface; it is filled with thousands of feet of marine and continental sediments. The basin extends south to Pueblo, north into Wyoming, and eastward nearly to Nebraska. At its deepest, near Denver, its layered sediments are 13,000 feet thick. It is an important petroleum province, with many oil fields. Oil and gas occur in Cretaceous rocks, where they fill spaces between sandstone grains, as well as joints and fissures in the Pierre shale and interlayered beds of limestone. On the western flank of the basin, oil was trapped in folds near the mountain front.

Two types of erosion and deposition characterize mountain uplifts. Gravel deposits may continue up onto edges of an eroded uplift, or pediments may be carved into an uplifted block. Pediments are commonly coated with thin sheets of gravel or sand.

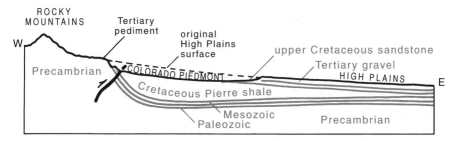

Diagrammatic section across the High Plains and Colorado Piedmont.

The worldwide extinction that marks the end of the Mesozoic Era left its record in the rocks of the sagging basins. In oil well cores a sudden absence of many kinds of microscopic pollen coincides exactly with an iridium-rich layer of clay known to mark the great extinction, apparently fallout from a meteor that plowed into Earth 65 million years ago.

Mountain building continued until around 40 million years ago, well into Eocene time, as did erosion. By 30 million years ago the mountains were half drowned in their own debris, which spread like a broad, sloping apron far into eastern Colorado, Nebraska, and Kansas. The apron was continuous with its erosional counterpart, a bedrock surface cut into the mountains themselves—the Tertiary pediment.

Mid-Tertiary Regional Uplift

Around 28 million years ago, after a long period of relative stability, the entire region between mid-Kansas and the deserts of Nevada began to rise. Volcanoes erupted in the area from the Rockies westward. Geologists have yet to find the cause of the broad uplift, often referred to as mid-Tertiary (or Miocene-Pliocene) regional uplift. Possibly the Pacific plate, as it was overridden by the North American plate, failed to plunge downward into the mantle and remained fairly shallow, resulting in two layers of crust on top of one another. Or perhaps the North American plate overrode a mid-ocean ridge, whose underlying convection currents might account for the uplift and for the fact that the crust in this region is warmer than average.

The uplift again jump-started erosion. Mountain streams cleaned and scoured the mountain slopes, eventually revealing the Tertiary pediment. Streams flowing eastward from the Rockies spread sand, gravel, and clay of the Ogallala formation across the High Plains. Many streams on the High Plains still flow east today.

Close to the mountains, however, downcutting eventually reached the line of upturned sedimentary rocks at the mountain edge. Controlled by the pattern of hard and soft rock layers, streams abandoned their old eastward paths and turned northward or southward, cutting into the soft shale at the base of the mountains. The South Platte River flowed north to its junction with the Cache la Poudre River; farther south, Cherry Creek and Fountain Creek abandoned their eastward paths as well. Together, these streams and their tributaries stripped away the High Plains veneer, cutting into early Tertiary and Cretaceous rocks to form the Colorado Piedmont.

Interstate 25
Wyoming—Denver
85 miles (137 km)

Just before entering Colorado, I-25 begins to descend from the escarpment that edges the High Plains, dropping gradually into the Colorado Piedmont. For several miles it crosses Tertiary sediments, stair-stepping down through a layer of gravelly sandstone to a lower layer of very fine-grained white to tan sandstone. This rock erodes easily into badlands or rock castles like those at the rest area at milepost 296. Northeast of here in the White River Badlands of Colorado, South Dakota, and Nebraska, many fossil vertebrate skeletons have been discovered in these sediments: saber-toothed tigers, rhinoceroses, camels, little three-toed horses, giant pigs, turtles, and strange rhinoceros-like titanotheres.

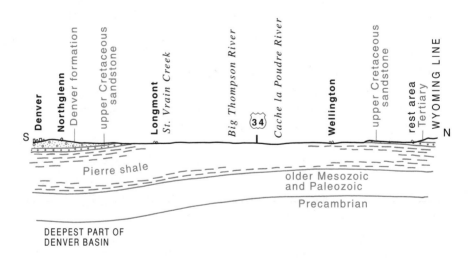

Section along I-25 between Wyoming and Denver.

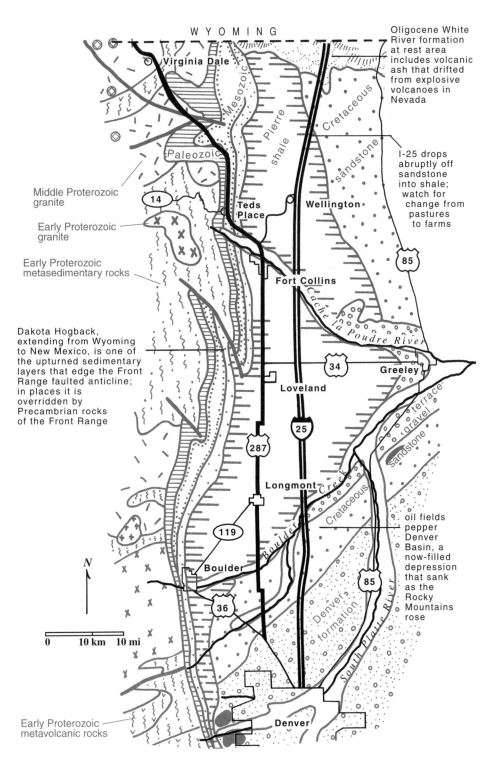

WYOMING

Virginia Dale

Oligocene White River formation at rest area includes volcanic ash that drifted from explosive volcanoes in Nevada

Mesozoic

Pierre shale

Cretaceous sandstone

Paleozoic

I-25 drops abruptly off sandstone into shale; watch for change from pastures to farms

Middle Proterozoic granite

14

Teds Place

Wellington

85

Early Proterozoic granite

Early Proterozoic metasedimentary rocks

Fort Collins

Cache la Poudre River

Dakota Hogback, extending from Wyoming to New Mexico, is one of the upturned sedimentary layers that edge the Front Range faulted anticline; in places it is overridden by Precambrian rocks of the Front Range

34

Greeley

Loveland

25

terrace gravel

sandstone

287

Longmont

Boulder Creek

Cretaceous

oil fields pepper Denver Basin, a now-filled depression that sank as the Rocky Mountains rose

N

119

85

Boulder

Denver formation

South Platte River

0   10 km   10 mi

36

Early Proterozoic metavolcanic rocks

Denver

Geology along I-25 between Wyoming and Denver.

At the edge of the High Plains, erosion has carved rugged sandstone figures in poorly consolidated Tertiary sedimentary rocks. —Jack Rathbone photo

Looking north from 8 or 10 miles south of the state border, you can see the High Plains escarpment clearly. The High Plains surface slopes steadily eastward in Wyoming in a long ramplike slope known as the Gangplank—only the uppermost peaks of nearby mountains stand above it. In 1869 the Gangplank was chosen as the route of the nation's first transcontinental railroad—a relatively easy way across the Rockies.

In Colorado there is no Gangplank: the broad valley of the Colorado Piedmont comes between the High Plains and the mountains. But the plains extended clear to the mountains here, too, until about 28 million years ago when regional uplift invigorated streams emerging from the mountains, enabling them to excavate the valley of the Colorado Piedmont. Three lines of evidence support this history:

- High Plains rock layers slope eastward as if they were outwash from the mountains.

- Stream courses on the High Plains flow eastward also, as if they originally continued on downslope from the mountains.

- An erosion surface on the mountains, the Tertiary pediment, slopes eastward as well, and if continuous eastward would merge with the High Plains. Ranging from 8,000 to 10,000 feet in elevation, this surface is not at all a perfect plain, but it is nevertheless easily discerned.

Near milepost 293, Interstate 25 leaves the Tertiary sediments of the High Plains and drops into the Colorado Piedmont. For about 12 miles it crosses a level, natural step of late Cretaceous sandstone, with a long fingerlike promontory of Tertiary sediments paralleling the eastern side of the highway. Sandstone under and near the highway was once beach and bar sand edging the retreating Cretaceous sea. Notice that it forms rather poor soil, suitable only for grazing.

At milepost 281 the highway descends to the top of the Pierre shale. Land use changes abruptly from grazing on the sandstone to farming on the shale. This change is so consistent it can be used in the piedmont area to map the formations, which are rarely exposed in outcrop but strongly influence soil chemistry and texture.

The route to Denver parallels the Front Range, whose northern end can be seen clearly from I-25. Like many other ranges in the Colorado Rockies, the Front Range is a huge upward-faulted block of ancient Precambrian granite and metamorphic rocks, longer in a north-south direction than it is wide. Typically, Mesozoic and Paleozoic sedimentary rocks are bent up along its edges—all that remains of the sedimentary blanket that extended westward before the mountains rose.

West of Fort Collins and Loveland lies one of the highest and most spectacular parts of the Front Range, Rocky Mountain National Park. The Front Range includes many 13,000-foot summits and four Fourteeners, among them 14,255-foot Longs Peak, easily recognized by its flat top. The interstate's elevation is about 5,000 feet, so the total topographic relief here is over 9,000 feet. However, geological or structural relief is much greater: the surface of Precambrian rocks at the summit of Longs Peak is 22,000 feet higher than the surface of Precambrian rocks in the deepest part of the Denver Basin, an upward displacement of more than 4 miles!

South of milepost 250, look westward to a jagged row of peaks south of Rocky Mountain National Park. Ogallala, Paiute, Pawnee, Shoshone, Arapaho—they are named after Native American tribes and are known collectively as Indian Peaks. The line of their summits marks the Continental Divide, the crest separating Pacific and Atlantic drainages.

The hilly yet basically level surface below Indian Peaks and about halfway down the mountain slope is the Tertiary pediment, the erosion surface once continuous with the High Plains.

Near milepost 225, the interstate leaves the Tertiary and Cretaceous sandstone and shale on which it has been traveling and climbs slightly onto the Denver formation, coarse gravelly Tertiary sandstone and conglomerate that fill the Denver Basin. The interstate remains on this formation and its approximate equivalent, the Dawson formation, for much of the way to Colorado Springs.

Curving west through Denver, I-25 follows the South Platte River. Now peppered with factories, roads, storehouses, railroad yards, and stadiums,

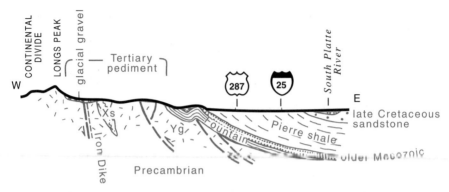

An east-west section from Longs Peak into the Colorado Piedmont shows that the total structural relief between Longs Peak and the piedmont is far greater than the topographic relief.

the river's floodplain was deluged by rushing, tumbling waters in June 1965. The flood crested rapidly and destroyed almost everything in its path. Debris piled up behind bridges, and bridge after bridge collapsed or washed out, effectively cutting Denver in two. The flood came from the south, where the river and its tributaries drain areas that received 14 inches of rain in one afternoon. Catchment dams now reduce the danger of such sudden calamities.

Rhinoceros-like titanotheres and other strange mammals inhabited northeastern Colorado in Oligocene time. Their fossilized skeletons are exhibited at the Denver Museum of Nature and Science. —Photo courtesy of Denver Museum of Nature and Science

Interstate 25
Denver—Colorado Springs
65 miles (105 km)

Between Denver and Colorado Springs, I-25 crosses Tertiary rocks that fill the center of the Denver Basin, mostly light-colored tan, white, and yellowish sandstone and conglomerate of the Denver and Dawson formations. These rocks often appear black in outcrops because black lichens thrive on them. Between some layers are thin volcanic flows and ash beds, but these are hard to spot at interstate speeds.

West of I-25, the Front Range rises like a gigantic ocean wave to crest above 14,000 feet at Mt. Evans, Mt. Bierstadt, and Pikes Peak. Summits between these high points are hidden from I-25 by the lower Rampart Range, in the foreground.

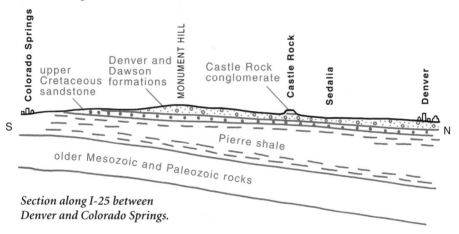

Section along I-25 between Denver and Colorado Springs.

The castle-shaped hill for which the town of Castle Rock is named wears a cap of hard Castle Rock conglomerate, somewhat younger than the Denver and Dawson formations and a good deal more resistant to erosion. This coarse, pebbly rock caps similar mesas and buttes farther south. Its sand and pebbles, derived from nearby mountains, may be a remnant of a huge alluvial fan formed by the ancestral South Platte River, which emerged from the mountains north of Pikes Peak.

As they reach a sharp decrease in slope, mountain streams deposit most of their sandy, rocky load in alluvial fans.

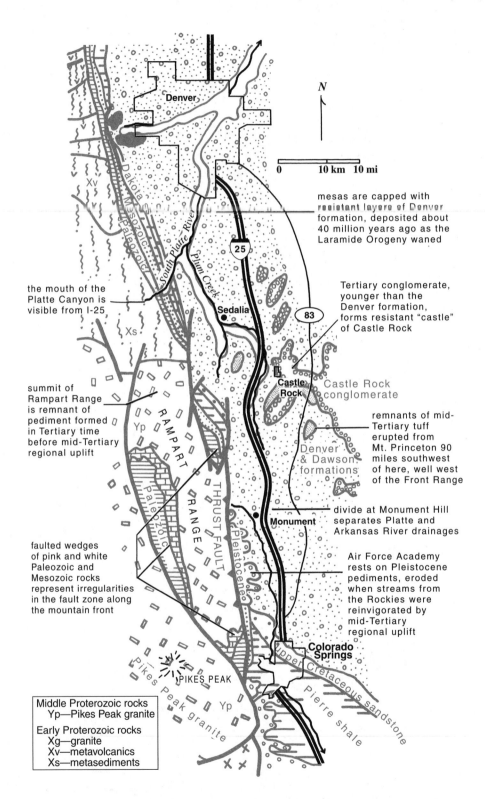

Geology along I-25 between Denver and Colorado Springs.

Castle Rock Butte is the type section of the Castle Rock conglomerate, the location for which the rock unit was named. —Halka Chronic photo

About 4 miles south of Castle Rock there is an excellent view of the Rampart Range. Evidence that this range is a faulted anticline can be seen between I-25 and the mountains, where isolated pink rocks of the Fountain formation stand out among the pines. They are late Paleozoic sedimentary rocks similar to those in Red Rocks Park near Denver and the Garden of the Gods at Colorado Springs, made of fine sand, mud, and gravel eroded from the Ancestral Rocky Mountains. Bent into an almost vertical position as the Front Range rose, these sedimentary rocks probably once extended much higher up the slope. On either side of them Tertiary sediments run right up to a fault at the edge of the Precambrian Pikes Peak granite that makes up the core of the mountains.

At Monument Hill, the high point between Denver and Colorado Springs, the interstate crosses the divide between Arkansas River and Platte River drainages. A prominent white tower standing out from the mountains, an erosional left-over of the same Tertiary sandstone that is exposed in roadcuts nearby, gives the town of Monument its name. Unlike the pink rocks farther north, which were tipped up as the mountains rose, these white rocks are nearly horizontal, showing they were deposited *after* rather than *before* uplift of the Rockies.

The Rampart Range, a lower block of the Front Range, stretches southward to Colorado Springs. Faults separate it from the Front Range and from the foothills. Its almost horizontal upper surface is a remnant of the Tertiary pediment eroded in the Pikes Peak granite. Much younger pediments

Carved by erosion, Tertiary sandstone giants line up at the edge of the forest near Monument. A layer of hard brown sandstone, resistant to erosion, forms their caps. —N. H. Darton photo, courtesy of U.S. Geological Survey

form the site of the U.S. Air Force Academy. Like their older counterpart they testify to long stable periods in the history of these mountains.

North of Colorado Springs and east of the Air Force Academy, the Western Museum of Mining and Industry presents an overview of Colorado's mining industry.

Scars on the mountain south of the Air Force Academy are quarries in hard Paleozoic limestone used for concrete aggregate and road material.

Colorado Springs originally nestled in the valley at the confluence of Monument Creek and Fountain Creek, protected by hills of Cretaceous and Tertiary sandstone to the east. Older parts of the city, near these streams, lie on Pierre shale, but a growing population has overflowed westward onto pediment gravels and into the mountains, as well as eastward over the sandstone hills.

Interstate 25
Colorado Springs—Walsenburg
96 miles (154 km)

Most of Colorado Springs, particularly the older part of town, is built on Pierre shale, Cretaceous rock that underlies the valley of Fountain Creek between the Rampart Range and whitish, cliffy sandstone hills east of town. Between Colorado Springs and Pueblo, I-25 stays close to the contact between the marine shale and the edge of Fountain Creek's floodplain. Pierre shale normally weathers into soil-covered slopes and is poorly exposed. In a number of roadcuts and in a few gullies and stream-cut banks, the thin-layered gray shale can be seen in greater detail. Running through it are pale yellow bands of bentonite, a fine, clayey substance formed from volcanic ash later modified by chemical changes. The bentonite tells us volcanoes erupted somewhere windward in Cretaceous time when the shale was being deposited.

To the west, all along the foot of Cheyenne Mountain, the Pierre shale bumps right into Pikes Peak granite and other Precambrian rocks, for a major fault separates the Precambrian mountain mass and the sedimentary rocks of the Colorado Piedmont. The granite has pushed nearly a mile eastward here, *over* some of the sedimentary rocks. This is a case where older rocks lie above, rather than below, younger rocks. The Front Range seemingly ends south of Cheyenne Mountain, but a gentle arch in sedimentary rocks continues southward for many miles, reflecting the range's uplift.

Fountain Creek, east of I-25, flows south toward the Arkansas River. In sight across the valley from mileposts 124 to 108, small conical bumps on the eastern skyline, known as tepee buttes, are mounds of relatively resistant limestone in the Pierre shale. Up to 50 feet high, they originally developed as reeflike mounds near springs on the floor of the Cretaceous sea, which

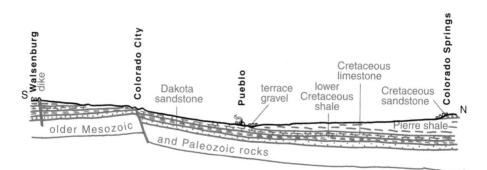

Section along I-25 between Colorado Springs and Walsenburg.

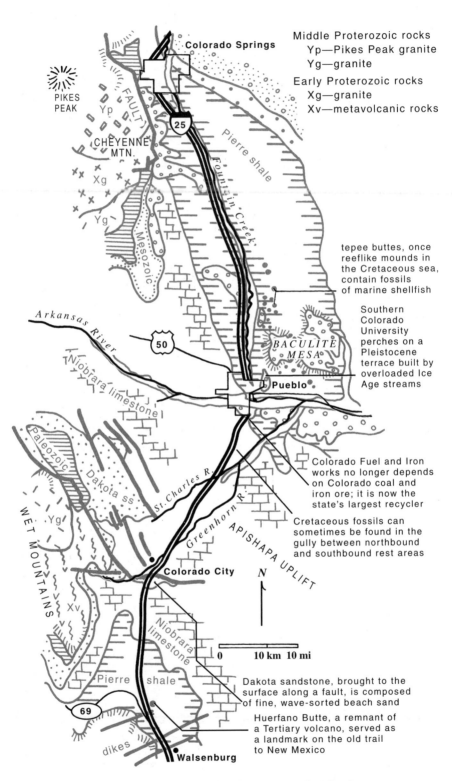

Colorado Springs

Middle Proterozoic rocks
Yp—Pikes Peak granite
Yg—granite

Early Proterozoic rocks
Xg—granite
Xv—metavolcanic rocks

PIKES PEAK

FAULT

Yp

CHEYENNE MTN.

Xg

Yg

Mesozoic

Pierre shale

Fountain Creek

tepee buttes, once reeflike mounds in the Cretaceous sea, contain fossils of marine shellfish

Arkansas River

50

Niobrara limestone

BACULITE MESA

Pueblo

Southern Colorado University perches on a Pleistocene terrace built by overloaded Ice Age streams

Paleozoic

Dakota ss

St. Charles R.

Colorado Fuel and Iron works no longer depends on Colorado coal and iron ore; it is now the state's largest recycler

WET MOUNTAINS

Yg

Greenhorn R.

APISHAPA UPLIFT

Cretaceous fossils can sometimes be found in the gully between northbound and southbound rest areas

Xv

Colorado City

N

0   10 km 10 mi

Niobrara limestone

Dakota sandstone, brought to the surface along a fault, is composed of fine, wave-sorted beach sand

Pierre shale

69

Huerfano Butte, a remnant of a Tertiary volcano, served as a landmark on the old trail to New Mexico

dikes

Walsenburg

Geology along I-25 between Colorado Springs and Walsenburg.

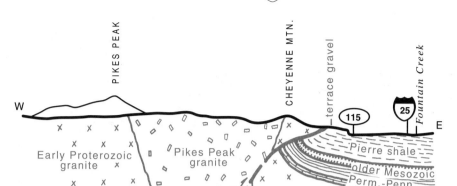

East-west section across Cheyenne Mountain shows Precambrian rocks thrust over younger sedimentary rocks. Half-arrows indicate direction of movement along the thrust fault.

was never very deep in the interior of the continent. The buttes contain marine fossils, notably little round clams called *Lucina*.

About 20 miles south of Colorado Springs, watch to the west for glimpses of the Sangre de Cristo Range 60 miles away behind the Wet Mountains. One of the youngest ranges in Colorado, the Sangre de Cristos extend in an unbroken rampart from Salida, Colorado, to Santa Fe, New Mexico, a distance of 235 miles. Their western side is sharply faulted and very steep, with a main fault that is much younger and more active than most others in this region. A branch of the fault along their western slope has moved within the last few hundred years—just yesterday by the geologic clock.

The large mesa to the east between mileposts 108 and 102 is called Baculite Mesa because on its flanks of Pierre shale are fossils of the straight-shelled Cretaceous ammonite *Baculites*. In this case a geographic feature has been given a fossil's name, rather than the other way around.

In the last few miles before its confluence with the Arkansas River, Fountain Creek cuts more deeply into the Pierre shale, which is rising as it approaches the southern end of the Denver Basin. The Pierre shale thins and almost disappears south of Pueblo.

Like many towns along the eastern edge of the Rockies, Pueblo has seen its share of flash floods. In early June 1921 heavy rains burst a dam and roared through town in waves 12 feet high. Levees designed to control flooding proved inadequate, and more than half the city was destroyed. Larger levees built in the 1950s and 1980s on Fountain Creek and the Arkansas River now protect the city.

As I-25 leaves Pueblo it passes the towering CF&I smelters and huge man-made mesas of slag, a lavalike waste product of iron smelting. The

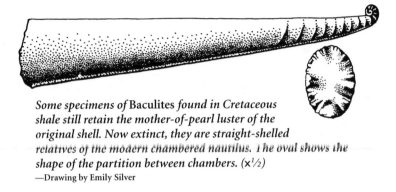

Some specimens of Baculites *found in Cretaceous shale still retain the mother-of-pearl luster of the original shell. Now extinct, they are straight-shelled relatives of the modern chambered nautilus. The oval shows the shape of the partition between chambers. (x½)*
—Drawing by Emily Silver

smelters were located here in the 1880s because abundant coal and lime were available nearby in Cretaceous rocks, water could be obtained from streams draining the Wet Mountains, and iron ore could be brought in by railroad from mines in the Sangre de Cristo and Mosquito Ranges. CF&I no longer uses coal-fired smelters or needs mines to supply it with raw materials. Instead it has become the state's largest recycler, using scrap metal to make railroad rails, pipe, rebar, wire, fencing, nails, and other products. The smelter is now fueled with natural gas, a much cleaner-burning fuel than coal.

Between Pueblo and Walsenburg, I-25 follows the old Taos Trail across a surface of Niobrara limestone, older Cretaceous rock that underlies the Pierre shale. The limestone rises southward onto a broad, gentle anticline that rims the southern end of the Denver Basin. Sloping limestone layers are well exposed on cuestas east of milepost 88. Derived from the Spanish word meaning "hills," cuestas are gently sloping mesas. In roadcuts and streambanks you can see that the impure, light gray limestone is interbedded with darker layers of shale. Fossil clams (*Inoceramus*) and oysters (*Ostrea*) are common in it. A good place to look for them is in the deep gully between the northbound and southbound rest areas at mileposts 82 and 81. Shark teeth have been found in a sandstone layer in the overlying Pierre shale just west of here. South of milepost 75 the limestone cuestas slope in the opposite direction, having arched across the crest of the anticline.

South of milepost 73 the oldest of Colorado's Cretaceous rocks comes to the surface—the Dakota sandstone, formed from beach sand left by the sea as its shoreline receded eastward across Colorado about 100 million years ago. Significantly harder than shales above and below it, it forms the famous Dakota Hogback, a ridge that can be traced along the edge of the mountains from Wyoming to New Mexico. Notice its tawny color and

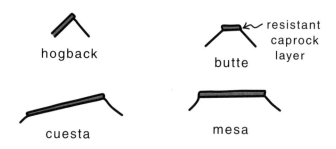

angular way of breaking. Easily recognized, it is Colorado's most widespread rock unit. Dinosaur footprints mark it in a few scattered places.

The Wet Mountains, directly west of the highway, are another faulted anticline. They are cored with 1.8- to 1.7-billion-year-old Precambrian metamorphic rocks, intruded by 1.4-billion-year-old granite. On their northeastern flank, large blocks of Paleozoic and Mesozoic rocks are faulted against the Precambrian rocks. At the south end of the range, upturned Paleozoic and Mesozoic rocks describe an arrowhead-shaped swath around the mountain tip.

The Huerfano Valley, southwest of the Wet Mountains, is floored with Tertiary sedimentary rocks. Many fossil mammals, including almost perfect skeletons of *Eohippus*, a tiny four-toed ancestor of the horse, have been found here.

Spanish Peaks, two large Tertiary intrusions, are surrounded by numerous radiating dikes, one of which shows clearly in this photograph. —Jack Rathbone photo

The twin domes of Spanish Peaks jut up to the south, huge masses of igneous rock whose magma pushed and melted its way upward in Tertiary time, possibly never reaching the surface. They are probably much reduced from their former height. Molten rock filling fissures and cracks that opened around the rising masses hardened into prominent radiating dikes from 1 to 100 feet wide. More resistant to erosion than surrounding sedimentary rocks, many of the dikes stand as vertical walls. One of the northernmost dikes crosses I-25 just south of milepost 56. Another, though not exposed along I-25, crosses the I-25 business loop just before it enters Walsenburg and the Cucharas River valley.

The lonely cone-shaped hill east of milepost 59 is a small volcanic neck, the last remnant of an early Tertiary basalt volcano. Named Huerfano, the "Orphan," by some poetic Spaniard, it marks the old trail from Denver to Taos, New Mexico. —Jack Rathbone photo

Interstate 25
Walsenburg—New Mexico
51 miles (82 km)

Between Walsenburg and Trinidad, I-25 parallels a high, juniper-covered escarpment of Cretaceous sandstone. As far as Trinidad the interstate is on gray Pierre shale, which is also Cretaceous but is under and therefore older than the sandstone. The shale and the light-colored, blocky sandstone intertongue to some extent, a situation not uncommon geologically. In doing so, they show that the shoreline moved back and forth with changes in sea level. Dark gray marine shales were deposited when the sea covered the area, and beach sand and sandy coastal bars were deposited as it receded. With each new retreat of the sea—each probably representing many thousands of years—more sandstone accumulated.

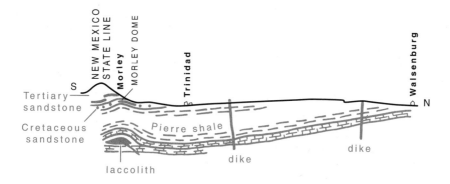

Section along I-25 from Walsenburg to New Mexico.

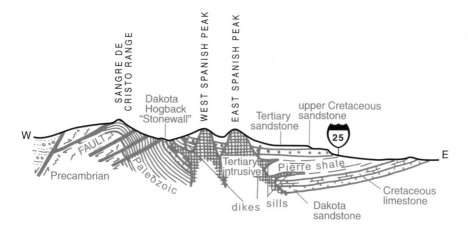

Section across I-25 near Aguilar.

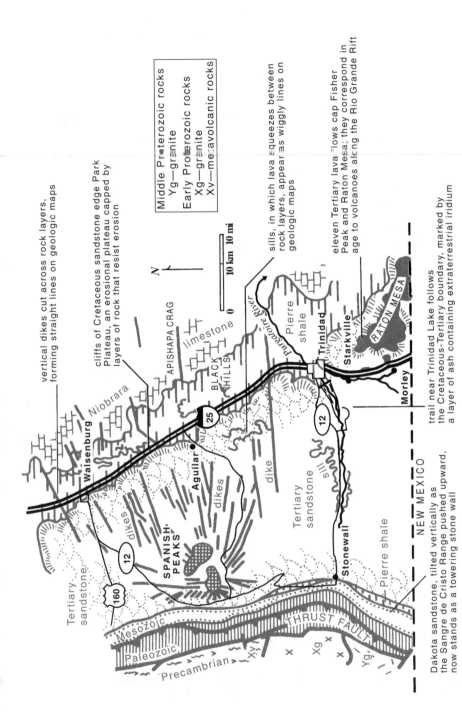

vertical dikes cut across rock layers, forming straight lines on geologic maps

cliffs of Cretaceous sandstone edge Park Plateau, an erosional plateau capped by layers of rock that resist erosion

Middle Proterozoic rocks
Yg—granite
Early Proterozoic rocks
Xg—granite
Xv—metavolcanic rocks

sills, in which lava squeezes between rock layers, appears as wiggly lines on geologic maps

eleven Tertiary lava flows cap Fisher Peak and Raton Mesa; they correspond in age to volcanoes along the Rio Grande Rift

N

0   10 km 10 mi

APISHAPA CRAG

limestone

Niobrara

BLACK HILLS

Purgatoire River

Pierre shale

Walsenburg

Trinidad

Starkville

RATON MESA

Morley

25

Aguilar

12

dikes

dike

Tertiary sandstone

sill

SPANISH PEAKS

dikes

12

160

Tertiary sandstone

Stonewall

Pierre shale

Mesozoic

Paleozoic

Precambrian

Xv

Xg

Yg

THRUST FAULT

NEW MEXICO

trail near Trinidad Lake follows the Cretaceous-Tertiary boundary, marked by a layer of ash containing extraterrestrial iridium

Dakota sandstone, tilted vertically as the Sangre de Cristo Range pushed upward, now stands as a towering stone wall

Geology along I-25 between Walsenburg and New Mexico.

Spanish Peaks, two large domes with centers of igneous rock, developed after the faulted anticlines of the Colorado Rockies had formed. Most likely, they are related to Tertiary rifting on the western side of the Sangre de Cristo Range. Such medium-size intrusions of igneous rock are called stocks. The molten rock, under tremendous pressure, forced its way through Tertiary sedimentary layers, doming them upward and baking a 900-foot-wide zone around each intrusion. Colorado 12, a scenic route over Cucharas Pass west of the Spanish Peaks, crosses part of the Spanish Peaks intrusions and passes over or tunnels through many dikes.

Between Walsenburg and Trinidad, I-25 crosses some of the many dikes that radiate from Spanish Peaks. The wall-like sheets of rock formed as magma penetrated vertical cracks and cooled. Some of the dikes are light colored, others are dark, as the composition of the magma varied. A particularly large dike, known locally as Apishapa Crag, forms the backbone of a prominent ridge east of milepost 30. South of it are the Black Hills, composed of Pierre shale but protected and made resistant by the hard rock of several igneous sills. Sills form in much the same way as dikes, except that instead of intruding vertical cracks, the magma forces its way

The Dakota sandstone, steeply tilted by the rising Sangre de Cristo Range, forms a stunning vertical rampart near the town of Stonewall. Colorado 12 slips through a gap in this formidable wall.
—W. T. Lee photo, courtesy of U.S. Geological Survey

between layers of sedimentary rocks, which here are horizontal or nearly so. These sills, like some of the dikes, are composed of andesite and are connected to the intrusive masses of Spanish Peaks.

Behind the cliffs that border the interstate is a broad plateau composed of Cretaceous and Tertiary sandstone and shale deposited in beaches, bars, and coastal swamps during and after retreat of the Cretaceous sea. The abundant fallen vegetation of the swamps now forms coal seams, of which there are many here and farther south. The town of Trinidad owes its existence to coal deposits in Cretaceous and Tertiary sandstone.

Deposition of sediment in this area was continuous from Cretaceous into Tertiary time. Geologists can put their fingers right on the boundary between the two time periods here. Where strata that similarly span this boundary are preserved, its precise position may be marked by a thin ash layer that coincides with the extinction of many Cretaceous life forms—including dinosaurs. The ash is unusual because it contains quite a bit of iridium, an element very rare on Earth, but common in some meteorites. The iridium-enriched layer is the clue that a very large meteorite struck Earth, its impact leading to the extinctions.

Near Trinidad Lake the clay layer at the Cretaceous-Tertiary boundary shows up as a scant inch of white claystone overlain by an inch of coal. Note pen for scale. —Charles Pillmore photo

The probable meteor impact site is a subsurface crater 110 miles across, lying partly in the Gulf of Mexico, partly on the Yucatán Peninsula. Geologists discovered that this crater, sketchily outlined on the surface by sinkholes and small lakes, lies right at the boundary between Cretaceous and Tertiary time and that the iridium-enriched clay layer thickens close to it. To create a 110-mile-wide crater, the meteor must have been about 6 miles in diameter.

A few outcrops of the clay layer along I-25 are difficult to get to. The easiest place to see it up close is west of Trinidad, near the southern shore of Trinidad Lake. A sample of the layer is displayed at the museum at Trinidad State Junior College.

Coal-bearing rocks extend along the foot of the mountains for many miles in both Colorado and New Mexico. Several large mines lie west of Trinidad along Colorado 12; others are south of Trinidad near I-25. Mines in this area closed down in 1995 and 1996, for economic reasons rather than for lack of coal.

Sandstone and shale, coal seams, and several dikes are clearly exposed in hills, railroad cuts, and large highway cuts, particularly near Morley, only a ghost town now, with ruins of a little church above the coal dumps. Notice how the shale and sandstone layers thicken and thin. Coal seams are discontinuous, too, as you would expect in a shifting beach-sandbar-lagoon setting. Frustrating for coal miners!

Morley is at the center of Morley dome. Strata dip away from the dome, which is 450 feet high in terms of its geologic structure. A small igneous intrusion pushed between rock layers and lifted overlying strata, creating the dome. An oil well drilled on the crest of the dome penetrated about 450 feet of igneous rock below the sedimentary layers—a thickness just equal to the structural height. An igneous mass of this type—flat bottom, round top, and doming up overlying beds—is a laccolith.

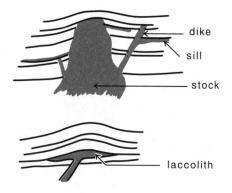

Stocks are medium-size intrusive masses that may or may not lift up overlying rocks. Laccoliths squeeze between rock layers like sills but are viscous enough to dome up overlying strata.

Fisher Peak and Raton Mesa rise dramatically northeast of Morley. Their layered Tertiary and Cretaceous sedimentary rock, the same sandstones and shales you have seen by the roadside, are protected by a many-tiered cap of late Tertiary basalt. Although you can't count all the volcanic flows from the highway, there are eleven of them, one on top of another. The lava came from small volcanic openings, or vents, near La Junta; remnants of the flows are scattered on mesa tops south and east of here.

The Sangre de Cristo Range, the long sharp-toothed line of peaks to the west, is part of a narrow, tilted fault block that extends from Salida, Colorado, to Sante Fe, New Mexico. Paleozoic sedimentary rocks tilted by the faulting reach clear to the top of this range, with Pennsylvanian strata forming many of the summits. Coarse red Permian sandstone and siltstone, formed from sand and gravel washed off the Ancestral Rockies, covers the eastern slope. The range is considerably younger than other ranges in Colorado; faults on its western side have displaced alluvial fans that formed within the past 10,000 years.

Raton Pass, at 7,834 feet, is a historic route into New Mexico, used by Native Americans, Spaniards, Frenchmen, and Union soldiers, and by ox-drawn wagon trains and swaying stagecoaches on the Bents Fort branch of the Santa Fe Trail. In 1878 a railroad was completed across this pass; a narrow highway built in 1922 still winds among the piñons and junipers of surrounding slopes. The pass gets its name from a local inhabitant, the furry-tailed pack rat.

Fisher Peak's multilayered volcanic cap protects underlying Tertiary and Cretaceous sandstone and shale. —Jack Rathbone photo

Interstate 70
Kansas—Limon
88 miles (142 km)

Between the Kansas border and Genoa, I-70 crosses the High Plains. They are not nearly as smooth as they seem at a distance; numerous small watercourses gully them, most of them dry except after rains.

This stretch of the highway rests on coarse Tertiary gravel of the Ogallala formation, which surfaces the central and southern High Plains. Also known as the Ogallala aquifer, the porous rock contains water—huge amounts of it—filling pore spaces between its sand grains and pebbles. Rainwater and snowmelt, as well as meltwater from Ice Age glaciers in the Rocky Mountains, have percolated down into the aquifer for many thousands of years, slowly flowing away from the mountains and out under the plains. The impermeable Pierre shale beneath the Ogallala formation prevents the water from sinking deeper.

West of the 100th meridian, the approximate dividing line between the moist mid-continental United States and the arid West, agricultural land benefits greatly from the Ogallala aquifer, which provides water for the big center-pivot sprinkler systems that circle many fields during the growing season. Each sprinkler system, pumping from several hundred feet below the surface, uses as much as 1,000 gallons a minute; sometimes one well supports two or three systems. However, this is "fossil water," most of it from Pleistocene snow and ice, and it is now pumped out twenty-five times as fast as it is replaced. Despite new pumping regulations and sprinkling methods established to lengthen the usefulness of the aquifer, many wells have already run dry. A return to dryland farming—a much more risky business—would reduce production to about a quarter of that made possible by sprinkler irrigation, and for dryland farmers, of course, there is the ever-present specter of drought. What if there were another twenty-four-year drought like that known to have taken place in the last years of the thirteenth century, or even a ten-year drought like the Dust Bowl of the 1930s? Can new technologies in agriculture and water conservation give this land a new lease on life?

The town of Genoa lies at the western rim of the High Plains, where the land drops abruptly away westward into the Colorado Piedmont. In clear weather the Front Range of the Rockies is visible, though still 80 miles away. At Genoa, I-70 winds down off the High Plains into the broad valley of Big Sandy Creek.

The High Plains escarpment is well defined at Cedar Point, north of this valley. A low cliff of Ogallala gravels, washed from the west since the Rocky Mountains rose, forms the prowlike point. Below it is the Pierre

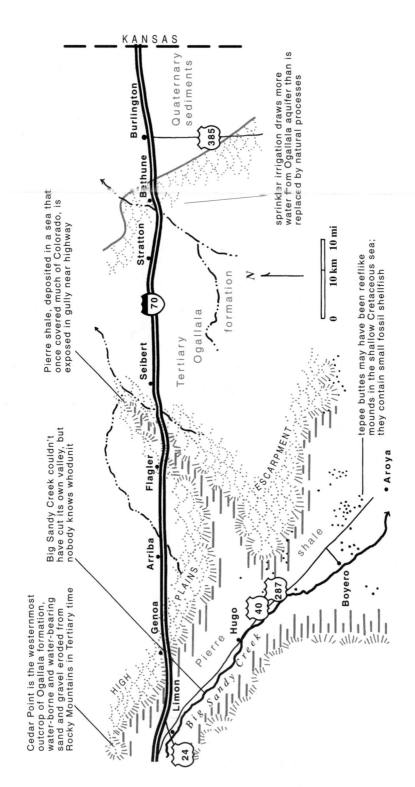

Geology along I-70 between Kansas and Limon.

Cedar Point is the westernmost outcrop of Ogallala formation, water-borne and water-bearing sand and gravel eroded from Rocky Mountains in Tertiary time

Big Sandy Creek couldn't have cut its own valley, but nobody knows whodunit

Pierre shale, deposited in a sea that once covered much of Colorado, is exposed in gully near highway

sprinkler irrigation draws more water from Ogallala aquifer than is replaced by natural processes

tepee buttes may have been reeflike mounds in the shallow Cretaceous sea; they contain small fossil shellfish

KANSAS

Burlington

Bethune

Stratton

Seibert

Flagler

Arriba

Genoa

Limon

Hugo

Boyero

Aroya

Quaternary sediments

Tertiary Ogallala formation

ESCARPMENT

Pierre shale

HIGH PLAINS

Big Sandy Creek

N

0   10 km   10 mi

Center-pivot sprinkler systems in eastern Colorado use groundwater pumped from the Ogallala aquifer. —Photo courtesy Butler Mfg. Co.

shale, deposited when the Cretaceous sea covered this area. Though the Pierre shale is easy to recognize by its muddy gray color and fine texture, it converts so readily to soil that good exposures are quite rare. Beautiful ammonite shells, with mother-of-pearl surfaces still shiny and lustrous, occur in this formation, as do large Cretaceous clams called *Inoceramus*. Skeletons and teeth from sharks and big swimming reptiles like mosasaurs and icthyosaurs are fairly common as well.

The wide valley of Big Sandy Creek is floored with river floodplain deposits: layered sand, mud, and gravel derived from both Tertiary and Cretaceous rocks. A few miles southeast of here, the valley widens out, and the Pierre shale is better exposed and spotted with small, conical tepee buttes, actually much flatter than tepees! These buttes were probably small patches of reef formed near springs on the floor of the shallow Cretaceous sea.

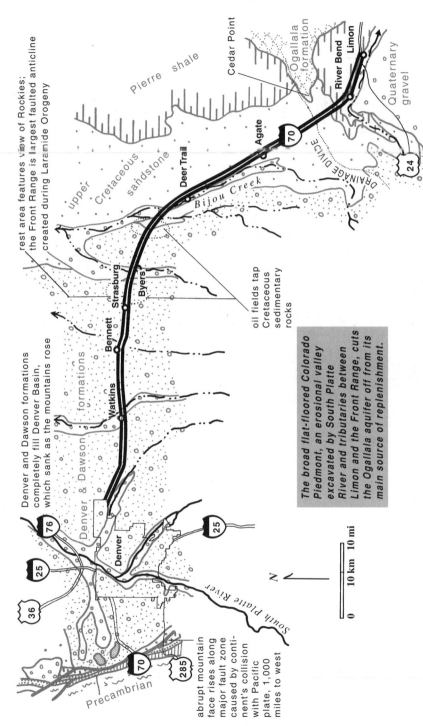

rest area features view of Rockies; the Front Range is largest faulted anticline created during Laramide Orogeny

Pierre shale

Cedar Point

Ogallala formation

River Bend

Limon

Quaternary gravel

upper Cretaceous sandstone

Deer Trail

Agate

70

24

DRAINAGE DIVIDE

Bijou Creek

Strasburg

Byers

oil fields tap Cretaceous sedimentary rocks

Denver and Dawson formations completely fill Denver Basin, which sank as the mountains rose

Bennett

Watkins

Denver & Dawson formations

76

Denver

25

25

36

70

285

abrupt mountain face rises along major fault zone caused by continent's collision with Pacific plate, 1,000 miles to west

South Platte River

Precambrian

N

0   10 km   10 mi

The broad flat-floored Colorado Piedmont, an erosional valley excavated by South Platte River and tributaries between Limon and the Front Range, cuts the Ogallala aquifer off from its main source of replenishment.

Geology along I-70 between Limon and Denver.

Interstate 70
Limon—Denver
89 miles (143 km)

Between Limon and Denver, I-70 crosses the Colorado Piedmont, travers-ing younger and younger rocks that fill the center of the Denver Basin. Just west of Limon, as the highway swings northwest toward River Bend, it crosses the divide between the Arkansas River drainage (Big Sandy Creek) and the drainage of the South Platte River (Bijou Creek). The South Platte itself is about 65 miles north of here. Most of its tributaries flow north, and most are sporadic or intermittent streams that run only in spring or after heavy rains.

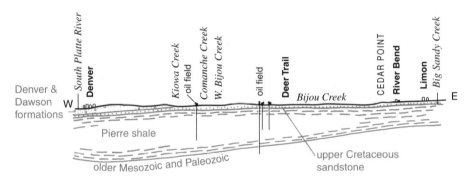

Section along I-70 from Limon to Denver.

As I-70 crosses the drainage divide it leaves the Pierre shale and passes onto higher, younger sandstone layers that were once beaches and sand-bars along the edge of the retreating Cretaceous sea. This sandstone forms poor soil and is used mostly for grazing land. Some of the sand-stone contains thin layers of coal, evidence of marshes and lagoons along the Cretaceous shore. Cedar Point, a prowlike prong of the High Plains, rises to the north, revealing the eroded western edge of the Ogallala for-mation of Tertiary age.

Between Deer Trail and Byers, I-70 reaches the Dawson formation, with sandstone and conglomerate layers that span the end of the Cretaceous period and the beginning of Tertiary time. When these rocks were depos-ited, the Denver Basin, bounded on the west by the rising Front Range, was already well defined; as it bowed down, sand and gravel of the Dawson and Denver formations kept filling it. Between mileposts 320 and 310, a few wells penetrate the Denver formation to tap oil in Cretaceous rocks be-neath it. More Denver Basin oil fields lie farther north.

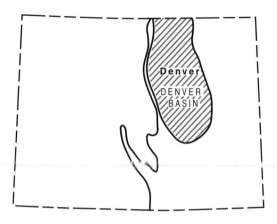

The Denver Basin is a geologic basin, with strata bowed down toward its center. It is filled with loosely consolidated debris from the Rockies and is not a topographic basin at present.

Much of the irrigation water in the piedmont area comes from shallow wells tapping Pleistocene gravel that in some places forms a veneer over older rocks. The gravel accumulated during glacial episodes in the mountains, when rivers were greatly overloaded with sand and rock dumped into them by melting ice.

Water rights are a major legal issue. The rights to surface water—water in streams and rivers—were established by law soon after the area was settled. Much later, when wells were drilled, farmers realized that pumping from wells often depleted surface flow. Such indirect interference with prior water rights has caused some knotty problems for Colorado courts, which often call on geologists to serve as expert witnesses.

If the weather is fine, you should be able to see the Rockies clearly. To pioneers coming from the east this rampart was the front of the mountains, so they called it the Front Range. To the northwest (straight ahead from milepost 288) Longs Peak rises to 14,255 feet in elevation, the highest point in Rocky Mountain National Park. Directly west beyond Denver, the highest peak is 14,264-foot Mt. Evans, with 14,060-foot Mt. Bierstadt close behind it. Seventy-five miles away to the southwest, you may be able to see the dome of 14,110-foot Pikes Peak. Ancient Precambrian rocks—gneiss, schist, and granite—form these peaks as well as the core of the entire Front Range.

You can also see the Tertiary pediment, an irregular horizontal surface about halfway up the mountains at about 9,000 to 10,000 feet elevation. Don't confuse it with timberline, the top of the dark evergreen forest, which is at about 11,000 feet and well highlighted by snow in winter.

Interstate 76
Julesburg—Fort Morgan
108 miles (173 km)

Interstate 76 enters Colorado on the floodplain of the South Platte River and closely follows the historic Overland Trail established in the 1800s. Low hills on either side are surfaced with Tertiary sedimentary rocks, loosely consolidated sand and gravel called the Ogallala formation or, because it carries abundant groundwater, the Ogallala aquifer. Soon after entering the corner of Colorado, I-76 rises onto these hills for 30 miles and then near milepost 134 passes almost imperceptibly into a very broad, shallow valley of older rocks, the Colorado Piedmont. In the distance to the north, tan bluffs of the High Plains escarpment become more and more distant from the highway; they roughly parallel the Nebraska boundary.

Between Crook and Hillrose, I-76 crosses an eroded surface of Cretaceous rocks, first late Cretaceous sandstone that formed as beaches and sandbars of the receding sea, then dark gray Pierre shale, frequently containing fossil shells of marine animals that lived on or sank to the muddy seafloor. Outcrops of these rocks are rare because this entire area is covered with sand hills, former dunes now stabilized by vegetation or in some recent "blowouts" by old automobile tires. The dunes developed during the Pleistocene Ice Age, when glacier-fed streams frequently overflowed their banks to cover their floodplains with sand and gravel, and strong winds sweeping down from mountain ice fields winnowed the sand and piled it into dunes. Some sand came from loosely cemented late Cretaceous sandstone as well. Patchy sand hills extend most of the way to Denver.

Just southwest of Hillrose at milepost 92, the interstate descends from the sand hills onto a broad river terrace 60 to 80 feet above the level of the South Platte River. This terrace, on which lie Brush and Fort Morgan, is composed of river-deposited gravel and sand. It is an old floodplain formed at about the same time as the dunes, during a long period when the river, heavily charged with rock debris, habitually overflowed its banks. In places several terrace levels are present, each representing a similar period of stabilized but nonetheless heavy river flow. These river-formed terraces can be correlated with others nearer to the mountains, and some of those in turn can be traced continuously into the mountains, where each terrace appears to end at a moraine left by a glacier. So we are fairly sure that terrace formation coincides with periods of colder, wetter climate, when glaciers occupied mountain valleys, torrential steams were overloaded with debris, and cold winds rushing down from the mountains picked up sand and piled it into dunes.

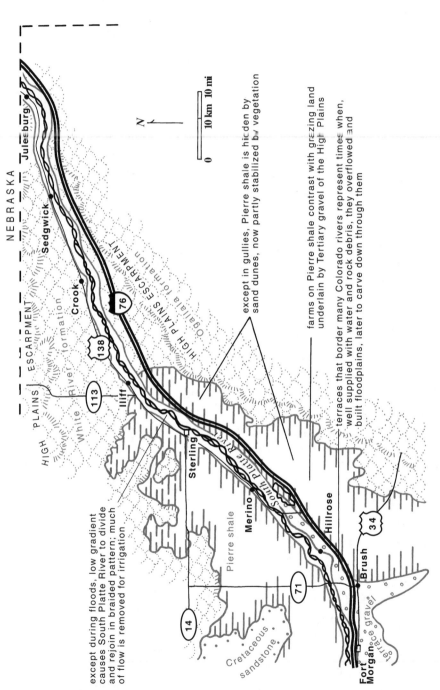

NEBRASKA

Julesburg

Sedgwick

Crook

76

138

113

Iliff

HIGH PLAINS   ESCARPMENT

White River formation

Ogallala formation

HIGH PLAINS ESCARPMENT

Sterling

South Platte River

Merino

Pierre shale

Hillrose

71

14

34

Brush

Cretaceous sandstone

Fort Morgan gravel terrace

Fort Morgan

N

0   10 km   10 mi

except in gullies, Pierre shale is hidden by
sand dunes, now partly stabilized by vegetation

farms on Pierre shale contrast with grazing land
underlain by Tertiary gravel of the High Plains

terraces that border many Colorado rivers represent times when,
well supplied with water and rock debris, they overflowed and
built floodplains, later to carve down through them

except during floods, low gradient
causes South Platte River to divide
and rejoin in braided pattern; much
of flow is removed for irrigation

Geology along I-76 between Julesburg and Fort Morgan.

Close to the mountains and the High Plains escarpment, a famous archeological locality known as the Lindenmeier Site is on one of these terraces. Artifacts found there tell us that sometime between 12,000 and 13,000 years ago, as the Pleistocene Ice Age waned, early North Americans camped on these terraces.

Interstate 76
Fort Morgan—Denver
84 miles (135 km)

West of Fort Morgan, I-76 travels along a South Platte River terrace for about 12 miles. It then leaves the river and climbs into sand-hill country. The sand hills formed during the Ice Age as wind from the mountains picked up fine sediment washed down by swollen glacial rivers and carried it eastward onto the plains, depositing it there as dunes.

For half of the distance between Fort Morgan and Denver, the sand hills cover Cretaceous rocks—Pierre shale deposited in a shallow sea and sedimentary rocks of beaches and bars that bordered the sea as it receded eastward. The Rocky Mountains had not yet begun to rise when the shale formed; it thickens westward in the direction of the Wasatch Range of Utah. Drilling data from oil wells tell us the Pierre shale is less than 3,000 feet thick near the Nebraska line, 7,000 feet thick in the deepest part of the Denver Basin, and 20,000 feet thick in North Park, west of the Front Range.

Both oil and gas have been produced from Cretaceous rocks of this area. You may notice a few oil wells along the highway. However, I-76 slips between the two areas of greatest production, south of Fort Morgan and north of Sterling.

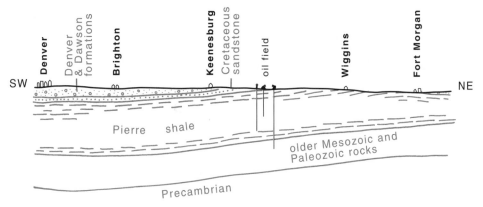

Section along I-76 from Fort Morgan to Denver.

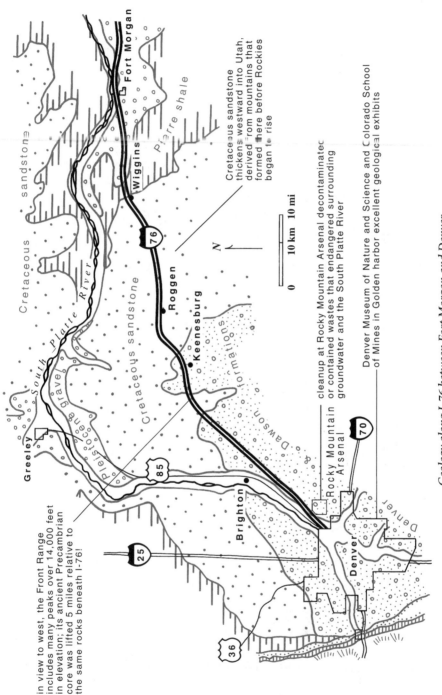

in view to west, the Front Range includes many peaks over 14,000 feet in elevation; its ancient Precambrian core was lifted 5 miles relative to the same rocks beneath I-76!

Cretaceous sandstone thickens westward into Utah, derived from mountains that formed there before Rockies began to rise

cleanup at Rocky Mountain Arsenal decontaminated or contained wastes that endangered surrounding groundwater and the South Platte River

Denver Museum of Nature and Science and Colorado School of Mines in Golden harbor excellent geological exhibits

N

0   10 km   10 mi

Geology along I-76 between Fort Morgan and Denver.

West of milepost 59 there are good views of the Front Range of the Colorado Rockies, depending of course on the weather and the time of day. Due west, the tallest peak is Longs Peak, elevation 14,255 feet. North of it are summits of the Mummy Range, usually snow-capped in spring and early summer and often with snow lasting year-round. South of Longs Peak the jagged ridge of Indian Peaks is profiled against the sky. A prominent, broadly pyramid-shaped peak west of Denver, just left of the highway's direction near milepost 51, is Mt. Evans, 14,264 feet above sea level. And far away to the south, seeming to stand east of the general mountain trend, is the rounded summit of Pikes Peak, 14,110 feet in elevation. All these peaks are built of ancient Precambrian rocks that make up the core of the Front Range.

Near milepost 44, I-76 passes onto the Denver formation. Here, tilled soils lose their sandiness and become clayey. The Denver formation, which is often concealed by soil or sand, contains late Cretaceous through early Tertiary sediments that washed into the Denver Basin from the newly risen Rocky Mountains. It is older than Tertiary rocks such as the Ogallala formation and other units that surface the High Plains to the east and north.

As you look at the mountains, see if you can distinguish the Tertiary pediment. It is at about 9,000 feet in elevation, halfway up the slope, a visible step on the front of the mountains. It is particularly evident northwest of Boulder and west of Denver.

Just northeast of Denver between mileposts 16 and 12, you will pass Rocky Mountain Arsenal. Now a national wildlife refuge, it was established in 1942 as a site for munitions and chemical production, testing, and disposal. DDT, mustard gas, napalm, nerve gas, rocket fuel, chlorinated benzenes, and other toxic chemicals were made, stored, and disposed of here. For forty years waste chemicals were routinely dumped into unlined evaporation ponds, where roughly 230 million gallons of solution sank into underlying sediments. Some of the most toxic wastes were pumped down a 12,045-foot deep injection well. In the 1960s so much waste was pumped down the well that it lubricated a deep fault below the Denver Basin, causing a series of minor earthquakes in the Denver area—their timing coinciding perfectly with fluctuations in the pumping.

Superfund cleanup of the area began in 1987. Roughly 10 percent of the surface area of the arsenal was contaminated, but beneath the arsenal, in underlying sediments, lay the largest problem. The Denver and Dawson formations are between 200 and 700 feet thick here. They are very porous, so groundwater fills and moves through them easily. By the 1970s toxic chemicals had begun to seep into well water north of the arsenal, contaminating much of the soil and ruining the farmland. Even the water in the South Platte River was at risk.

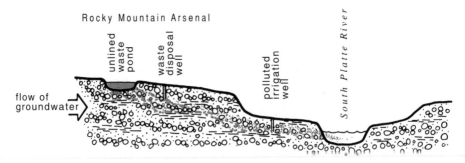

In porous, permeable rock like the Dawson and Denver formations, groundwater flows slowly through spaces between sediment grains. Contaminants from the Rocky Mountain Arsenal percolated into the groundwater and were carried slowly north, where they seeped into neighboring water wells and the South Platte River.

When a well is pumped for water, surrounding groundwater flows toward the well, so several rows of new wells were drilled on and north of arsenal land and used to pump out the contaminated water with its polluting chemicals. A new water system has been installed for areas where wells were contaminated. Chemicals that remained in unsealed ponds were transferred to sealed containers or incinerated. Even the dust, dangerous to breathe, was stabilized. The injection well was sealed off. Wastes still remain in deep subsurface rock layers because there is no known way of cleaning them up.

U.S. 24
Limon—Colorado Springs
71 miles (114 km)

Between Limon and Colorado Springs, U.S. 24 crosses from the High Plains to the Colorado Piedmont, paralleling the drainage divide that separates Platte River tributaries to the north from those of the Arkansas River to the south. Just west of Limon the highway climbs out of the valley of Big Sandy Creek, a tributary of the Arkansas, onto a corner of a huge isolated mesa capped with Pleistocene gravel.

The wide valley of Big Sandy Creek, with its diminutive trickle of water, presents us with a puzzle: How could such a stream, usually dry or nearly dry, scour out a valley 8 to 15 miles wide and nearly 100 feet deep? It is likely that for a time during uplift of the Rockies the valley was occupied, deepened, and widened by a larger, more powerful river—probably one

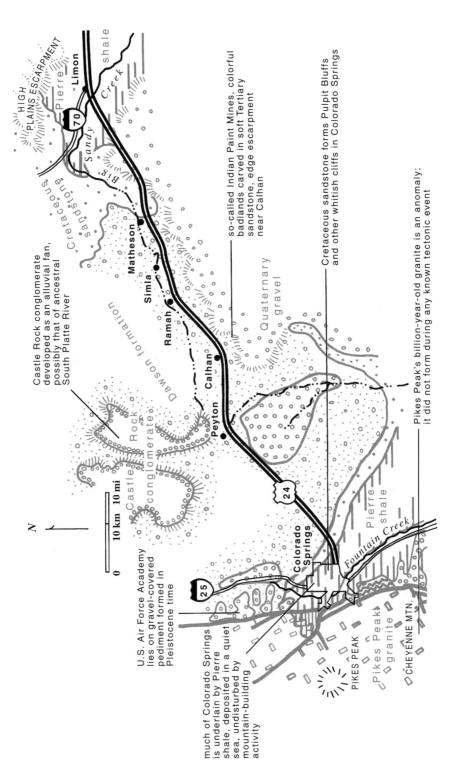

Geology along U.S. 24 between Limon and Colorado Springs.

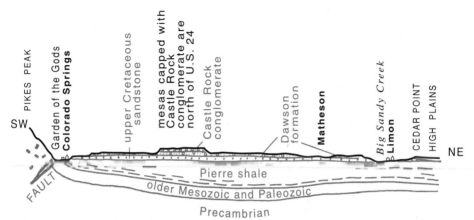

Section along and north of U.S. 24 between Limon and Colorado Springs.

draining the mountains west of Colorado Springs, an area that now drains south into the Arkansas.

From the high area west of Limon, there is a good view northwestward toward Cedar Point, a western prong of the High Plains. You can see distinctly the sharp cliff edge of the High Plains escarpment and the Tertiary rocks that form the High Plains surface.

For 6 miles, U.S. 24 crosses the gravelly mesa surface; then it descends to Matheson. This part of the piedmont is fairly high: Simla, almost on the Platte-Arkansas drainage divide, is over 6,100 feet in elevation. The piedmont here is surfaced with Dawson formation, thick layers of Tertiary sandstone that fill the center of the Denver Basin. Most sandstone is made primarily of quartz grains, but this rock also contains abundant grains of feldspar, common in granite. Feldspar minerals readily break down into clay; their presence here indicates that the sand didn't travel very far from its original source—the granite of the Front Range—before being deposited.

Looking northwestward, notice the flat-topped, forested mesas and buttes on the skyline. They are capped with a coarse, resistant rock, the Castle Rock conglomerate, also derived from the nearby mountains. This pebble-filled conglomerate occurs in a limited area about 30 miles across and may be part of a large alluvial fan deposited before the South Platte River changed course to flow northward toward Denver.

Between Matheson and Calhan, U.S. 24 skirts a steep escarpment, at places only a mile or two southeast of the road. At Calhan, erosion of this escarpment has bared several little badland areas with gaily colored ravines, gullies, and standing rock pillars known locally as Indian Paint Mines. Native Americans frequented this area—many arrowheads and spear points have been found—but there is no direct evidence that they used the colored clays. Most of the yellow, red, and purple rock is pigmented with iron oxides; gypsum crystals whiten some layers. Around 1888 clay was dug here for fire bricks, and some has been used for making pottery.

As U.S. 24 approaches the mountains, every cliff and crag and slope of Pikes Peak stands out sharply, in clear weather anyway. Two miles west of Calhan, the view to the southwest also includes the sawtooth crest of the Sangre de Cristo Range 100 miles away. The country between Calhan and Colorado Springs opens out into typical Colorado Piedmont—rolling, farm-strewn hills divided by intermittent streams.

Pikes Peak, lone sentinel of the southern end of the Front Range, towers 14,110 feet above sea level. Although not the tallest mountain in Colorado, it rises nearly a mile and a half above its base, more than any other peak in the state. The Pikes Peak massif—a word used by geologists to indicate the whole of a mountain mass that behaves geologically as a single unit—includes Cheyenne Mountain, the flat-topped smaller mountain to the south. The Rampart Range, stretching in a long, low, horizontal line northward, is composed of the same type of 1.1-billion-year-old granite as Pikes Peak, set off by a giant northwest-southeast break, the Ute Pass fault. The flat, almost horizontal tops of both Cheyenne Mountain and the Rampart Range are parts of the Tertiary pediment, an erosion surface formed before the most recent uplift of this region.

In this aerial photograph, the Tertiary pediment shows up as a horizontal surface immediately below the mountains on the skyline. Here and there, the pediment is trenched by stream canyons. —T. S. Lovering photo, courtesy of U.S. Geological Survey

As U.S. 24 nears Colorado Springs it drops into sandstone laid down on beaches, bars, and coastal floodplains of the retreating Cretaceous sea. Below the sandstone is fossil-bearing marine shale of the Pierre formation, deposited before the sea's retreat. Lower parts of the city are built on this gray shale; higher parts lie on the sandstone or on terraces of sandstone topped with younger gravels. The terracelike hills close to the mountains are true pediments, levels carved back into the bedrock of the mountain. They also are topped with gravel.

U.S. 50
Kansas—La Junta
88 miles (142 km)

U.S. 50 enters Colorado on the floodplain of the Arkansas River near Colorado's lowest point, 3,300 feet above sea level. A fairly sizable river for this part of the country, the willow- and cottonwood-bordered Arkansas drains much of southeastern Colorado as well as the area between the Front Range and the Sawatch Range farther west. For most of the distance to La Junta, U.S. 50 remains in sight of the river, following the old wagon route of the Santa Fe Trail across sand and gravel of the river floodplain or dark gray Cretaceous shale bordering the floodplain. Low hills north of the river are surfaced with Cretaceous limestone, a little younger than the shale but hardly differing in appearance from shale slopes to the south. Thin beds of light-colored limestone can be seen in some roadcuts.

In this area, most of the Arkansas River's tributary streams are intermittent: they flow only seasonally or after heavy rains. The Arkansas itself no longer flows as heavily as it used to, as much of its water is now drawn off in irrigation ditches and canals. Upstream reservoirs store spring runoff for summer use.

The irrigated floodplain produces sugar beets, soy beans, corn, alfalfa, and garden crops, whereas non-irrigated slopes north and south of the floodplain are used for grazing. Sand hills and small dune areas edge these slopes. A hundred years ago immense herds of cattle, driven north from Texas over a then fenceless land, crossed the Arkansas River here.

Colorado's oldest and most widespread Cretaceous rock, the Dakota sandstone, is well exposed in river bluffs below John Martin Dam. Older than the gray shale, this sandstone is more resistant than surrounding sedimentary rocks, so it tends to form ledges, cliffs, and narrow canyons. It is responsible here for narrowing the Arkansas River's channel, thereby creating a good site for the dam. Along the road to the recreation area be-

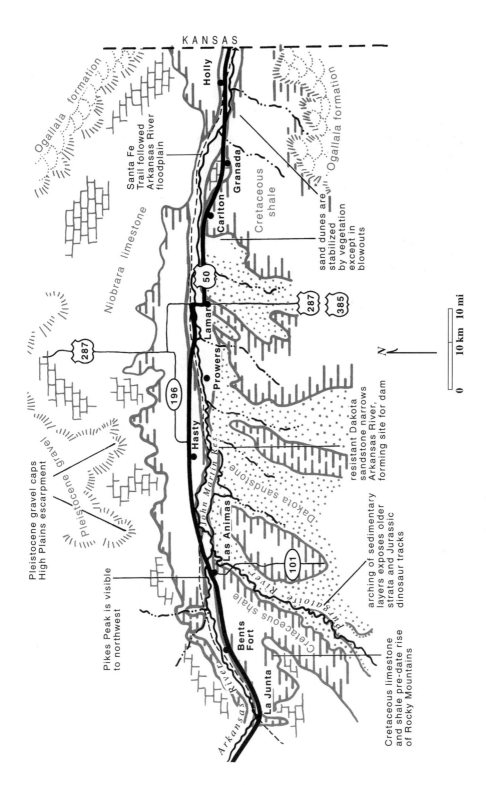

Geology along U.S. 50 between Kansas and La Junta.

Alternating layers of limestone and shale represent rapid and repeated changes from muddy water, in which dark clay was deposited, to clear water, in which limestone was deposited. In the Cretaceous sea, similar changes apparently extended more than 2,000 miles in a north-south direction. Note figure for scale. —P. T. Voegeli photo, courtesy of U.S. Geological Survey

low the dam, the sandstone is crossbedded with diagonal laminae that record currents that sorted the sand during deposition. In early Cretaceous time the sea rapidly covered the land, then receded briefly, leaving the Dakota sandstone to mark its receding shoreline.

All the rock layers, except the young gravels topping terraces in the distance to the north, dip very gently northward into the great sag of the Denver Basin. The basin, now filled completely with sediment, formed in late Cretaceous and early Tertiary time as the mountains were thrust upward.

The Purgatoire River, one of the area's few permanent streams, enters the Arkansas River near Las Animas. Fed from the Sangre de Cristo Range to the west and the high country around Mesa de Maya to the south, the Purgatoire drains a large and unusually colorful region. With its tributaries it cuts across a broad but gentle anticline of sedimentary rocks ranging in age from Permian to Cretaceous, where miniature Grand Canyons form a twisting maze of pink sandstone cliffs and dark red shale slopes. This gentle arch, called the Apishapa uplift, rose during the Laramide Orogeny, at the same time as the Rockies; it had earlier been lifted during Pennsylvanian time, when Frontrangia and Uncompahgria were rising farther west. Rocks as old as Permian age are exposed in its canyons, and 1.8-billion-year-old gneiss and 1.4-billion-year-old granite like those in the high ranges of the Colorado Rockies are known from numerous exploratory wells. The uplift extends in a broad crescent from here to the mountains.

Few roads lead into the deeply incised wonderland south of U.S. 50. Remoteness has probably been good protection for one of Colorado's best dinosaur footprint sites. Deep within the canyons a dinosaur trackway marks a sandy layer in the Jurassic Morrison formation. The site can be visited by four-wheel-drive vehicle by prearrangement with the U.S. Forest Service in La Junta. Collecting is not permitted.

In the valley of the Purgatoire River, footprints of several Apatosaurus-*like sauropods lead across an outcrop of the Morrison formation. The animals that left the tracks were walking side by side along a lakeshore, suggesting that this species of dinosaur lived in family groups or herds.* —Photo courtesy of Teresa Kesterson, Santa Fe Trail Detours

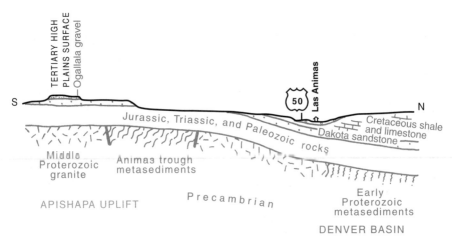

Section across U.S. 50 at Las Animas shows the structure of the Apishapa uplift.

West of Las Animas the slopes north and south of the Arkansas River are composed of dark gray shale capped with light gray Niobrara limestone, both Cretaceous in age. Here, the rocks dip westward. Eventually the shale disappears below the surface of the river floodplain, and cliffs and ledges of overlying limestone converge on the highway. Thin beds of limestone alternate with layers of darker shale, as many as 30 alternations within 20 feet. Geologists think widespread rapid fluctuations in sea level or climate caused the alternating pattern.

U.S. 50
La Junta—Canon City
103 miles (166 km)

U.S. 50 remains close to the Arkansas River all the way from La Junta to Pueblo; west of Pueblo it is never more than a few miles from the river.

Near La Junta light-colored Niobrara limestone of Cretaceous age, interbedded with thin layers of shale, covers slopes north and south of the river. The limestone, which accumulated in the broad, shallow sea that covered much of the continent 90 to 80 million years ago, dips gently northward.

Fossils of many Cretaceous shellfish occur in the Niobrara limestone and its interbedded shale: tiny beadlike shells of one-celled foraminifera; much larger ammonites, relatives of the modern octopus and chambered nautilus; and *Inoceramus*, a large clam with a corrugated shell.

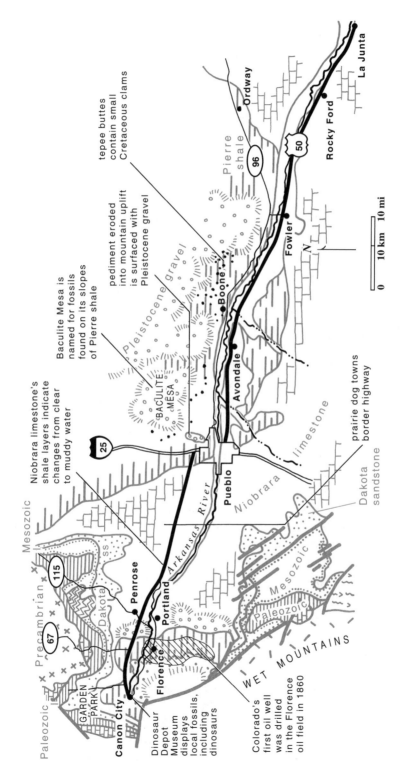

tepee buttes
contain small
Cretaceous clams

pediment eroded
into mountain uplift
is surfaced with
Pleistocene gravel

Baculite Mesa is
named for fossils
found on its slopes
of Pierre shale

Niobrara limestone's
shale layers indicate
changes from clear
to muddy water

prairie dog towns
border highway

Colorado's first oil well
was drilled in the Florence
oil field in 1860

Dinosaur Depot
Museum displays
local fossils,
including
dinosaurs

Ordway

La Junta

Rocky Ford

Fowler

96

50

Boone

Avondale

Pierre shale

Pleistocene gravel

BACULITE MESA

25

Pueblo

Niobrara limestone

Dakota sandstone

Mesozoic

Paleozoic

WET MOUNTAINS

Mesozoic

Paleozoic

Arkansas River

Portland

Florence

Penrose

115

67

Canon City

GARDEN PARK

Precambrian

Paleozoic

Dakota ss.

Mesozoic

N

0   10 km   10 mi

Geology along U.S. 50 between La Junta and Canon City.

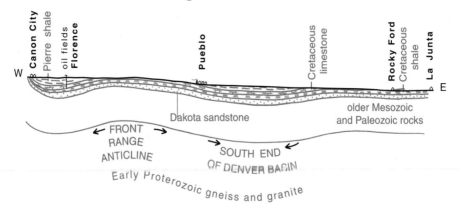

Section along U.S. 50 from La Junta to Canon City.

Just east of Rocky Ford, watch for your first glimpse of Pikes Peak (if you're headed west) just to the right of the highway's bearing.

N. H. Darton, writing in 1916 about the geology along the Santa Fe Railroad, wrote about Rocky Ford's famous cantaloupes. Lettuce, onions, sugar beets, corn, fruit, and other garden crops are grown here too.

Near Fowler we encounter a younger rock formation, the Pierre shale, which is rarely well exposed. The gray rock, deposited as sea-bottom mud, appears in low slopes across the river. The contact between the Niobrara limestone and the Pierre shale outlines the southern end of the Denver Basin, and the Pierre shale thickens rapidly northward and westward.

Avondale perches 30 to 40 feet above the Arkansas River on an Ice Age terrace composed of gravel, sand, and small boulders washed from the Front Range when the river was overloaded with glacial debris. The terraces are remnants of old floodplains and represent fairly stable periods in the development of the river, when downcutting was minimal and the river swung back and forth across its floodplain.

Southern Colorado University, at Pueblo, stands on a higher mesa. Northeast of it is still higher Baculite Mesa, whose gray slopes of Pierre shale contain fossils of *Baculites,* a straight-shelled relative of the modern octopus and chambered nautilus. Coarse Pleistocene gravel caps these mesas.

West of Pueblo, U.S. 50 climbs out of the Pierre shale onto the Niobrara limestone—beds of fossil-bearing white limestone alternating with thin shale. More resistant than thick shale layers above and below it, the limestone tops many gently sloping cuestas; its best exposures are in ravines between the cuestas. West of Pueblo it rises northward onto a gentle upward fold or anticline.

Across the Arkansas River south of Penrose, near the town of Portland (named after Portland, England, original home of Portland cement),

Small, scattered cones of harder rock material rise above the shale surface east of Avondale. Known as tepee buttes, the cones resist erosion because they contain more limestone than the surrounding soft shale. These reeflike aggregates, built on the floor of the shallow Cretaceous sea, are good places to look for fossils, particularly small clams named Lucina. *Hundreds of the buttes are scattered north of the river near Avondale, just east of Pueblo, and north and east of Canon City.*
—G. R. Scott photo, courtesy of U.S. Geological Survey

Niobrara limestone has been quarried for use in cement. Cretaceous rocks furnish clay for bricks near Canon City and coal south of Florence. Colorado's first oil field was discovered at Florence in 1860.

Pikes Peak dominates the landscape to the north. Both this peak and its lower neighbor Cheyenne Mountain are composed of Pikes Peak granite, a huge mass of igneous rock that intruded this area about a billion years ago. The flat top of Cheyenne Mountain, as well as the broad shelf extending westward from Pikes Peak, are parts of the Tertiary pediment, the erosion surface once continuous with the High Plains. The Wet Mountains to the south display the same nearly horizontal surface.

Gray shales in roadcuts and on mesa slopes east of Canon City are the Pierre shale. Often beveled across their dipping bedding planes, they are thinly surfaced with Pleistocene gravel.

At the south end of the Front Range, Cretaceous rocks as well as older strata are caught up in the structure of the mountains and bend up dramatically at their edge. The steeply tilted rock layers form hogbacks that curl around the end of the Precambrian mountain mass. We can imagine that sedimentary layers once arched as a titanic anticline over the top of the range, but in fact they probably eroded away about as fast as the mountains rose. You can see the tipped-up rocks, particularly the resistant layers of Dakota sandstone, north of U.S. 50. A short side-trip up Colorado 67 toward Phantom Canyon or up the Garden Park Road will take you right through the hogbacks for a closer view. Except where Cretaceous rocks are absent because of faulting, the Dakota Hogback extends southward past the New Mexico line.

Directly north of Canon City the hogbacks bend sharply and circle south past town. The most prominent of them, the Dakota Hogback, forms the western wall of the penitentiary grounds and the ridge traversed by Skyline Drive—a good place for a look around. The hogbacks continue southeast along the flank of the Gore Hills and Wet Mountains.

In the great curve of exposed, upward-bent sedimentary rocks known as the Canon City Embayment, Paleozoic and Mesozoic rocks drape and fold around the end of the Front Range and the side of the Wet Mountains. The Canon City Embayment attained paleontological fame in 1877 when Jurassic dinosaur bones were found in the Morrison formation a few miles north of Canon City. Within months of the discovery, two competitive

A Stegosaurus *skull, part of a nearly complete skeleton, was unearthed in 1992 near Canon City; amazingly, a fossilized piece of knobby skin from the animal's throat was still attached. Above the skull are two scutes, or armor plates, from the animal's back. The fossil is on display at the Denver Museum of Nature and Science.* —Felicie Williams photo

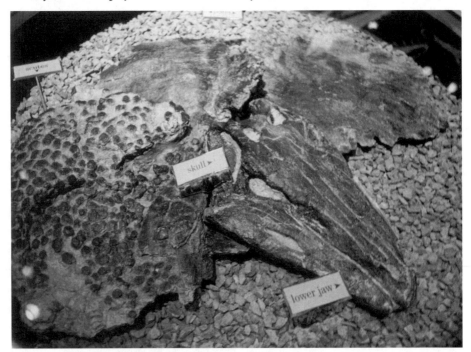

paleontologists, Edward Cope of Philadelphia and O. C. Marsh of Yale's Peabody Museum, opened up rival quarries in Garden Park and discovered many kinds of dinosaurs and other Jurassic animals. Visit Dinosaur Depot, Canon City's dinosaur museum, and check on plans for an active quarry in Garden Park. Dinosaur skeletons are still found here. Federal law protects fossils on public lands, so don't pick up or move any bones you find.

Devonian arthropods and fossil tracks, as well as impressions of Ordovician marine animals, have been found in the Canon City area. In 1887 fish scales of some of the oldest vertebrates then known were discovered here; recently some early sharks have been found.

Colorado 115
Colorado Springs—Penrose
36 miles (58 km)

Cheyenne Mountain rises steeply west of the junction of Colorado 115 with I-25. To the southeast the poorly exposed greenish gray Pierre shale, deposited in a Cretaceous sea, extends across the valley of Fountain Creek. Faults at the foot of Cheyenne Mountain have cut off Paleozoic and older Mesozoic rocks, so the Pierre shale butts directly against Precambrian Pikes Peak granite. The great granite massif of Pikes Peak and Cheyenne Mountain was wedged upward between outward-slanting thrust faults and slid eastward 4,000 to 6,000 feet, coming to rest above the Pierre shale.

Near the mountains the Pierre shale is beveled by a sloping pediment that probably developed in Pleistocene time; it is covered with the remains of a bouldery rockslide that extends like fingers to the southeast.

Small conical tepee buttes jut up here and there above the Pierre shale east of Colorado 115. The buttes contain numerous little fossil clam shells and remains of other marine animals that may have gathered around underwater springs.

Toward the southern end of Cheyenne Mountain, between mileposts 37 and 36, the Dakota Hogback juts up between the highway and the mountain slopes. Here, the faults that edge the east side of the Front Range die out. The highway soon crosses older and older strata, passing through Cretaceous limestone, hogback-forming Dakota sandstone, and the Jurassic Morrison formation, whose Easter-egg colors make it easy to recognize.

Near milepost 36 the highway passes into the Permian Lyons sandstone, the same salmon-colored dune sandstone that appears in the Garden of the Gods in Colorado Springs. South of it is the Pennsylvanian Fountain

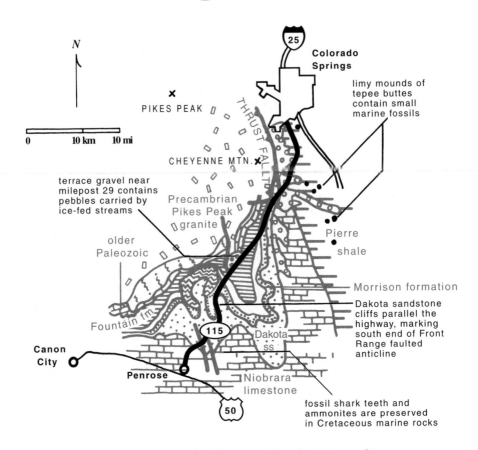

N

PIKES PEAK

THRUST FAULT

Colorado
Springs

limy mounds of
tepee buttes
contain small
marine fossils

0   10 km   10 mi

CHEYENNE MTN.

terrace gravel near
milepost 29 contains
pebbles carried by
ice-fed streams

Precambrian
Pikes Peak
granite

Pierre

shale

older
Paleozoic

Morrison formation

Dakota sandstone
cliffs parallel the
highway, marking
south end of Front
Range faulted
anticline

Fountain fm.

115

Dakota
ss.

Canon
City

Penrose

Niobrara
limestone

50

fossil shark teeth and
ammonites are preserved
in Cretaceous marine rocks

Geology along Colorado 115 between Colorado Springs and Penrose.

formation, darker red than the Lyons sandstone and with interspersed layers of conglomerate and deep red shale.

Between milepost 35 and U.S. 50 near Penrose, Colorado 115 crosses back and forth among these rock layers. Close to the mountains these rocks are steeply upturned, and the resistant ones—the Lyons, Fountain, and Dakota formations—form well-defined hogbacks. Farther from the mountains, where they do not slope as steeply, they cap cuestas.

To the southwest, beyond the valley of the Arkansas River, are the Wet Mountains, part of another faulted anticline. Blue with distance is the sharply toothed skyline of the Sangre de Cristo Range. The eastern slope of these mountains, which you see from here, consists mostly of east-dipping Pennsylvanian and Permian sedimentary rocks; the west face is rugged and cliffy, abruptly cut off by downfaulting of the San Luis Valley.

Between mileposts 25 and 21, a cuesta capped with Dakota sandstone parallels the highway. Well around on the south side of Pikes Peak, at milepost 20, younger Cretaceous sedimentary rocks begin to appear above the Dakota sandstone, notably the cuesta-capping Niobrara limestone, on which Colorado 115 descends toward Penrose and U.S. 50. Sharks teeth, ammonites, and other marine fossils have been found in this marine limestone.

*What a frightening barrier Colorado's majestic mountains must have seemed to cov-
ered-wagon travelers! Following the great western routes—the Santa Fe and Oregon
Trails—they turned aside, preferring low passes of New Mexico and Wyoming to threat-
ening Colorado peaks. Even now only a handful of highways challenge the high passes.*
—Halka Chronic photo

*The central cores of most Colorado ranges are composed of hard Precambrian igneous
and metamorphic rock. Relatively smooth uplands probably date from erosion before
regional uplift. Pleistocene glaciers carved horseshoe-shaped cirques and deep canyons.
South Park, a broad high-altitude intermountain valley, is at the upper left.* —T. S. Lovering
photo, courtesy of U.S. Geological Survey

IV
Ranges Folded and Faulted

Studies of rocks in and along the flanks of the dozen or so ranges of the Colorado Rockies reveal that most of them are like giant wrinkles in Earth's crust, formed where areas of Precambrian rocks—the ancient foundation of the continent—were squeezed sideways, folding and breaking into long north-south slivers that were wedged many thousands of feet up and over neighboring rocks. Younger layers of sedimentary rock that lay above the old foundation were folded and faulted, too. More flexible, they stretched and draped like rippled blankets across the broken edges of the mountain cores. Most of the sedimentary layers, however, eroded from the mountain summits even as uplift—a long drawn-out process—continued. Less noticeable but equally important, the land in front of the uplifted ranges warped downward into basins, almost like the troughs before ocean waves.

Let's review the tectonic events that led to the mountains as we know them today, remembering that each time mountains rise, erosion immediately sets about obliterating them, depositing their debris close at hand or spreading it far and wide over surrounding land or carrying it to some distant sea. Think of processes we know today—slow invisible movements in Earth's crust, sudden earthquakes that are an expression of that movement, volcanoes that rise over centuries and blow themselves apart in a blink of an eye, rocks dissolved or disintegrating, landslides, floods, frost cracking rocks apart year after year or night after night, wind and wave and muddy river—players in the never-ending drama between build-up and break-down. Somewhere along the geologic timeline the destructive and protective effects of plants and animals should be added to the equation. Keep in mind that peaceful periods of erosion have through geologic time tended to be far longer than episodes of mountain building.

Mountain-Building Events in Colorado
- The first mountains we know of in Colorado formed 1.8 to 1.7 billion years ago when an island arc collided with and plunged beneath the craton called the Wyoming Province. Melting rock rose to form a chain of volcanoes along the craton's margin. Sediments that washed from

the craton and volcanic lavas that accumulated on the seafloor mingled to form the oldest rocks now known in most of the state.

- Around 1.7 billion years ago when volcanic arcs formed in New Mexico, the forces that created them compressed the crust in Colorado, causing metamorphism of older rocks and intrusion of granite—the granite of the Routt plutonic suite.

- Another round of compression came from the southeast 1.4 billion years ago as crust was added to Texas and Oklahoma. In Colorado it brought about intrusion of scattered masses of granite, the Berthoud plutonic suite, into the older rocks.

- The Pikes Peak batholith pushed upward around 1.1 billion years ago, its tectonic cause as yet a mystery.

- About 300 million years ago, after a break of nearly a billion years, the Ancestral Rockies developed: notably two island ranges we call Frontrangia and Uncompahgria. These ranges seem to have formed in response to the collision of North America with Africa and Europe.

- The Laramide Orogeny 72 to 40 million years ago established the present ranges of the Colorado Rockies. Beginning in California in Jurassic time when an offshore subcontinent (now parts of California, Idaho, Nevada, and Utah) collided with the west coast of North America, a wave of mountain building moved gradually eastward, reaching and affecting Colorado in late Cretaceous and early Tertiary time.

- Relatively late in Colorado's dynamic history, Oligocene volcanism left its mark in the western part of Colorado.

- Mid-Tertiary regional uplift 28 to 5 million years ago domed up a large area of the West, including Colorado, to its present elevation. Plains that had been near sea level rose to 4,000 feet or more; mountains 5,000 feet high became 10,000 feet high, those over 9,000 feet became Fourteeners. As uplift proceeded, the Colorado Plateau (largely in Arizona and Utah but extending into western Colorado) broke away as a solid block from the craton to the east, leaving the long, narrow sliver of the Rio Grande Rift to mark the fracture between them.

The Rocky Mountains stretch from Alaska to New Mexico, part of a chain of mountains that runs almost unbroken along the entire length of the Americas. Most of Colorado's share of the Rockies consists of a cluster of long, narrow faulted anticlines of which the largest, the Front Range, is typical. Its core is a block of hard 1.8- to 1.1-billion-year-old granite, gneiss, and schist, a segment of the Precambrian rock that underlies North America. Numerous faults edge both the eastern and western sides of the Front Range, many of them are thrust faults along which the block of Precambrian rocks moved upward and outward over surrounding sedimentary rocks, like a wedge of soap squeezed between your hands. The upward and outward

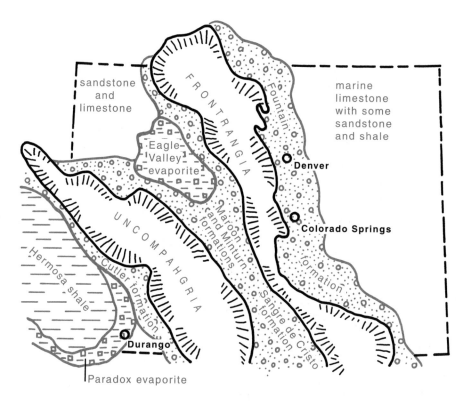

The mountains ranges of Frontrangia and Uncompahgria rose above the sea in Penn-sylvanian time. As erosion gnawed into them, the mountains became encircled with alluvial aprons, now the Fountain, Maroon, Sangre de Cristo, and Cutler formations, all deposited above sea level. In all-but-landlocked basins west of each range, gypsum and salt were left behind as seawater evaporated.

movement accomplished a sideways shortening of the crust to relieve pres-sure from the west. As far as we know, the thrust faults curve, their steepness increasing downward. We are not sure how deep they go.

The Front Range is complicated by thousands of faults too small to put on the map. Rocks of the mountain core are so intensely faulted that it's difficult to distinguish the boundary between igneous and metamorphic rocks, or to trace gold-bearing veins for any distance.

Most of the mines in the Front Range lie along the Colorado Lineament, a Precambrian fault zone that intersects the Front Range between Boulder and Dillon. Magma rose in many places along this zone during and after Laramide mountain building; mineral-rich solutions escaping from the magma seeped into old cracks that reopened as the mountains rose. Min-eralized veins, whether or not they contain economically valuable ores, are

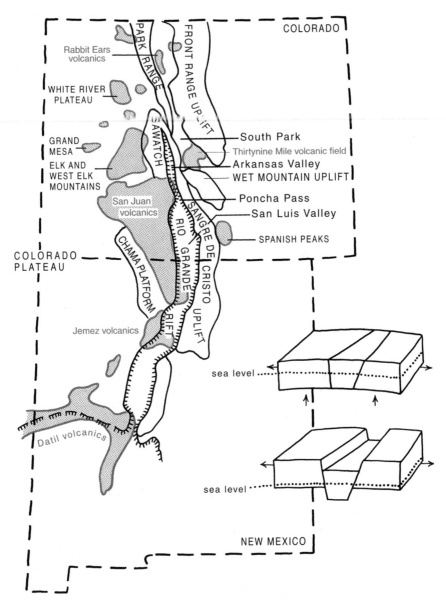

The San Luis Valley and the Arkansas Valley are the northern end of the Rio Grande rift, a down-dropped wedge of crust that stretches from central Colorado to southern New Mexico. Like rift valleys elsewhere, it marks a pulling apart of Earth's crust and is bordered by areas of volcanism, shown in red.

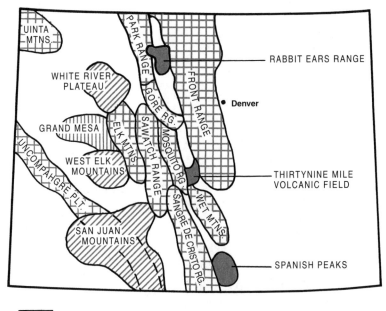

faulted anticlines

plateaus covered with volcanic rocks

volcanic mountains

domes covered with volcanic rocks

The Colorado Rockies consist of about a dozen individual, geologically distinct ranges. In casual and official usage each range may be divided again: the Front Range includes the Mummy Range, Indian Peaks, the Rampart Range, and the Tarryall and Kenosha Mountains; part of the Sawatch Range is called the Collegiate Range; and the Culebras are part of the Sangre de Cristos. Your road map may use these and other local names.

usually stained near the surface by iron oxide in rusty shades of yellow and red, colors that led observant prospectors to rich orebodies. Such staining shows up in many highway cuts.

Faults seem to be particularly abundant in the Colorado Mineral Belt. Is the amount of mineralization related to the local abundance of faults? Or have we simply found more faults in mining areas because of the detailed probing that goes on in the search for minerals?

Patterns of erosion in the mountains are varied. Rivers and rivulets carve steep-walled, V-shaped canyons. Storm-swollen mountain streams tumble giant boulders and undercut tall cliffs. Avalanches, mudflows, and rockslides

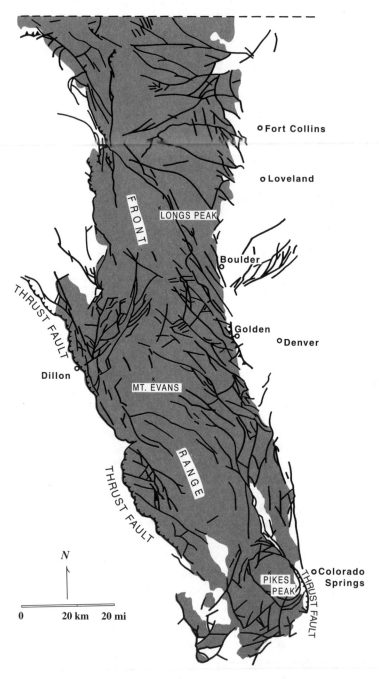

Complex arrays of faults border and cut across the Front Range.
Thrust faults are shown by teeth on the upper plate. Thousands of
smaller faults are not shown.

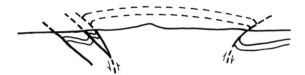

A section across the Front Range, drawn with no vertical exaggeration, shows the hypothetical faulted anticline in overlying strata. Sedimentary rocks erode far more easily than hard granite and metamorphic rocks of the mountain core, and there was probably never a time when they arched completely across the mountains.

scar and reshape timbered slopes. And with patient ferocity, water seeps into tiny cracks and freezes, its expansion forcing the cracks apart. Such frost wedging is one of the most potent agents of erosion in the mountains.

Another active erosional process is chemical weathering, particularly in granite. Decomposition of granite occurs mostly because mica crystals absorb moisture and change chemically into clay minerals. They increase in volume, too, loosening adjacent quartz and feldspar grains. The process rounds exposed granite into knobs and boulders. Day-night and winter-summer temperature changes help the process along.

During the Pleistocene Ice Age, and even to some degree more recently, glaciers helped carve the peaks and ridges. Glaciers are nourished by snowfall. When winter snow exceeds summer melting, snow deepens with each passing year. Gradually recrystallizing, it forms tiny ice granules—you can find them on any summer snowfield—that slowly fuse into solid ice under the weight of added snow. When masses of ice become heavy enough to flow (100 feet or more of packed snow and ice) glaciers are born. Their birthplaces may be on the leeward sides of high ridges where winter winds drift snow on north-facing, shadowed slopes. Or they may originate on broad, fairly level mountain uplands where deep snow accumulates, and where ice fields flow slowly in all directions from the highest point. Such upland ice masses become ice caps; along their margins they may finger down into valley glaciers.

At least three times glaciers scoured the Colorado Rockies. There may have been earlier glacial episodes, but if so, proof of their existence was erased by one or another of the last three. All three episodes are part of the last of four great glacial stages known in North America—the Wisconsin glaciation of Pleistocene time.

As they pluck rock from the heads of valleys or gouge it from floors and walls of canyons, glaciers leave large, clear fingerprints. They excavate semi-circular cirques on high mountain peaks and ridges. Gripping rocks for

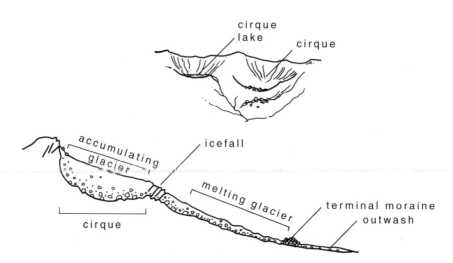

Valley glaciers originate in high-altitude snowfields as snow compacts into ice. In Colorado they flowed downward to about 8,000 feet in elevation, where their lower ends melted, leaving rocky terminal moraines.

cutting tools, glaciers widen, straighten, and deepen their pathways, giving them U-shaped profiles quite unlike V-shaped, stream-cut canyons. Like powerful rasps, they groove, grind, and striate rocks over which they pass, and like conveyor belts, they dump their loads in moraines as they melt.

Glacial features show up in Colorado's ranges above 8,000 feet in elevation. Below 8,000 feet, where glacier tongues melted faster than they were resupplied from above, evidence of glaciers appears in glacial outwash and broad, stepped terraces shaped by streams swollen with meltwater.

Tiny glaciers existing today in the Front Range and the Park Range are not simple remnants of Ice Age glaciers. They formed within the last few thousand years, after a warmer period when their predecessors melted completely. These little glaciers have been shrinking over the last few hundred years.

Colorado's water resources are greatest on the western side of the mountains, an accident of prevailing westerly winds. Water needs are greatest in the band of cities east of the mountains. As you drive through the mountains, you will see many reservoirs, water conduits, and tunnel portals that are parts of the huge network of water projects that increase eastern slope supplies.

Resistant Dakota sandstone at the eastern end of the I-70 highway cut (right side of photo), *with its distinctive black coaly layers, protects soft purple, green, and brown shales of the Morrison formation below it. The Morrison formation is covered with soil where it has not been sliced open by highway construction.* —Jack Rathbone photo

Interstate 70/U.S. 6
Denver—Dillon
66 miles (106 km)

There is a developed geologic site where I-70 cuts through the Dakota Hogback on the outskirts of Denver. A cross-section of Mesozoic rocks is particularly well displayed, with layers tilted steeply upward by uplift of the Front Range. Just east of the highway cut, I-70 crosses the Golden fault zone, the main line of faulting that edges the eastern side of the range. The fault is hidden, however, by relatively soft rock weathered into soil.

The hard Precambrian rocks that border I-70 for 25 miles as it climbs into the mountains originally formed around 1.8 billion years ago as volcanic and sedimentary rocks in an island arc. They show obvious signs of the pushing and wrenching, melting and squeezing to which they have been subjected during repeated periods of mountain building. In many places,

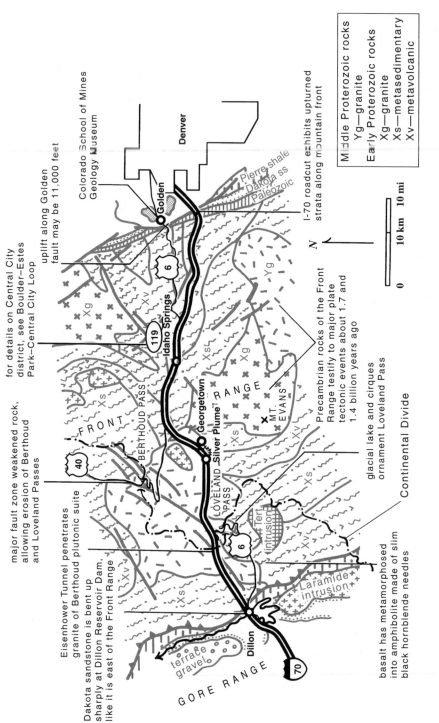

Geology along I-70 between Denver and Dillon.

Middle Proterozoic rocks
Yg—granite
Early Proterozoic rocks
Xg—granite
Xs—metasedimentary
Xv—metavolcanic

I-70 roadcut exhibits upturned strata along mountain front

N

0   10 km   10 mi

Denver

Colorado School of Mines Geology Museum

Golden

Pierre shale
Dakota ss
Paleozoic

for details on Central City district, see Boulder–Estes Park–Central City Loop

uplift along Golden fault may be 11,000 feet

6

119

Idaho Springs

Xv

Xg

FRONT

BERTHOUD PASS

Georgetown

Silver Plume

RANGE

Xs

Xg

Xs

×
MT. EVANS

Precambrian rocks of the Front Range testify to major plate tectonic events about 1.7 and 1.4 billion years ago

major fault zone weakened rock, allowing erosion of Berthoud and Loveland Passes

40

Eisenhower Tunnel penetrates granite of Berthoud plutonic suite

LOVELAND PASS

Xs

Xv

glacial lake and cirques ornament Loveland Pass

Continental Divide

Dakota sandstone is bent up sharply at Dillon Reservoir Dam, like it is east of the Front Range

6

Tert. intrusion

Xv

Xs

Laramide intrusion

basalt has metamorphosed into amphibolite made of slim black hornblende needles

Dillon

70

terrace gravel

GORE RANGE

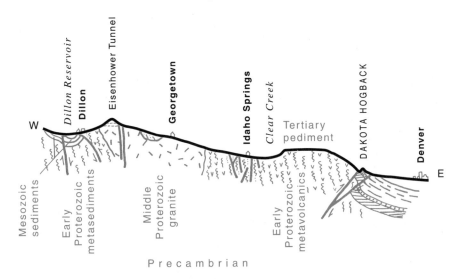

Section along I-70 from Denver to Dillon. Half-arrows show direction of movement along faults.

as near milepost 257, they are crisscrossed by light-colored pegmatite dikes with unusually large quartz, feldspar, and mica crystals. Pegmatites form from fluids escaping into cracks as surrounding rock is heated during mountain building.

The interstate crosses the Colorado Mineral Belt near Idaho Springs and Georgetown. Numerous veins containing gold and other metals began as cracks in the ancient rocks, which acted as passageways for mineral-rich fluids rising from cooling magma between 59 and 35 million years ago. As liquids and vapors rose through the cracks, the lessening of pressure and temperature caused minerals to precipitate on their walls, much as vapor precipitates on the outside of a glass of ice water. Many of these veins have been cut or blasted completely away; this was one of Colorado's principal mining districts from 1860 to the late 1880s. More than $400 million worth of gold, lead, zinc, and copper were produced from this area.

Just west of Idaho Springs a frontage road on the north side of I-70 gives safe access to some of the mineralized rocks of this area. You will see dark Precambrian biotite gneiss intruded by light-colored granite and cut by numerous veins, some of which contain small amounts of metallic ore minerals.

A few miles west of Idaho Springs, at about 8,000 feet in elevation, I-70 enters glaciated territory. The lowest well-defined, hilly terminal moraine is near the junction with U.S. 40. Above it, the valley assumes the U-shaped profile of valley glaciation, though rock slides, avalanche debris, and dense forest sometimes obscure its characteristic shape.

At Georgetown, condominiums on the south side of the valley lie dangerously close to paths of floods and avalanches. Avalanche tracks mark the slopes above them, and big boulders between the buildings arrived by crashing down the mountainside.

Patches of light gray Silver Plume granite begin to appear near Georgetown. Named for a nearby town and a waterfall since destroyed by highway construction, the granite is 1.45 billion years old, part of the Berthoud plutonic suite but locally called the Silver Plume granite. Irregular bands of metamorphic rocks alternate with slender granite fingers that branch from the edges of the main granite mass.

A wide avalanche track at milepost 218 has been a headache for highway maintenance crews for decades; a ridgelike berm now protects the highway.

As I-70 approaches Loveland Basin, either continue straight ahead through the Eisenhower Tunnel or follow U.S. 6 over scenic Loveland Pass. Both routes are bordered by alternate patches of gray granite like that at Silver Plume and gneiss and schist like that at Idaho Springs. Along U.S. 6, good outcrops are rare because the rock is badly shattered near a broad fault zone. Erosion exploited the broken, weakened rock, determining the position of Loveland Pass. The same fault zone extends northeast through Berthoud Pass, also on the Continental Divide. East of the divide, streams flow to the Gulf of Mexico; those on the west flow west toward the Colorado River and the Pacific.

Viewed to the south from Loveland Pass, three cirques—scoop-shaped basins gouged by glaciers—frame Arapaho Basin ski area. To the northwest, a larger cirque curves around Loveland Basin. The smoothly contoured uplands were rounded by Pleistocene ice that covered them completely, but Pettingell Peak to the north, contrastingly sharp and craggy, jutted up

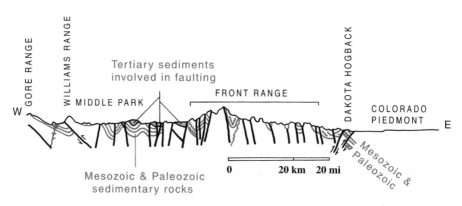

During the boring of tunnels that now bring western slope water to cities along the eastern slope, geologists discovered that faulting within the Front Range was more complex than formerly suspected.

*Proterozoic gneiss
displays typical bands of
light crystalline quartz
and feldspar alternating
with dark, fine-grained,
platy biotite and
needlelike hornblende.*
—T. S. Lovering photo, courtesy
of U.S. Geological Survey

through the ice cap. Pass Lake, just south of Loveland Pass, occupies another glacial cirque. Huge talus cones form where rocks, loosened from the cliffs by freezing and thawing, tumble toward the lake.

Interstate 70 and U.S. 6 descend toward Dillon through glaciated valleys, but many glacial features along I-70 are hidden beneath landslide debris. The slides pose serious problems for highway engineers. Near milepost 211, where the toes of two large slides reach the highway, broken and tilted trees show that despite engineering efforts the slides are still moving.

Along U.S. 6 several terminal moraines can be identified by the appearance of jumbled, unsorted rock, sand, and silt in highway cuts. The hummocky, irregular moraines are hard to discern because they usually support a dense forest of pines and Douglas-fir, trees that like a loose, well-drained foothold.

Along both highways Cretaceous sedimentary rocks surface east of Dillon Reservoir Dam—steeply west-dipping Pierre shale and Dakota sandstone. In places the formations are overturned and seem to dip east. They are overridden from the east by Precambrian metamorphic rocks (milepost 208 on I-70), which dragged the sedimentary formations with them, folding them back on themselves.

Dillon Reservoir stores western slope water for Denver. Water from the reservoir is pumped 23 miles through one of the longest water tunnels in the world, passing under the Continental Divide. Although there is considerable resistance to further "theft" of western slope water, the needs of eastern slope cities have increased western slope opportunities for water sports and lakeside recreation. The western slope is part of the Colorado River drainage basin, where water diversion and use is strictly regulated to protect downstream users between here and the Gulf of California.

Interstate 70
Dillon—Dotsero
76 miles (122 km)

Below Dillon Reservoir Dam, I-70 crosses the Blue River, a tributary of the Colorado River, as it flows northwest along the faulted syncline between the Front Range and the Gore Range. Curving across a broad, boulder-strewn moraine, the highway enters Tenmile Canyon between the Gore and Tenmile Ranges. These two ranges are geologically continuous, as faults and Precambrian-Paleozoic contacts along their boundaries can be traced right across Tenmile Canyon.

This narrow defile opens the heart of these ranges to view, gouging deeply into their Precambrian core along the line of a particularly large fault. The smooth rock surface on the east wall of the canyon just south of milepost 196 is an actual fault surface. Smooth, shiny areas are slickensides produced by high-pressure grinding, rasping, and sliding of rock against rock. The

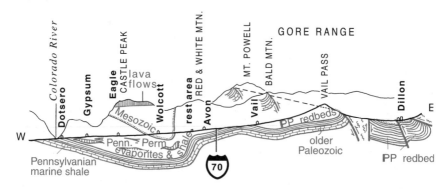

Section along I-70 from Dillon to Dotsero.

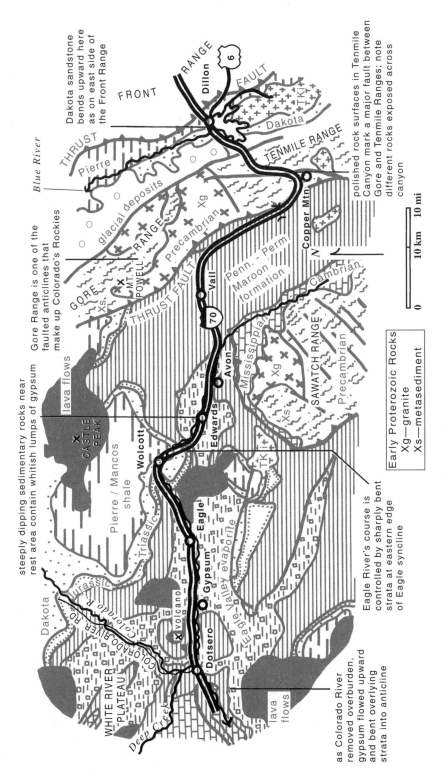

Geology along I-70 between Dillon and Dotsero.

Dakota sandstone bends upward here as on east side of the Front Range

Gore Range is one of the faulted anticlines that make up Colorado's Rockies

polished rock surfaces in Tenmile Canyon mark a major fault between Gore and Tenmile Ranges; note different rocks exposed across canyon

steeply dipping sedimentary rocks near rest area contain whitish lumps of gypsum

Eagle River's course is controlled by sharply bent strata at eastern edge of Eagle syncline

as Colorado River removed overburden, gypsum flowed upward and bent overlying strata into anticline

FRONT RANGE

Dillon

THRUST FAULT

Pierre

Blue River

glacial deposits

GORE RANGE

Precambrian

MT. POWELL

Xs

Xg

THRUST FAULT

TENMILE RANGE

Penn. - Perm. Maroon formation

Copper Mtn

Dakota

Tᴋi

Vail

Cambrian

70

Avon

SAWATCH RANGE

Mississippian

Precambrian

Xg

Xs

Edwards

lava flows

CASTLE PEAK

Pierre / Mancos shale

Wolcott

Triassic

Eagle

Tᴋi

Eagle Valley evaporite

Gypsum

volcano

Dotsero

Jurassic

Dakota

COLORADO RIVER RD.

Colorado R.

WHITE RIVER PLATEAU

Deep Creek

lava flows

N

Early Proterozoic Rocks
Xg—granite
Xs—metasediment

0   10 km
0   10 mi

wall is composed of metavolcanic rock, dark gray hornblende gneiss with a few pegmatite veins.

Precambrian rocks on the west side of the canyon, west of the fault, were once a series of sedimentary strata and volcanic flows; they are beautifully exposed in natural outcrops and huge highway cuts such as the one at milepost 199. The black-and-white bands in the gneiss represent bedding inherited from 1.8 billion years ago. Dozens of large and small veins cut the gneiss.

Just opposite Copper Mountain, I-70 passes from banded gneiss to reddish brown granite. To the west hummocky reddish glacial deposits and landslide debris cover all the older rocks and hide the great thrust fault along which the granite pushed westward over younger rocks.

The younger strata, however, appear at the surface along the approach to Vail Pass. The red sedimentary rocks resemble those of the Fountain formation at Red Rocks Park near Denver; here they are the Maroon formation, made of debris washed from the western slopes of Frontrangia. Like their eastern counterparts, they are composed of dull red sandstone and conglomerate interlayered with dark red shale. Thousands of feet thick, they border I-70 for nearly 18 miles. Their exact age is a mystery for they contain no fossils by which they can be dated. But we assign them a Pennsylvanian-Permian age because they are sandwiched between early Pennsylvanian and Triassic rocks. You can get a close look at them from the Vail Pass rest area.

Beneath the red rocks are Pennsylvanian brown and gray sandstone and shale deposited in a basin west of Frontrangia just as it began to rise. Well

Snow highlights bedding on 13,205-foot Jacque Peak. Notice the cirque east (to the left) of the peak. Farther east, through the low saddle where a thrust fault crosses the ridge, is 14,264-foot Quandary Peak. View looking south from Vail Pass rest area. —Felicie Williams photo

exposed between mileposts 172 and 193, the shale layers tend to slump and slide, especially when they are soaking wet. They are responsible for all the bumps in I-70 near Vail. They also form steep slopes above the Vail ski resort; in 1997 a rockslide carried limestone boulders as big as cars 1,800 feet down the side of the valley, where they crashed into several condominiums.

Between Vail and Dotsero, I-70 crosses first an anticline and then a broad syncline in these sediments. Strata at Vail rise westward, flex sharply downward along the Eagle River between Edwards and Wolcott, and then rise again toward Dotsero. Crossing these flexes you can also see a gradual change in the rocks; the redbeds give way westward to light gray shale containing a lot of gypsum and eventually to dark gray true marine shale. These changes mark the increasing distance from the western flank of Frontrangia, where alluvial fans gave way to shallow saltwater basins. The many minor flexes and small faults make interpretation difficult, but watch for changes in dip from westward-dipping rocks at Vail Pass to eastward-dipping ones at Vail, and then to westward-dipping ones near Edwards.

At Avon and Gypsum, bluffs along the Eagle River contain sediments particularly rich in gypsum. White stringers and large lumps of white gypsum bodies are visible from the highway. Gypsum often occurs with salt, both minerals left behind when seawater evaporates, but here the salt has washed away and only the less soluble gypsum remains. The Eagle Valley evaporite accumulated more than 280 million years ago when a shallow sea in a nearly landlocked basin west of Frontrangia dried up.

The mineral gypsum is lighter than most rocks and flows slowly even in a solid state, like glacial ice or Silly Putty. As the river removes overlying layers, the gypsum flows toward the river and upward. Along the sides of the valley, overlying rocks drop downward as gypsum moves out from beneath them. This ongoing process helps to widen the valley.

Wolcott lies within the deepest part of the syncline; west of it the highway encounters older and older strata:

- Cretaceous limestone forming a yellowish surface on hills across the river from milepost 157;

- black Benton shale, also Cretaceous, on slopes below the limestone;

- resistant Cretaceous Dakota sandstone, with coaly beds, capping hills at milepost 156;

- varicolored purple and green shale of the Jurassic Morrison formation, across the river from milepost 156;

- blocky, peach-colored Entrada sandstone, also Jurassic, at milepost 154;

- bright red Triassic Chinle formation at milepost 153;

Near the town of Gypsum, Pennsylvanian strata are folded and contorted by upward flow of gypsum layers where overlying rock layers have been removed by the Eagle River. —T. S. Lovering photo, courtesy of U.S. Geological Survey

Just west of milepost 136, black rock in sagebrush on either side of I-70 is a basalt flow coming from a gulch in the bluffs north of the highway. The flow erupted from a fissure at the base of a cinder cone—the dark mass above the center of this photograph. Colorado's youngest volcano, a little over 4,000 years old, is quarried for its foamy, lightweight cinders, which are used for concrete blocks and landscaping. —Felicie Williams photo

- Pennsylvanian Eagle Valley evaporite. Where the valley widens, shale and gypsum are more than 1,000 feet thick.

Pennsylvanian gypsum and shale beds are as contorted as marble cake, with gray, white, and brown swirls and folds, all caused by flowing gypsum. Contortions in these rocks, not at all like the normal parallel bedding of sedimentary rocks, show up particularly well on the north bank of the river east and west of Gypsum.

In this area gypsum is mined for manufacture of drywall. Notice that weathering of the gypsum-rich shale produces monotonously poor soil on which few plants grow.

Gypsum and salt, more buoyant than other rocks, tend to push upward against overlying strata, just as a helium balloon might push up against an awning or the roof of a circus tent. West of the confluence of the Eagle and Colorado Rivers, where the Colorado has removed much of the overlying weight, the Eagle Valley evaporites moved upward more easily, bending up the dark Pennsylvanian Belden shale now exposed on either side of the river. Thin beds of low-grade coal in the Belden shale began as densely vegetated swamps edging a marine embayment.

The road up the Colorado River from Dotsero makes an interesting and informative side trip. The river cuts deeply through Pennsylvanian and Mesozoic sedimentary rocks as it crosses the syncline between the Gore Range and the White River Plateau to the west. Two miles up this road is Deep Creek, and 1.5 miles up Deep Creek along a dirt road and trail, Devonian and Mississippian rocks contain many fossils.

Interstate 70
Dotsero—Rifle
43 miles (69 km)

West of Dotsero, I-70 enters Glenwood Canyon, a 15-mile-long gorge excavated by the Colorado River. Most of the canyon-cutting took place in Pleistocene time, when runoff was far greater and floods much more common than now, but you can't see the muddy river in spring without realizing that the process is still going on. The position of the river was governed originally by the contact between hard Mississippian limestone and soft Pennsylvanian shale, but now the canyon has cut through older Paleozoic layers and into Precambrian granite and gneiss, following vertical joints in the ancient rock.

Paleozoic rocks are fairly easily recognized as they appear along the highway. From east to west, they are:

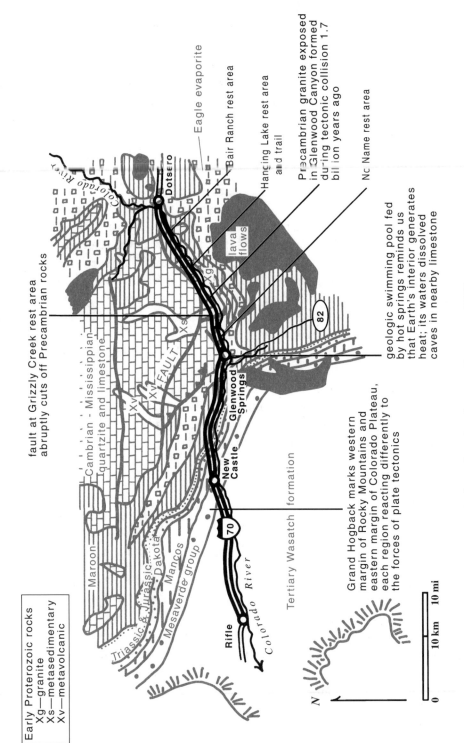

Early Proterozoic rocks
Xg—granite
Xs—metasedimentary
Xv—metavolcanic

fault at Grizzly Creek rest area
abruptly cuts off Precambrian rocks

Eagle evaporite

Bair Ranch rest area

Hanging Lake rest area
and trail

Precambrian granite exposed
in Glenwood Canyon formed
during tectonic collision 1.7
billion years ago

Nc Name rest area

geologic swimming pool fed
by hot springs reminds us
that Earth's interior generates
heat; its waters dissolved
caves in nearby limestone

Colorado River

Dotsero

lava
flows

Cambrian - Mississippian
quartzite and limestone

Maroon

Triassic & Jurassic

Dakota

Mancos

Mesaverde group

Tertiary Wasatch formation

Grand Hogback marks western
margin of Rocky Mountains and
eastern margin of Colorado Plateau,
each region reacting differently to
the forces of plate tectonics

Glenwood
Springs

New
Castle

Rifle

Colorado River

82

70

N

0   10 km   10 mi

Geology along I-70 between Dotsero and Rifle.

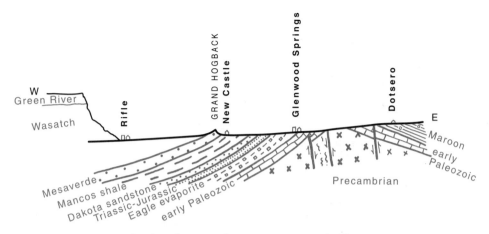

Section along I-70 between Dotsero and Rifle.

- Mississippian Leadville limestone, massive and cliff-forming, about 260 feet thick, high up to the west at the entrance to the canyon;
- Devonian thin-bedded Chaffee formation, limestone and shale forming greenish gray slopes and ledges below the Mississippian limestone;
- Ordovician Manitou dolomite and interbedded shale in brownish cliffs near I-70 at the Garfield County line, just east of Bair Ranch rest area;
- Cambrian Sawatch quartzite, with light and dark purplish banding, forming the 400- to 600-foot cliff at milepost 130;
- coarse pink Precambrian granite around 1.7 billion years old near milepost 128.

At milepost 128 a fault brings Pennsylvanian rocks down to roadside level, and the sequence starts all over again.

At milepost 125 Hanging Lake rest area provides access to Hanging Lake, a beautiful travertine lake up a 1.2-mile trail. Rest area buildings are made of Sawatch quartzite, the rock that forms the banded cliffs above. Along the trail east of the rest area, boulders of schist, pegmatite, and fine-grained granite are typical of Precambrian rocks in Glenwood Canyon. Small tight folds mark the schist, set off by coarse grains of black biotite and white muscovite. Deposited originally as sedimentary rock, the schist was metamorphosed and intruded by the granite during the Early Proterozoic Orogeny. Pegmatite veins decorate both schist and granite.

Hanging Lake Trail crosses some loose rock debris that hides the unconformity between Precambrian and Cambrian rocks. The lower part of the Sawatch quartzite is streaked with purple iron-rich minerals and pink feldspar eroded from the underlying granite. A small cave by the trail

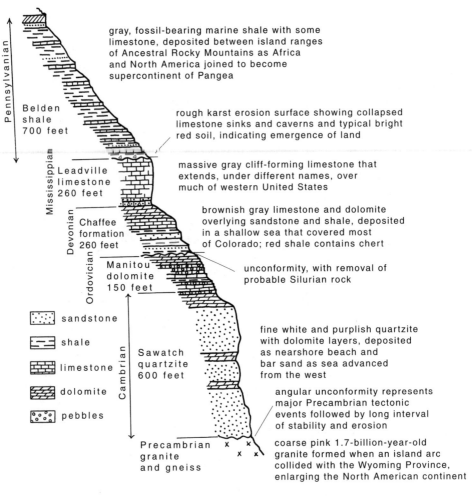

Pennsylvanian

gray, fossil-bearing marine shale with some limestone, deposited between island ranges of Ancestral Rocky Mountains as Africa and North America joined to become supercontinent of Pangea

Belden shale 700 feet

rough karst erosion surface showing collapsed limestone sinks and caverns and typical bright red soil, indicating emergence of land

Mississippian

Leadville limestone 260 feet

massive gray cliff-forming limestone that extends, under different names, over much of western United States

Devonian

Chaffee formation 260 feet

brownish gray limestone and dolomite overlying sandstone and shale, deposited in a shallow sea that covered most of Colorado; red shale contains chert

Ordovician

Manitou dolomite 150 feet

unconformity, with removal of probable Silurian rock

sandstone

shale

limestone

dolomite

pebbles

Cambrian

Sawatch quartzite 600 feet

fine white and purplish quartzite with dolomite layers, deposited as nearshore beach and bar sand as sea advanced from the west

angular unconformity represents major Precambrian tectonic events followed by long interval of stability and erosion

Precambrian granite and gneiss

coarse pink 1.7-billion-year-old granite formed when an island arc collided with the Wyoming Province, enlarging the North American continent

Stratigraphic column of rocks exposed in Glenwood Canyon.

reveals the soft greenish shale layers that cause the quartzite's thinly-bedded weathering pattern.

Along Dead Horse Creek, light tan travertine coats and cements boulders and occasionally forms mounds, presaging the huge travertine mound that supports Hanging Lake. Travertine forms by evaporation of calcium-laden water, usually water that has passed through limestone or dolomite. The wall behind the lake is mostly quartzite and contains thin layers of porous dolomite through which rain and snowmelt have apparently etched out channels. The amount of travertine here tells us that there are extensive solution cavities within the mountain. Bridal Veil Falls cascades lightly over the edge of the lake and down the mound.

Cambrian, Devonian, and Mississippian strata rise above coarse pink Precambrian granite in Glenwood Canyon. The beveled surface of the granite is halfway up the photo. —Felicie Williams photo

Shoshone Dam diverts Colorado River water to a power plant 2 miles downstream from the dam. Near the dam the abrupt contact between the gray granite, garlanded with pink and white veins, and overlying Sawatch quartzite marks a 1.2-billion-year gap. By the end of this time mountains that had existed here, products of at least three Precambrian orogenies, were eroded to a featureless, sea-level surface. More than twice as much time is represented by this gap, known as the Great Unconformity, as by all the sedimentary rocks above it.

West of the power plant Paleozoic sediments dip westward; the highway progresses through older to younger rocks, the reverse of the progression outlined earlier.

Weakened by fault zones, the valley widens near both Grizzly Creek and No Name rest areas. Near milepost 119, a fault abruptly cuts off Mississippian

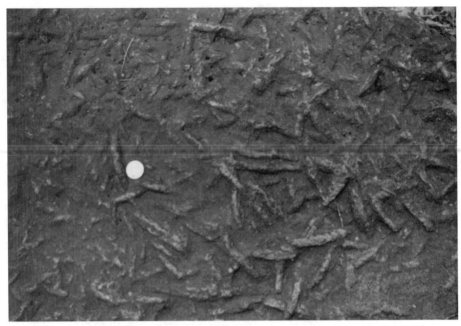

A sign of life from Cambrian time, these strange patterns in the Sawatch quartz-
ite were left by unknown invertebrate animals that lived 510 million years ago.
Quarter shows scale. —Felicie Williams photo

Hanging Lake lies on a travertine mound deposited in a steep-walled ravine
by dolomite-laden water emerging from cliffs above. —Felicie Williams photo

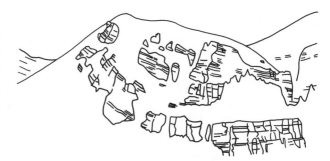

South of Grizzly Creek rest area, a gentle fold appears where the Sawatch quartzite has dropped down west of a north-south fault.

rocks and brings up Precambrian granite once more. West of the highway tunnel, which passes through the granite, I-70 crosses the Precambrian-Cambrian contact and travels through older to younger rocks.

As this part of Colorado, near the equator in Pennsylvanian time, rose above sea level, a karst surface developed on the Mississippian limestone. Watch the top of the gray limestone cliff near mileposts 119 and 120 for evidence of old solution caves and sinkholes, and for reddish soil full of broken limestone blocks. Similar features develop today in warm, humid regions underlain by limestone.

Hot springs at Glenwood Springs flow from shattered rocks in a fault zone along the Colorado River, but their exact site is difficult to see from I-70. Total flow of all the springs approaches 2,750 gallons per minute, with water temperature around 123 degrees Fahrenheit. The spring water is salty; it probably flows through Pennsylvanian salt layers on its way to the surface. There has been talk of removing salt from the Colorado River at Glenwood Springs for the sake of downstream users, but so far salt-extraction processes have been too expensive to be practical.

Massive gray rock north of I-70 just west of Glenwood Springs is the Leadville limestone. West of it are contorted gypsum-bearing Pennsylvanian shales, and then the dark red Maroon formation, 3,000 feet of red rocks deposited along the eastern flank of Uncompahgria in Pennsylvanian and Permian time.

West of Glenwood Springs and north of the highway, a 1994 fire destroyed groundcover on shaly slopes of Storm King Mountain. Before reseeded plants developed enough roots to bind the soil, heavy rains caused mudflows here.

West of the burn the interstate enters Triassic, Jurassic, and Cretaceous rocks that edge the west side of the mountains, with red Chinle shale, light brown Entrada sandstone, varicolored slope-forming Morrison formation, and the Dakota sandstone. The last forms a hogback that is the geological mirror image of the Dakota Hogback east of the mountains.

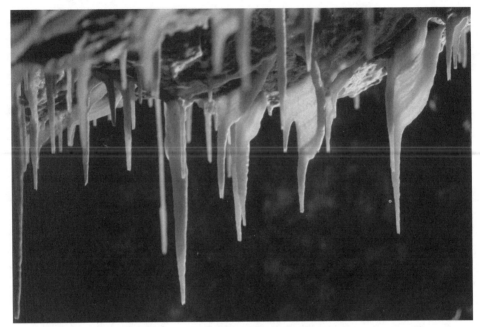

Delicate foot-long stalactites formed as water, carrying calcite dissolved from Mississippian limestone, dripped from the cave ceiling at Glenwood Caverns, formerly called Fairy Cave. When the floor of the valley was higher, hot spring water flowed up through cracks in the limestone, dissolving the cave chambers. Guided tours of the caverns are available. —Felicie Williams photo

About 9 miles west of Glenwood Springs, I-70 swings northward across a valley in gray Mancos shale. Deposited on the floor of the shallow Cretaceous sea, the Mancos shale was originally continuous with the Pierre shale east of the Rockies; geologists named the two units separately before they recognized their similar origin. West of the Mancos shale rises a particularly prominent ridge, the Grand Hogback, topped by resistant sandstone of the Mesaverde group. This series of thick nearshore Cretaceous sandstones contains many coal layers and coaly shales.

Several old coal mine tunnels and dumps mark the Grand Hogback near New Castle. Red streaks on slopes below the Mesaverde cliff are shale baked brick red by burning coal seams. Coal was mined here beginning in 1886, but this particular coal releases gases that ignite easily, and mine explosions caused several tragic accidents. Between 1896 and 1918 burning mines were sealed. Snow falling here melts unusually rapidly, showing that underground coal is still smoldering.

A narrow gap just west of New Castle passes through the youngest of the Cretaceous strata, shoreline sandstone and conglomerate deposited as the sea receded eastward. Coarsening debris in these rocks indicates that mountains were beginning to rise as the sediment was deposited.

West of the Grand Hogback, I-70 passes into flat-lying Tertiary rocks, the Wasatch and Green River formations. The pinkish gray Wasatch formation, poorly cemented sandstone and shale, represents sand and silt washed from the Rockies and the Uncompahgre Plateau into the basin between them. The town of Silt lies on silt derived from the Wasatch formation. Eventually a lake filled the basin, and lakebed deposits, now topping cliffs to the north, make up the Green River formation. Dark brown ledges of shale within the lakebeds are oil shale, an immense and as yet untapped petroleum reserve. Notice how poorly consolidated these Tertiary sedimentary rocks are—not nearly as durable as Paleozoic ones farther east that were hardened by burial and cemented with groundwater-borne minerals.

Between 1921 and 1977 mills at Rifle processed vanadium and uranium ore mined near Rifle Falls. Tailings were dumped near the Colorado River without considering environmental damage. Cleanup of the site began in 1988. Radioactive tailings and unsafe mill structures have been removed and safely stored far from the river and populated areas.

From Rifle, Colorado 13 and Colorado 325 lead through Rifle Gap to Rifle Box Canyon, a narrow oasis carved in the Leadville limestone with remarkable travertine terraces. The road guide **Colorado 13: Craig—Rifle** in Chapter VI covers this side trip.

Like the Dakota Hogback east of the mountains, the Grand Hogback outlines the Colorado Rockies for many miles, marking the boundary between the Rockies and the Plateau Country. —Jack Rathbone photo

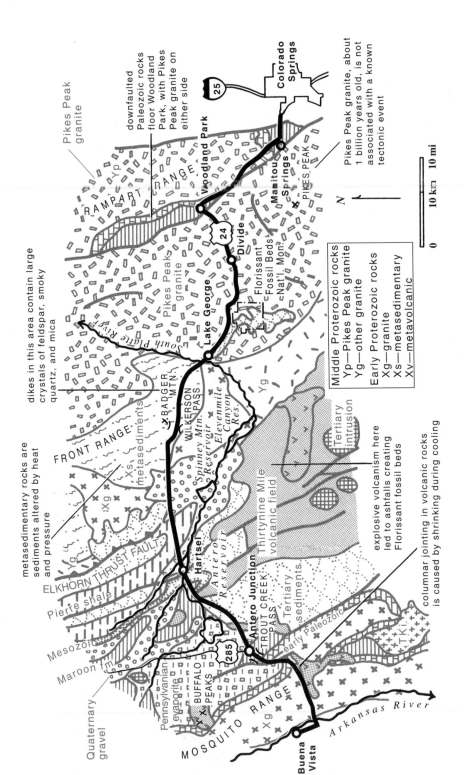

Geology along U.S. 24 between Colorado Springs and Antero Junction.

U.S. 24
Colorado Springs—Antero Junction
81 miles (130 km)

Plunging into the mountains, U.S. 24 follows very closely the line of the Ute Pass fault, one of the major faults edging the east side of the Front Range. Slicing through the mountains, this fault separates the Rampart Range from the rest of the Front Range and can be traced for about 60 miles, from Cheyenne Mountain north beyond Woodland Park. Rocks of the Rampart Range are raised more than 1,000 feet higher than those to the west in the main part of the Front Range. Except at Pikes Peak and a few other high points, erosion of the Tertiary pediment has evened out the difference. The fault zone is quite wide; its broken, fractured rocks erode easily, establishing the route of Fountain Creek.

Beeline section paralleling U.S. 24.

Near Manitou Springs and Woodland Park, Paleozoic rocks still lie in place on the Precambrian granite of the Rampart Range, represented by Cambrian sandstone, Ordovician to Mississippian limestone, and the coarse red sandstone and conglomerate of the Pennsylvanian Fountain formation. At Woodland Park these Paleozoic rocks form a long, narrow syncline trending northwest along the fault.

The Rampart Range consists of Pikes Peak granite, as does the area southwest of it. The granite is part of an ancient batholith about a billion years old. Typically pink, with chunky crystals of pink feldspar, glassy quartz, and biotite, it is cut in many places by pegmatite bands with large crystals and by white veins of light-colored muscovite and milky quartz.

Half a mile west of Woodland Park, U.S. 24 swings southwest and crosses Ute Pass fault. Sheared and broken granite along the fault has weathered deeply.

By milepost 280, U.S. 24 is on the 8,000- to 10,000-foot Tertiary pediment. The granite here has been weathering for at least 28 million years, and in places along the highway has turned into bouldery gravel and coarse sand called grus. Glimpses of Pikes Peak to the southeast show broad, smoothly contoured, glacially scoured uplands where glacial erosion and wind have removed these products of weathering.

Near the town of Florissant, roadside rocks change suddenly to thinly bedded, fine-grained, light gray shale. These are the Florissant lakebeds, formed in Oligocene time when volcanic ash from the Thirtynine Mile volcanic field to the west showered into a mountain lake. The ashfalls brought down many insects, leaves, and even birds, some of them now preserved as fossils in the thin gray beds.

From milepost 268 near Lake George, Crystal Peak is in sight in the distance to the north. Large blue-green feldspar known as amazonite, as well as smoky quartz crystals of museum quality, occur in pegmatite veins on this peak. Most of them are on private property, so permission is needed to collect mineral specimens.

On U.S. 24 west of Lake George, metamorphic rocks appear, often with large granite blobs injected into them, as at mileposts 255 to 256. These rocks were sediments near the margin of the continent 1.8 billion years ago; they were metamorphosed 1.7 billion years ago. Both then and 1.4 billion years ago, granitic magma melted its way between bands of gneiss and schist. These granites are far older than the 1.1-billion-year-old Pikes Peak granite.

Right at Wilkerson Pass, Precambrian rocks are concealed by Tertiary volcanic flows from the Thirtynine Mile volcanic field southwest of here, one of many Tertiary volcanic centers in Colorado. A Forest Service information center at the pass explains some of the geology of the region.

From Wilkerson Pass look westward over South Park, one of Colorado's four big intermountain valleys, to the Mosquito Range, and beyond it to the high peaks of the Sawatch Range. South Park is a lopsided faulted syncline. Its eastern side is bordered by the Elkhorn thrust fault along which Precambrian rocks were thrust westward over Cretaceous and early Tertiary rocks of the valley floor during the Laramide Orogeny. On the west, South Park is edged by tilted, faulted sedimentary rocks of the Mosquito Range.

From the center of South Park the dark double summits of Buffalo Peaks are in view to the west. They contain thick layers of Tertiary lava and volcanic ash composed chiefly of feldspar and dark minerals. Near Hartsel a hogback of Cretaceous Dakota sandstone cuts South Park in two, with Pierre shale in the valley to the east and older rocks to the west—the same pattern seen east of the Front Range.

The Spanish called South Park "Valle Salada" or "Bayou Salado," recognizing the salty white deposits on flat parts of the valley floor. The white coating on these Tertiary lake deposits is a mixture of salts, primarily common salt (NaCl, or halite) and a mineral resembling Epsom salt ($MgSO_4$). Taste it if you doubt this.

At Antero Junction, U.S. 24 joins U.S. 285. See the road guide for U.S. 285 from Fairplay to Poncha Springs.

Florissant Fossil Beds National Monument
5.5 miles (9 km) from U.S. 24

Fossil leaves, insects, fish, and even a marsupial have been exquisitely preserved in fine layers of volcanic ash that fell into ancient Lake Florissant nearly 35 million years ago. The lake formed when lava flows dammed a mountain valley. Swampy in places, it attracted an abundance of plant and insect life. Graceful palms and towering sequoias growing by its banks harbored birds and small animals almost like those we know today.

Volcanoes to the west in the Thirtynine Mile volcanic field and the Sawatch Range periodically filled the air with clouds of volcanic ash. Carried on the wind, some of the ash fell into Lake Florissant, killing fish, bringing

This petrified sequoia stump is one of several fossil tree stumps at Florissant Fossil Beds National Monument. A grove of these trees buried in volcanic ash became petrified as groundwater deposited silica in cells of the original wood. —U.S. National Park Service photo

down thousands of insects, and burying drifting leaves and bits of wood. The ash settled gently and quickly, preventing decomposition of plant and animal material. More volcanic ash, falling in thick unstable layers on surrounding hills, washed down in mudflows that buried trees and small mammals. Like an early Pompeii, the area was then covered with more ash.

Rainwater and groundwater filtering through the ash removed much of the plant and animal material, leaving only thin films of carbon—just enough to preserve the delicate traces of wings, leaf veins, even the soft parts of some plants and animals. Such perfect fossils are extremely rare. Since their discovery in 1874 the lakebeds have yielded thousands of

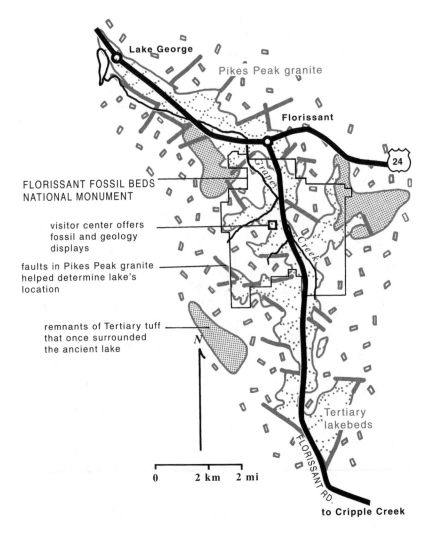

Geology of Florissant Fossil Beds National Monument.

specimens to museums and study collections all over the world. More than 1,100 species of insects, including dragonflies, butterflies, ants, flies, beetles, and bees; fossil leaves from birch, willow, maple, beech, and hickory; needles from fir trees; and fronds of giant sequoias have been collected. Petrified tree stumps, preserved by silica derived from the volcanic ash, show that some of these trees were as large as great redwoods growing today in California. The fossils indicate a warm climate.

Regional uplift of the Southern Rockies led to increased erosion that gradually exposed the old lake sediments. Only those that are on downfaulted blocks of Pikes Peak granite were preserved. The original lake may have been larger than the 12-mile-long, 2-mile-wide crescent that exists now.

Fossil collecting is not permitted within the monument. The delicate fossils do not survive long if they are exposed to weather, so newly uncovered fossils are removed immediately for preservation.

Like fossils of leaves and other insects, this bee is preserved as a fine film of carbon darkening light-colored volcanic ash at Florissant Fossil Beds National Monument. —U.S. National Park Service photo

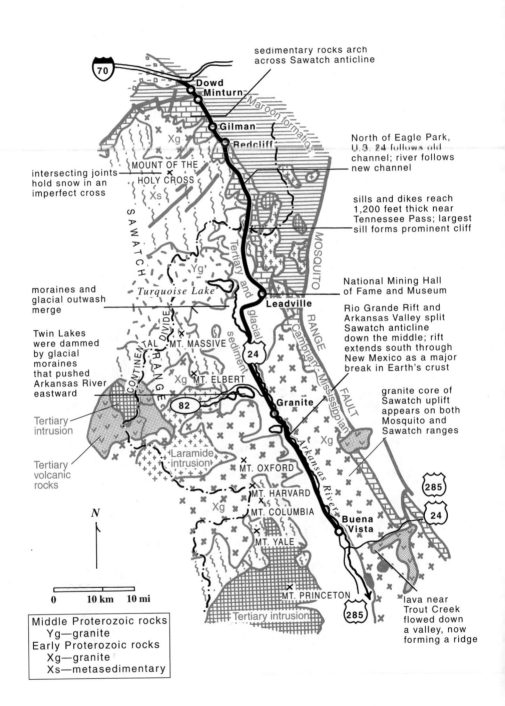

Geology along U.S. 24 between Buena Vista and I-70.

U.S. 24
Buena Vista—Interstate 70
72 miles (116 km)
Includes alternate route via Climax on Colorado 91

Between Buena Vista and Leadville, U.S. 24 ascends the Arkansas Valley, with the Mosquito Range on the east and the Sawatch Range on the west. In sight are five Fourteeners of the part of the Sawatch Range known as the Collegiate Range: Mounts Princeton, Yale, Columbia, Harvard, and Oxford. Mt. Elbert, at 14,433 feet Colorado's tallest, lies farther north. Big alluvial fans below these peaks were created by torrential Ice Age streams that dropped loads of rock and sand as they reached the valley floor. High cirques, U-shaped valleys between the peaks, and glacial moraines are further manifestations of an icy past.

Though glaciers from side valleys rarely reached into the Arkansas Valley, their meltwater contributed to the several hundred feet of Pleistocene stream and lake deposits on the valley floor. Visible in roadcuts, stream deposits are layered (stratified) collections of smooth, rounded cobbles and sand, mostly glacial outwash. Moraines deposited directly by ice are a completely unsorted hodgepodge of angular rock fragments, sand, and clay.

The Arkansas Valley is the northern end of the Rio Grande Rift, a long, thin wedge that subsided as land to the east and west rose 5,000 feet or so during mid-Tertiary regional uplift. Sedimentary rocks in the Mosquito Range dip east; those in the Sawatch Range dip west. Together the ranges form a single anticline, sliced lengthwise by the downfaulted rift valley.

High summits of the Collegiate Range tower beyond Buena Vista. Most of the peaks west and northwest of Buena Vista are composed of Precambrian gneiss, schist, and granite. —Jack Rathbone photo

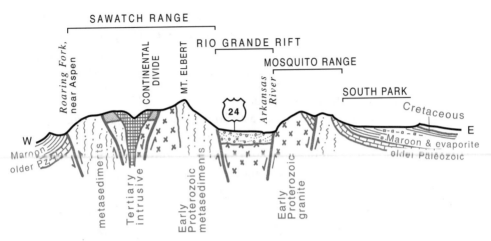

Section across U.S. 24 south of Leadville.

From Mt. Yale north, peaks of the Sawatch Range are composed of Precambrian gneiss and schist derived from even more ancient sedimentary and volcanic rocks deposited along the continent's margin about 1.8 billion years ago. They were altered into metamorphic rock by the immense heat and pressure of the Early Proterozoic Orogeny. Pinkish granite near the town of Granite intruded at about the same time, possibly formed from the complete melting of crustal rocks.

The southern Mosquito Range, much lower than the Sawatch Range, was not glaciated; it lacks the U-shaped valleys, terminal moraines, and large outwash fans of the Sawatch Range. Near Granite, debris from melting glaciers in the Sawatch Range forced the Arkansas River to the eastern side of the valley, where it carved down into solid granite. The river's earlier channel is buried beneath moraine and outwash gravels.

Immense gravel terraces on both sides of the valley north of milepost 187 merge upstream with glacial moraines. Each terrace level, once a river floodplain, formed during a relatively stable period when rivers were heavily choked with glacial debris. Later dissected as the river, freed of its ice-fed burden, deepened its channel, they record the on-again-off-again history of glaciation.

Mt. Elbert and Mt. Massive dwarf slag heaps near Leadville, waste from the town's mills and smelters. Mines pock steep slopes east of the picturesque metropolis. The Leadville Mining District, active up until 1998, yielded over $500 million in silver, lead, zinc, gold, and other metals. Mining began in the 1860s when placer gold was discovered along the Arkansas River. Years later, heavy gray material that settled out with the gold was identified as lead sulfide laced with silver.

The minerals were deposited well after the Laramide Orogeny when rich solutions from intruding mid-Tertiary magmas penetrated faults in Devonian and Mississippian limestone, which provided the right chemical environment for mineral precipitation. Faults in the mining area bring enriched veins to the surface stairstep style all the way up the slope above Leadville.

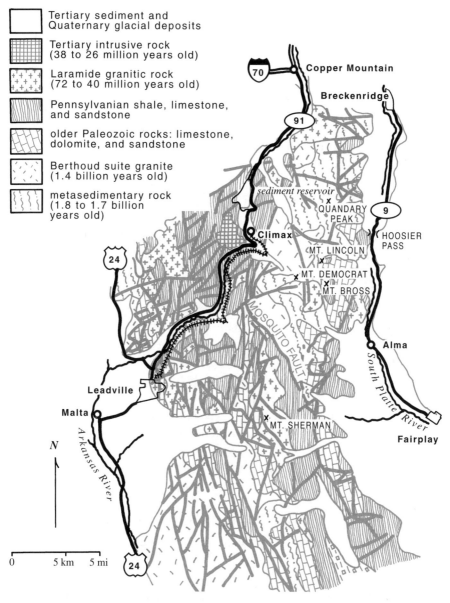

Faults that honeycomb the Leadville area provided routes for mineralizing fluids.

Across the Mosquito Range, mines at Breckenridge and Fairplay yielded similar gold placers and silver ores, though neither was as fabulously rich as Leadville.

Leadville's peacefulness now belies the amount of activity here during its heyday when hundreds of miles of underground tunnels probed beneath and east of the city and seventy-five mills and seventeen smelters were processing ore.

Little care was taken to contain lead and other toxic metals. Groundwater interfered with mining, so a drainage tunnel was excavated to carry it off, releasing acid and around 200 tons per year of heavy metals, including lead, into nearby streams. In winter lead-rich smelter slag was used to sand roads. In 1983 mining companies and the government began extensive testing and cleanup, including treatment of mine drainage.

The National Mining Hall of Fame and Museum describes the town's mining history and displays many mineral specimens. Mineral hunting on mine dumps is fun, but remember that old mines are extremely dangerous. The Leadville area is pocked with collapsed mine tunnels!

North of Leadville, U.S. 24 crosses hilly terminal moraines of three converging glaciers. Then it skirts the flat bed of a one-time lake dammed by glacial moraines, and climbs toward Tennessee Pass. Hills east of the highway contain a complicated mosaic of fault blocks in which Precambrian, Paleozoic, and Tertiary rocks are sliced and jumbled together. The fault zone parallels the highway across Tennessee Pass, where fault movement has ground the rocks into a fine, powdery, clayey material, setting the stage for erosion of the pass.

The Rio Grande Rift ends as the Mosquito and Sawatch Ranges converge around the northern end of their anticline. Northeast of the pass, a thick cliff-forming sill of Laramide intrusive rock is sandwiched between Pennsylvanian rock layers, forming a high ridge. Scores of dikes complicate the geology just north of the pass.

The valley widens at Eagle Park, where a Precambrian shear zone, trending northeast across the ranges, weakened the rocks. Dammed by a glacial moraine, the mountain basin contains thick lake, stream, and glacial deposits. On cliffs across Eagle Park, Precambrian rocks are overlain by Paleozoic sedimentary rocks that dip gently southeast, continuing the Mosquito Range trend.

Northwest of Eagle Park the Eagle River enters a rocky canyon cut when glaciers diverted the stream from its normal course. U.S. 24 follows a preglacial channel farther west.

At milepost 154, Paleozoic strata come into view, with a Cambrian quartzite cliff above the high bridge and Devonian and Mississippian limestone above that. At Gilman, Precambrian and Cambrian rocks lie below the town,

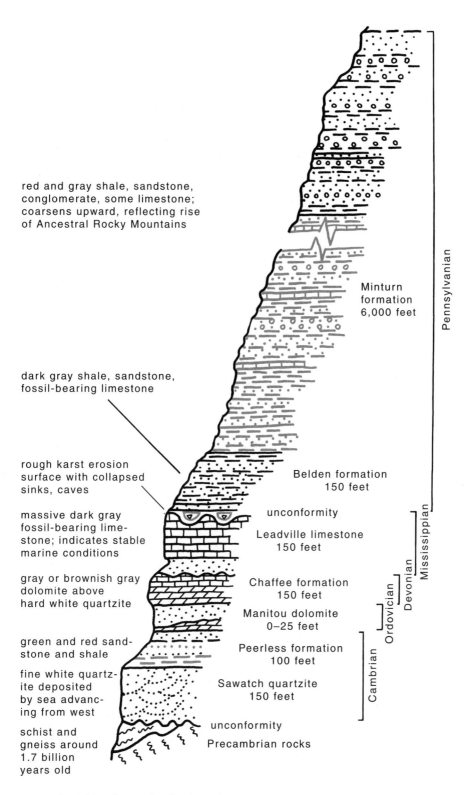

red and gray shale, sandstone,
conglomerate, some limestone;
coarsens upward, reflecting rise
of Ancestral Rocky Mountains

Minturn
formation
6,000 feet

dark gray shale, sandstone,
fossil-bearing limestone

Belden formation
150 feet

rough karst erosion
surface with collapsed
sinks, caves

unconformity

massive dark gray
fossil-bearing lime-
stone; indicates stable
marine conditions

Leadville limestone
150 feet

gray or brownish gray
dolomite above
hard white quartzite

Chaffee formation
150 feet

Manitou dolomite
0–25 feet

green and red sand-
stone and shale

Peerless formation
100 feet

fine white quartz-
ite deposited
by sea advanc-
ing from west

Sawatch quartzite
150 feet

unconformity

schist and
gneiss around
1.7 billion
years old

Precambrian rocks

Pennsylvanian

Mississippian

Devonian

Ordovician

Cambrian

Stratigraphic column of rocks along U.S. 24 between Tennessee Pass and Minturn.

the town and its mines are in Devonian and Mississippian strata, and Pennsylvanian rocks show up on steep slopes east of the town, above and below the highway.

The Gilman-Redcliff Mining District was for a time Colorado's main source of zinc; its mines closed in 1983. As at Leadville, most of the ores were in Devonian and Mississippian limestone that helped to precipitate ore minerals. Zinc occurs in fingerlike orebodies, possibly deposited in caves and caverns, near the top of massive Mississippian limestone. Extending from the ends of the fingers down through the Mississippian and Devonian layers are vertical chimneys of copper-silver-lead ores. And gold- and silver-bearing veins cut Cambrian and Precambrian rocks below. This is a classic example of mineral zoning—different ores at different levels—a common feature of metal mines.

When the mines were active, pumps ran night and day to prevent flooding. Since then, hazardous materials have been removed, and the mines have been sealed and allowed to flood. Waste, collected in one pile, is capped to keep water from leaching out poisonous metals. Water still seeping from the mining area is treated before it reaches the river.

Tall banded cliffs above Minturn were once debris washed from the Ancestral Rockies in Pennsylvanian time, with limestone layers formed by marine organisms. Where these rocks—the Minturn formation—come close to the highway farther north, you'll see massive limestone, rich in fossils, deposited as fringing reefs around Frontrangia, a reminder that Colorado lay in the tropics in Pennsylvanian time.

Climax Mine

An alternate route from Leadville to I-70 follows Colorado 91 through Paleozoic rocks past Climax, the world's largest molybdenum mine. Exhibits near the parking lot explain the geology of the huge orebody, which lies like three superimposed caps over three successive intrusions.

Ore at Climax is low in metal content—only ⅓ of 1 percent is molybdenum—but the large size of the orebody and efficient mining methods made Climax a profitable mine. For many years it produced over half the world's molybdenum. Mothballed now, the mine still contains lots of ore. Molybdenum was rare and little-used when Climax was discovered, so mine owners experimented with it, finding it improved the quality and hardness of steel. As demand grew, the mountaintop was mined completely away. In a "glory hole" like this, ore broken by blasting is funneled downward, then mined in underground workings.

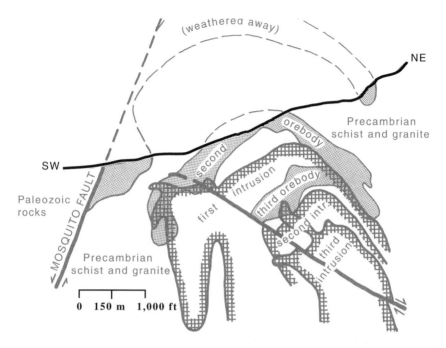

The three Climax orebodies, in cross section here, developed deep underground as superheated vapors escaped from three successive Tertiary intrusions. Pressure caused by escaping vapors fractured the rocks intensely. Where pressure and temperature were just right, the vapors reacted chemically with the shattered rocks, leaving caps of molybdenum-bearing minerals.

U.S. 34
Loveland—Granby via
Rocky Mountain National Park
92 miles (148 km)

The mountain front west of Loveland shows the complex folding and faulting that in many places edges the Rockies. Loveland lies on dark gray Cretaceous Pierre shale that weathers into soil and is rarely exposed. A few miles west of town, older Cretaceous rocks come to the surface, turned up along the edge of the Front Range.

Near milepost 88 the Dakota sandstone swoops up into a prominent hogback, with a valley of Morrison shale followed by a hogback of pink Lyons sandstone closer to the mountains. Then the Dakota sandstone turns down again—vertically, so it appears as a vertical wall locally known as the Devils Backbone. Farther west the same rocks bend up to the surface again, repeating the sequence. Finally, less steeply tilted, the Pennsylvanian

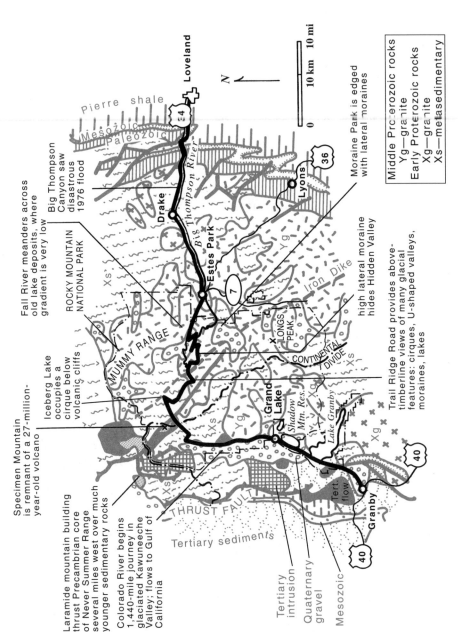

Geology along U.S. 34 between Loveland and Granby.

Specimen Mountain is remnant of a 27-million-year-old volcano

Fall River meanders across old lake deposits, where gradient is very low

Big Thompson Canyon saw disastrous 1976 flood

Pierre shale

Mesozoic

Paleozoic

Loveland

34

Big Thompson River

Drake

Estes Park

ROCKY MOUNTAIN NATIONAL PARK

Laramide mountain building thrust Precambrian core of Never Summer Range several miles west over much younger sedimentary rocks

Iceberg Lake occupies a cirque below volcanic cliffs

MUMMY RANGE

Xs

7

Lyons

36

Moraine Park is edged with lateral moraines

Iron Dike

Yg

LONGS PEAK

CONTINENTAL DIVIDE

Xs

high lateral moraine hides Hidden Valley

Colorado River begins 1,440-mile journey in glaciated Kawuneeche Valley; flows to Gulf of California

Xs

Yg

Grand Lake

Shadow Mtn. Res.

Xs

Xg

Lake Granby

Trail Ridge Road provides above-timberline views of many glacial features: cirques, U-shaped valleys, moraines, lakes

40

Tert. flow

Granby

40

THRUST FAULT

Tertiary sediments

Tertiary intrusion

Quaternary gravel

Mesozoic

N

0   10 km   10 mi

Middle Proterozoic rocks
Yg—granite
Early Proterozoic rocks
Xg—granite
Xs—metasedimentary

Fountain formation surfaces by the river. Just beyond it, across a fault, Precambrian rocks form the canyon wall of the Big Thompson River.

Before 1976 Big Thompson Canyon sheltered hundreds of pleasant, pine-shaded homes, summer cabins, motels, and picnic areas. Fed by high-country streams from Rocky Mountain National Park and the region north of it, the river rose during the night of July 1, 1976, when heavy rains fell on

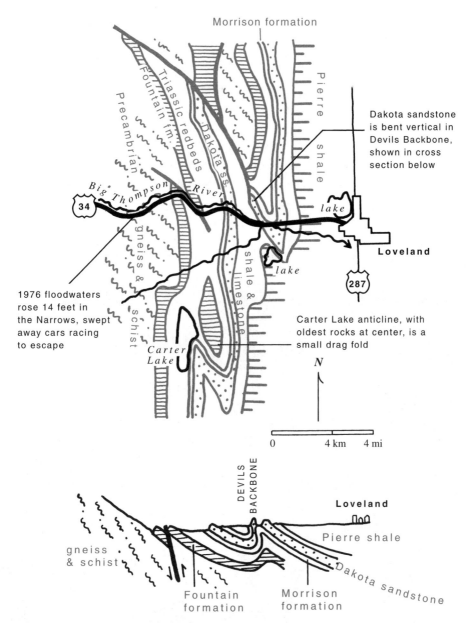

Detail of geology west of Loveland along U.S. 34.

its headwaters. Houses, cabins, trees, cars and trucks, bridges, a small dam, and U.S. 34 were wiped out by a flash flood so savage it moved 20-foot boulders. At least 139 people perished, many of them while trying to outrun the flood by driving down the canyon. At the Narrows near the canyon mouth, where it didn't even rain, the stream rose 14 feet in just a few minutes. Evidence of the rising water's rampage still appears in the canyon, a reminder that nature has not finished shaping these mountains.

The canyon walls display the oldest rocks in this part of Colorado: Precambrian banded gneiss and shiny schist formed from sediments deposited along the edge of the Wyoming Province 1.8 billion years ago. Later compressed and heated the during the Early Proterozoic Orogeny, they are marked by pegmatite veins, some with large mica crystals.

Between the Narrows and Drake the valley is wider and fairly straight, as the river follows a fault. A few miles west of Drake the canyon walls become granite, part of a batholith of the Routt plutonic suite, dating back about 1.7 billion years. Notice that granite weathers into rounded boulders and domes, distinct from craggy cliffs of metamorphic rocks.

The power of a flash flood is hard to imagine. This piece of a dam was carried 200 yards downstream in a matter of minutes by the Big Thompson Flood of 1976. —Felicie Williams photo

Dark red granite encountered a few miles east of Estes Park is part of the Berthoud plutonic suite, 1.4 billion years old. This rock is present over much of Rocky Mountain National Park, forming most of the scenery for which the park is famous. Its red color is due to slow radioactive decay of potassium in the feldspars.

Though Estes Park with its border of high granite domes looks like a glacier-carved valley, it is below the lower elevation limit of Ice Age glaciation. It is really a deep V-shaped, stream-eroded canyon like the one you have driven through, but it is filled in with thick deposits of gravel and rock.

Lake Estes, a reservoir, serves as a staging point for western slope water brought from Grand Lake via a 19-mile tunnel under Rocky Mountain National Park and the Continental Divide. The tunnel portal is visible across the valley. From there, over 13 more miles of tunnels carry the water to cities east of the mountains, generating electricity as the water speeds downhill.

West of Estes Park, U.S. 34 is once more confined within a narrow canyon carved in Precambrian granite. Almost precisely at the 8,000-foot

Granite domes like this one near Estes Park formed because granite expands slightly when freed of overlying rock, causing joints that parallel large exposed surfaces. Water seeps into the joints and freezes repeatedly, spalling off curved slabs of rock. —Halka Chronic photo

elevation—the lower limit of Pleistocene glaciation in Colorado—the canyon opens into a glacial valley. Three glacial episodes are represented.

U.S. 34 enters the valley across the lowest, oldest terminal moraine, so deeply weathered and decayed that it is hardly recognizable. Just below the entrance station is the terminal moraine from the second glacial episode, covered with grayish brown soil and dark with pines and spruce. The youngest moraine is half a mile beyond the entrance station, where U.S. 34 curves and begins to climb. Sandy soil and fresh-looking boulders on this moraine tumbled from the toe of the glacier less than 10,000 years ago. Horseshoe Park, above the moraine, became a lake as the glacier melted; it is now floored with lake-deposited mud and sand.

The old Fall River Road at the head of Horseshoe Park is an alternate route, with a nature guide available where it begins to climb. Routes rejoin at Fall River Pass.

The Lawn Lake flood in 1982 built an instant alluvial fan across the lower part of Fall River Road. The flood resulted from washout of an earthen dam 5 miles above. The arrow points to a large power shovel. —Halka Chronic photo

Sediments deposited in a moraine-dammed lake flatten the floor of Horseshoe Park. View from Trail Ridge Road. —Halka Chronic photo

Beyond the junction with U.S. 36, U.S. 34 becomes Trail Ridge Road, following an old Ute trail across the mountains. Views of Horseshoe Park and of Moraine Park to the south show that the trail climbs a narrow ridge between two broad glaciated valleys. Picture these green parks 10,000 years ago, filled with creeping tongues of ice cut by crevasses and streaked with rock and sand, terminating at high-piled, rocky moraines. Frigid winds blew downslope from icy peaks to the west, and little vegetation—only low-growing tundra plants and willows—grew here.

Looking down on Fall River in Horseshoe Park, you can see that it meanders in tight loops on the flat valley floor; there isn't much gradient to tell it which way is downhill. For more than a mile Horseshoe Park is hidden from view by a long lateral moraine that also hides Hidden Valley.

The core of the Front Range exposed along Trail Ridge Road is mostly 1.7-billion-year-old gneiss and schist intruded by irregular bodies of 1.4-billion-year-old granite. Younger than the granite by about 100 million years, a zone of dark, iron-rich vertical dikes cut northwestward across the range. The road crosses one of them a half mile after Many Parks Curve.

Longs Peak towers to the south in an area that is mostly granite. Some geologists reason that this peak's flat top is a remnant of the Great

Unconformity, the widespread erosion surface formed at the end of Precambrian time, 650 million years ago. Elsewhere in Colorado such surfaces are covered directly by Cambrian strata. Or the summit surface may have formed when the Ancestral Rocky Mountains eroded away 275 million years ago.

As Trail Ridge Road climbs higher, many glacial features become visible: a broad upland smoothed by an ice cap and edged by glacier-carved cliffs; cirques, cirque lakes, and hanging valleys; some small glaciers, distinguished from snowfields by the liplike piles of their terminal moraines.

From Forest Canyon Overlook, look down 2,500 feet into a steep-walled, glacially gouged trough. Before glaciation this upper part of the Big Thompson River probably followed as tortuous a course here as the lower part does today. But glaciers have trouble negotiating turns, so they straighten the valleys in which they flow, grinding off spurs and smoothing curves.

Above timberline, even today's climate is arctic. Freezing nights and frigid winters leave shattered rocks and barren, jagged surfaces. Trees cannot grow, and short-seasoned midsummer wildflowers toss in chilly winds. Growth is so slow that the remains of this old Ute trail are still visible near Trail Ridge Road.

—Halka Chronic photo

In high mountain lakes like Poudre Lake, biologic processes accomplish geologic ends. Dammed by a moraine and a small beaver dam, the lake is filling with marsh grass and other vegetation and will eventually become a mountain meadow. —Halka Chronic photo

Across the canyon from the overlook, tiny lakes stairstep down hanging valleys, filling hollows scooped from bedrock by glaciers. At the head of Hayden Creek (named for a pioneer American geologist) an ice field nestles in a small ice-carved cirque.

From Fall River Pass, look into the cloverleaf cluster of three cirques that heads Fall River Canyon. Exhibits in the visitor center explain many alpine geologic and biologic features.

Northwest across the Cache la Poudre Valley from Medicine Bow Curve is red-tinted Specimen Mountain, a volcano active around 27 million years ago. At that time it was much higher and perhaps more conical. Soft yellowish ash from its explosions is visible in roadcuts near Poudre Lake.

At Farview Curve look down into the Kawuneeche Valley, where the young Colorado River meanders through a long, straight valley carved by a succession of glaciers. Fed by glacial meltwater, the stream gathered strength and sped southwest, joining other streams and rivers to carve Glenwood Canyon, Glen Canyon, and ultimately Grand Canyon in Arizona.

Across the valley, decomposing igneous rocks add a touch of brilliance to the Never Summer Range. Mid-Tertiary intrusions and late Tertiary lava

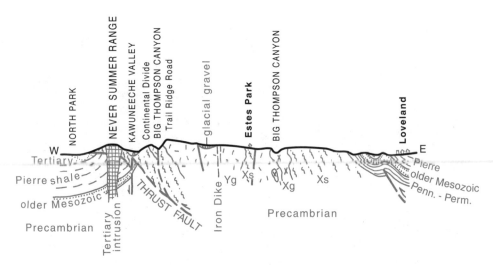

Section across the Front Range through Rocky Mountain National Park.

flows paint the range with iron oxide colors. Beneath the range is an immense thrust fault along which Precambrian rocks pushed westward, overriding Mesozoic and younger rocks by 6 miles or more, clearly reversing the usual rule of "youngest on top."

Grand Lake is naturally dammed by the lateral moraine of the big Kawuneeche Valley glacier and the terminal moraine of a smaller glacier from Paradise Creek. Shadow Mountain Reservoir and Lake Granby, also a reservoir, store Colorado River water that is eventually tunneled to Lake Estes and eastern slope cities. Mountains west of the river here are topped with Tertiary volcanic rocks that, along with broad, high river terraces, bury most of the band of upturned strata we would otherwise expect to find on this side of the Front Range.

U.S. 40
Empire—Kremmling
73 miles (117 km)

Between Empire and Winter Park, U.S. 40 zigzags up to Berthoud Pass and down again into the valley of the Fraser River, having crossed the Continental Divide and a large area of 1.7-billion-year-old Precambrian rocks. This is avalanche country; steep slopes and deep snow are a precarious mix. When packed snow breaks loose and whooshes down the mountainside, it gathers up more and more snow and grows more and

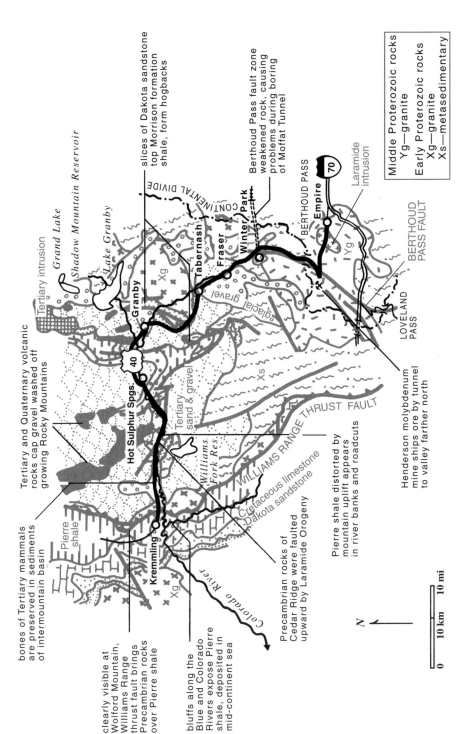

Geology along U.S. 40 between Empire and Kremmling.

Middle Proterozoic rocks
Yg—granite
Early Proterozoic rocks
Xg—granite
Xs—metasedimentary

slices of Dakota sandstone
top Morrison formation
shale, form hogbacks

Berthoud Pass fault zone
weakened rock, causing
problems during boring
of Moffat Tunnel

Laramide
intrusion

BERTHOUD
PASS FAULT

LOVELAND
PASS

BERTHOUD PASS

Empire

70

Winter Park

Fraser

Tabernash

CONTINENTAL DIVIDE

Granby

Xg

Lake Granby

Grand Lake

Shadow Mountain Reservoir

Tertiary intrusion

Tertiary and Quaternary volcanic
rocks cap gravel washed off
growing Rocky Mountains

Hot Sulphur Spgs.

40

glacial gravel

Xs

Tertiary
sand & gravel

WILLIAMS RANGE THRUST FAULT

Williams
Fork Res.

Cretaceous limestone
Dakota sandstone

Pierre
shale

Kremmling

Xg

Colorado River

Henderson molybdenum
mine ships ore by tunnel
to valley farther north

Pierre shale distorted by
mountain uplift appears
in river banks and roadcuts

Precambrian rocks of
Cedar Ridge were faulted
upward by Laramide Orogeny

bluffs along the
Blue and Colorado
Rivers expose Pierre
shale, deposited in
mid-continent sea

clearly visible at
Wolford Mountain,
Williams Range
thrust fault brings
Precambrian rocks
over Pierre shale

bones of Tertiary mammals
are preserved in sediments
of intermountain basin

N

0   10 km
0   10 mi

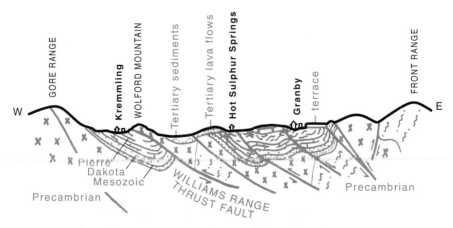

Section from the Front Range to the Gore Range, through Granby. As rocks slide over one another along faults, the crust is foreshortened, and pressure caused by plate collisions is relieved. Some of the faults probably merge at depth.

dangerous. Broad swaths overgrown with stands of young trees mark old avalanche paths. The Colorado Highway Department watches and measures snowpack and dynamites potential avalanches before they endanger the highway.

Several glacial moraines cross the U-shaped valley of the West Fork of Clear Creek. As highway switchbacks climb toward Berthoud Pass, look across the West Fork at two high, scoop-shaped cirques on Woods Mountain. Little glaciers that gouged out these cirques were tributaries to a larger glacier in the main valley. Rough, rocky moraines below each cirque show that the small glaciers survived there for some time after they no longer reached the main glacier.

Berthoud Pass marks the line where a fault zone crosses the divide, weakening the hard Precambrian granite so it erodes more easily. The same fault crosses Loveland Pass, 11 miles southwest of here. Seven miles to the north it caused collapse problems in boring Moffat Tunnel by which main-line trains tunnel through the Continental Divide.

The crest of the pass is a narrow strip of metasediment, with gray granite on both sides. These Precambrian rocks haven't changed much since 1.4 billion years ago, when ancient schist and gneiss were intruded by Berthoud suite granite. In roadcuts you can see intense, chaotic fracturing near the fault. On the northwest side of the pass, the Fraser River heads in this fractured zone.

The Fraser Valley is floored by glacial moraines, with humps and bumps enclosing small ponds and marshy depressions. Near Winter Park, where

Cirques from Pleistocene glaciers frame Berthoud Pass in this aerial photograph. James Peak is at the upper right; Fraser River valley lies beyond the pass.
—T. S. Lovering photo, courtesy of U.S. Geological Survey

terminal moraines cross the valley, notice their unsorted, angular rock fragments randomly mixed with sand and clay.

North of Winter Park the valley is underlain with Tertiary deposits concealed by stream-deposited glacial debris. Some of the Tertiary sediments are exposed in ridges across the valley—muddy brown sandstone that contains a good deal of volcanic ash, some pure volcanic ash layers, and several mud and gravel layers. High, flat-topped terraces in the valley near Fraser and Tabernash consist of glacial outwash deposited by streams below melting glacial tongues.

Swinging west, the road leads through hills of Tertiary sediment interspersed with purple granite. Older than the granite on Berthoud Pass by 300 million years, this purple rock dates from the Early Proterozoic Orogeny. Careful geologic mapping shows that the ancient granite is thrust westward over Mesozoic rocks, and that they in turn are thrust over Tertiary strata. Precambrian rocks also slid several miles westward here, as they did farther north in the Never Summer Range. Glaciers and rivers have removed a great deal of the overthrust rock, leaving only islands of

Precambrian granite on a sea of sedimentary strata. As would be expected, rocks beneath the thrust were intensely folded and faulted. So many faults go in so many directions that little blocks of this and that pop to the surface where you least expect them. Besides Precambrian granite, Cretaceous Dakota sandstone is the most recognizable unit, often adjacent to green and purple Jurassic Morrison shale.

The high terrace east of Granby is outwash from a large glacier that crept south down the valley of the Colorado River, swollen with ice from tributary glaciers in Rocky Mountain National Park and the Never Summer Range. West of Granby, U.S. 40 enters a narrow canyon bordered by cliffs of dark mid-Tertiary volcanic rocks.

At Hot Sulphur Springs, the springs rise where water heated well below the surface comes up rapidly along a fault. The water is highly mineralized and has built a large mound of travertine. Of all the minerals, the smelly hydrogen sulfide is the most noticeable. Dakota hogbacks east and west of town are small fault blocks.

West of the town, Jurassic Morrison shale lies between the Dakota sandstone and Precambrian rocks. This region was part of the Ancestral Rockies, and Paleozoic rocks were washed away before the Jurassic shale was deposited.

Just west of Hot Sulphur Springs, the Colorado River enters a canyon that exposes Precambrian granite on the north, faulted over Tertiary sediments

Near Kremmling the Pierre shale includes the sandstone unit that caps these bluffs. Marine fossils, particularly the straight-shelled ammonite Baculites, *have been found in shale slopes below the bluffs.* —G. A. Izett photo, courtesy of U.S. Geological Survey

on the south. The thrust fault is at the base of the coarse red granite north of the road. Across the river, another block of granite forms a ridge with the Morrison and Dakota formations leaning up against it.

Emerging from the canyon, the river and highway are at last out of the complexly faulted area and on the relatively featureless surface of much younger Tertiary sediments that floor Middle Park.

Because of all the faulting along its fringes, Middle Park is not a very coherent geographic area. Its margins are ragged, and various hills and ridges poke above Pleistocene terraces or Tertiary sediments of its floor.

Tertiary sediments in Middle Park, brown sand and gravel and stream-rounded boulders washed off the surrounding mountains, are so poorly consolidated you can hardly call them rocks. Sedimentary rocks harden with burial, compaction, and the slow cementing that accompanies the flow of groundwater through them. These Tertiary rocks are as yet only lightly glued together, so they erode easily.

Just east of Kremmling, U.S. 40 crosses the Williams Range thrust fault, which defines the western edge of the Front Range for more than 60 miles. East of the fault there are scattered outcrops of granite; west of it the Cretaceous Pierre shale appears at the surface. Accumulated at the bottom of an extensive but shallow Cretaceous sea, the Pierre shale is particularly well exposed in bluffs near Kremmling.

The Colorado River turns southwestward at Kremmling and cuts through the hard Precambrian core of the Gore Range–Park Range uplift; its deeply incised canyon is visible to the south. Both ranges are one large faulted anticline, but the Gore Range is unusual in that fragments of some of the sedimentary rocks that arched across its core are still in place on top of the range.

U.S. 40
Kremmling—Steamboat Springs
52 miles (84 km)

North of Kremmling, U.S. 40 traverses a shallow valley underlain by Cretaceous limestone and shale. The highway more or less parallels the Williams Range thrust fault, which brings ancient Precambrian rocks west over much younger shale. The line of the thrust fault is east of the highway, in most places concealed by soil and Tertiary sediments.

Near and north of Wolford Mountain, this shale contains marine fossils: high-spired snails, big corrugated clams called *Inoceramus,* and ammonites that are ancient relatives of the modern chambered nautilus. These marine

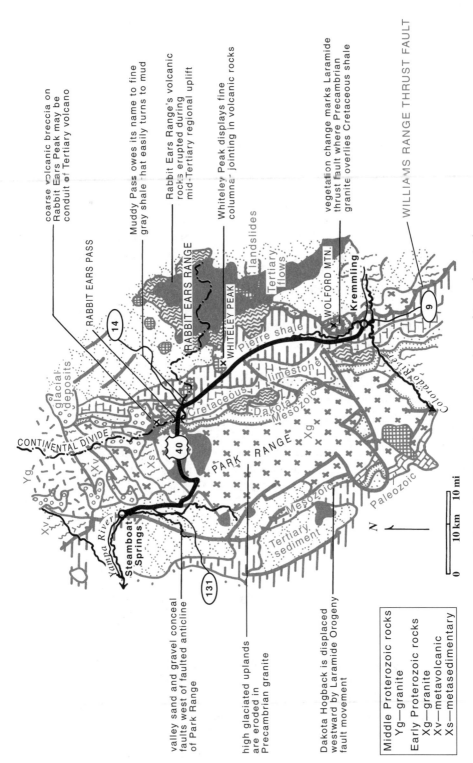

coarse volcanic breccia on Rabbit Ears Peak may be conduit of Tertiary volcano

Muddy Pass owes its name to fine gray shale that easily turns to mud

Rabbit Ears Range's volcanic rocks erupted during mid-Tertiary regional uplift

Whiteley Peak displays fine columnar jointing in volcanic rocks

vegetation change marks Laramide thrust fault where Precambrian granite overlies Cretaceous shale

WILLIAMS RANGE THRUST FAULT

RABBIT EARS PASS

14

RABBIT EARS RANGE

landslides

Tertiary flows

WHITELEY PEAK

Pierre shale

WOLFORD MTN.

Kremmling

9

CONTINENTAL DIVIDE

glacial deposits

Cretaceous

Dakota

limestone

Mesozoic

Colorado River

Yg

Xs

40

Xv

Xg

PARK RANGE

Steamboat Springs

Xv

Yampa River

131

Mesozoic

Paleozoic

Tertiary sediment

N

0   10 km   10 mi

Geology along U.S. 40 between Kremmling and Steamboat Springs.

valley sand and gravel conceal faults west of faulted anticline of Park Range

high glaciated uplands are eroded in Precambrian granite

Dakota Hogback is displaced westward by Laramide Orogeny fault movement

Middle Proterozoic rocks
  Yg—granite
Early Proterozoic rocks
  Xg—granite
  Xv—metavolcanic
  Xs—metasedimentary

animals inhabited the far-reaching sea that extended from Canada to the Gulf of Mexico in Cretaceous time.

Whiteley Peak, a prominent pointed peak about 15 miles north of Wolford Mountain, is capped with Tertiary basalt, with the columnar jointing that often characterizes lava flows. As lava cools and shrinks, it often cracks into polygonal columns. Shrinkage cracks develop from the bottom up and from the top down, so there often appear to be two flows when there is really only one. Tertiary volcanism occurred 20 to 30 million years after the Rocky Mountains rose, as mid-Tertiary regional uplift raised and stretched Colorado and adjacent states, opening pathways for lava and volcanic gases.

At Muddy Pass, U.S. 40 swings west to climb toward Rabbit Ears Pass, at the top of the Park Range. Beaver dams just beyond Muddy Pass are reminders that beavers are active geologic agents—damming streams, reducing floods, and eventually creating new mountain meadows as their ponds fill in with vegetation and soil.

The bumpy, irregular terrain in this area is the surface of a large glacial ground moraine. A melting glacier left big chunks of ice and unsorted rock material on the surface; as the ice chunks melted away, they left depressions, some becoming kettle lakes. Landslides of uncemented moraine make the surface even more hummocky.

The Williams Range thrust fault shows up clearly on Wolford Mountain, where trees grow in coarse Precambrian granite on the top half of the mountain but don't take root in Cretaceous shale below. —Jack Rathbone photo

Flat uplands of the Park Range supported several ice caps in Pleistocene time, with fingers of ice reaching down the slopes and into the canyons. North of here in the Mt. Zirkel Wilderness Area, small perennial snowfields still exist but do not move enough to be classed as glaciers. Geologists and climatologists think they started from scratch about 4,000 years ago, after a warmer climate cycle that completely melted the Ice Age glaciers. Valleys on both sides of the Park Range head in steep-walled cirques and terminate in extensive moraines.

As U.S. 40 climbs toward Rabbit Ears Pass, the Dakota sandstone appears and extends to the summit of the range. From the highway there are good views of the Park and Gore Ranges, North and Middle Parks, and the Front Range in the distance to the east. On clear days you can spot light-colored sand dune fields in North Park.

Rabbit Ears Peak, with its cluster of three summits, contains rough, coarse volcanic breccia, with fragments of broken volcanic material that accumulated in or near a volcanic vent and cemented together with reddish lava. On the sloping surface west of the pass, basalt flows cap outcrops of weathered Precambrian rock.

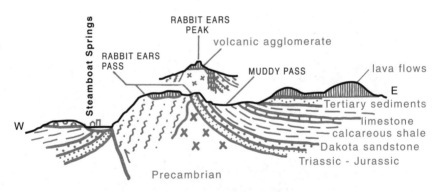

Section across the Park Range at Rabbit Ears Pass.

West of milepost 152 on Rabbit Ears Pass is a thin band of 1.7-billion-year-old dark red granite. Nearby, roadcuts sparkle with biotite in yet older gneiss and schist that core this part of the Park Range. Decoratively swirled like marble cake with light and dark bands and laced by a network of pegmatite dikes, these ancient rocks reached temperatures and pressures high enough to bring them to the verge of melting, probably 7 miles beneath the surface and at temperatures of 500 degrees Celsius. Only under such conditions can pegmatite dikes develop. The gneiss and schist probably originated as sedimentary rocks like sandstone and shale; they still retain the chemical composition of these sedimentary types.

The western side of the Park Range is steep and dramatic. Faults that edge it north and south of Steamboat Springs are hidden beneath glacial outwash and the sand and gravel of the Yampa River valley.

Hot springs at Steamboat Springs emerge from both banks of the Yampa River, some mere seeps and others active bubbling pools. One of them used to sound like the chugging of a steamboat, giving the town its name, but the sound ceased when a railroad cut was excavated nearby. The hot springs issue from the Dakota sandstone where water heated by hot rocks far below the surface rises rapidly along faults.

Along the banks of the Yampa, highly mineralized hot water charged with calcium carbonate dissolved from limestone cools and evaporates at the surface, leaving its calcium carbonate behind in deposits of dull gray travertine.

U.S. 50
Canon City—Poncha Springs
62 miles (100 km)

West of Canon City, U.S. 50 skirts the southern end of the Dakota Hogback, then runs north beside it. Mesozoic formations dip steeply here. For a better look at them, drive (carefully) up 2.5-mile Skyline Drive onto the crest of the hogback. From there you can see that the hogback comes up from the south, passes Canon City, and curves east around Garden Park. Resistant Cretaceous limestone layers form parallel but less pronounced ridges. Some sandstone layers in the Dakota formation are patterned with ripple marks, mudcracks, and tracks and trails of Mesozoic animals, left on beaches and sandbars bordering the Cretaceous sea.

To the west, quarries visible from Skyline Drive are in Ordovician rock, the Harding sandstone and Fremont dolomite, deposited in shallow equatorial seas. In the Harding Quarry in 1887, paleontologist Charles Walcott discovered bony plates of extinct armored fish called *Astraspis,* previously known only from Devonian rocks. Walcott had found the world's oldest known vertebrate animals—a record that held until 1977, when older fossil fish were discovered in Antarctica and Wyoming. Recently, skin denticles of sharklike but jawless fish were found in the Harding sandstone.

U.S. 50 crosses a gentle syncline in this sandstone as it curves westward a few miles north of Canon City. Thin Ordovician Manitou limestone beneath it lies right on coarse pink Precambrian granite, as in roadcuts between mileposts 273 and 272. This granite dates back to the Early Proterozoic Orogeny 1.7 billion years ago, when surrounding 1.8-billion-year-old rocks

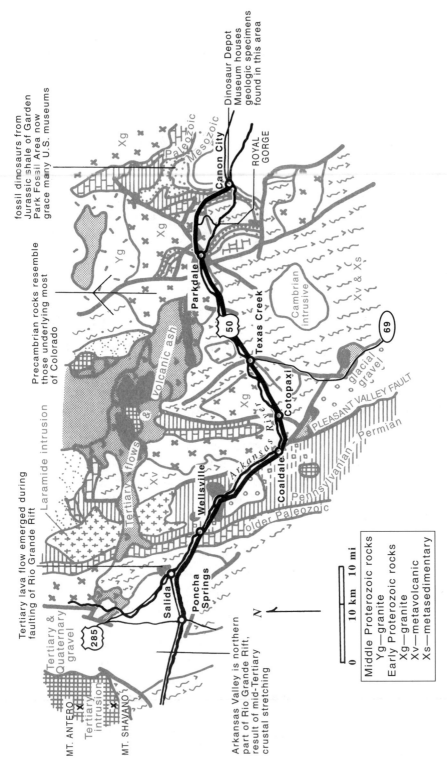

fossil dinosaurs from
Jurassic shale of Garden
Park Fossil Area now
grace many U.S. museums

Precambrian rocks resemble
those underlying most
of Colorado

Dinosaur Depot
Museum houses
geologic specimens
found in this area

Laramide intrusion

Tertiary lava flow emerged during
faulting of Rio Grande Rift

Tertiary &
Quaternary
gravel

Tertiary intrusion

MT. ANTERO
MT. SHAVANO

Arkansas Valley is northern
part of Rio Grande Rift,
result of mid-Tertiary
crustal stretching

Canon City

ROYAL
GORGE

Paleozoic
Mesozoic

Xg

Yg

Xg

volcanic ash

Parkdale

Texas Creek

Cambrian
intrusive

Xv & Xs

50

69

glacial
gravel

Cotopaxi

Xg

Tertiary flows &

Arkansas River

PLEASANT VALLEY FAULT

Pennsylvanian - Permian

Older Paleozoic

Coaldale

Wellsville

Xv

Salida

Poncha
Springs

285

N

0   10 km   10 mi

Middle Proterozoic rocks
  Yg—granite
Early Proterozoic rocks
  Xg—granite
  Xv—metavolcanic
  Xs—metasedimentary

Geology along U.S. 50 between Canon City and Poncha Springs.

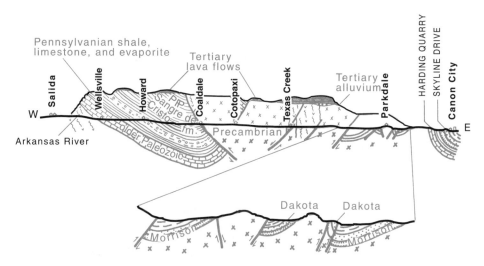

Section along U.S. 50 between Canon City and Salida.

were pushed up into mountains and metamorphosed by a plate collision along the southeast coast of the young continent.

A 5-mile entrance road leads to Royal Gorge, one of the deepest canyons in Colorado, 1,200 vertical feet from rim to river. The road to the gorge crosses Precambrian granite and then a band of brownish gray metamorphic rock where many large pegmatite dikes contain unusually large crystals of quartz, feldspar, and muscovite. Pegmatites develop when watery fluids escape from extremely hot igneous or metamorphic rock and collect in shrinkage cracks in the parent rock—a process similar to the way that sugary fluids sometimes seep from newly made chocolate pudding as it cools and shrinks. Crystals grow exceptionally large with slow cooling and lots of elbow room. "Books" of mica 4 or 5 inches across have been quarried

A specimen of Precambrian granite shows large, knobby crystals of pink feldspar (often blocky or rectangular), smaller glassy quartz and gray feldspar grains, and black biotite. Slow radioactive decay of potassium causes the salmon pink color of the feldspar. (x ⅔)

here; big chunks stand out in lane markers near Royal Gorge Bridge. Feldspar crystals, masses of snow white quartz, and large crystals of beryl, black tourmaline, and garnet occur here, too, as does graphic granite—intergrown quartz and feldspar crystals that look like a mysterious form of writing.

At Royal Gorge there is a curious mixture of gneiss, schist, and granite in which the granite magma appears to have squeezed along flaky cleavage planes in the gneiss and schist, forming granite masses hundreds of feet thick or only half an inch thick. Geologists refer to this type of rock as injection gneiss.

Royal Gorge cuts through a wedge of Precambrian granite and metamorphic rock that extends south from the Pikes Peak massif. The Arkansas River established its course on Cretaceous and Tertiary rocks. With mid-Tertiary regional uplift, it cut deeper and deeper, finally trenching a spectacular canyon in underlying hard Precambrian rock. —M. R. Campbell photo, courtesy of U.S. Geological Survey

The flat rim of Royal Gorge is part of an ancient peneplain, perhaps originally worn down during the lengthy erosion interval at the end of Precambrian time, perhaps eroded further as the Ancestral Rockies were worn away.

Near Parkdale, U.S. 50 enters a small patch of broken, intensely faulted rocks where wedges of red and yellow Mesozoic formations tilt at crazy angles, some vertical and some even upside down. There are no Paleozoic rocks here because this region was part of Frontrangia, the Ancestral Rocky Mountain highland, and erosion stripped them away.

West of Parkdale, Precambrian rocks are cut by black dikes and white dikes. Near milepost 257, injections of pink granite weather into characteristic bare, rounded slopes. North across the river the granite hills are capped with lava flows from the Thirtynine Mile volcanic field at the southern end of South Park. Uranium has concentrated in a few places in stream deposits below these volcanic rocks, probably leached from volcanic tuff above it. In the canyons Precambrian metamorphic rocks are common, especially black hornblende gneiss that probably originated as basalt. Its abundance in this area led geologists to map these rocks as metavolcanics.

Half a mile west of milepost 257, muscovite schist gives the roadcut a satinlike sheen. When clayey sediments are strongly metamorphosed, clay becomes muscovite. The rock's knobby texture is caused by another metamorphic mineral, cordierite, created when the rocks were buried at least 2 miles deep and heated to 500 degrees Celsius. Now exposed to weathering, the muscovite is slowly changing back to clay again. Hold a weathered piece in your hand, breathe on it, and quickly smell the clay. —Felicie Williams photo

Just east of Coaldale a striking change in scenery is brought about by the Pleasant Valley fault, which extends far northward and separates Precambrian rocks from dark red sandstone and shale of Permian and Pennsylvanian age. The steeply dipping rocks developed from rock debris washed off the western slope of Frontrangia; they contain coal and gypsum. Fine grains of iron oxide, the mineral hematite, color the rocks red.

Where they first appear (milepost 238) the red sediments are dragged upward and overturned by upward movement of granite along the Pleasant Valley fault. For a few miles the Arkansas River runs along the fault line.

About 20,000 feet of these redbeds, measured perpendicular to the layering, occur here. They may be doubled up by faulting, but if so the faults parallel the layers and are virtually impossible to detect. Redbeds rarely contain fossils, but fossil shells from limestone layers tell us these rocks are Permian at the top (east) and grade westward into Pennsylvanian.

Upriver near milepost 232 the shales become thicker and more abundant, and sandstone and conglomerate are supplanted by gray marine limestone. Here, the rocks are definitely Pennsylvanian. A few interspersed coaly layers were deposited in nearshore lagoons and swamps. Repeated sequences of sandstone-shale-sandstone-shale, as at milepost 231, or limestone-shale-limestone-shale show that the environment in which they were deposited must have fluctuated repeatedly. All over the world such cycles characterize Pennsylvanian rocks, possibly because sea level varied

West of Howard, thick Pennsylvanian-Permian redbeds bend upward, lifted along the flanks of the Sangre de Cristo Range farther west. —Felicie Williams photo

regularly with variations in the pace of seafloor spreading, or possibly in response to repeated glacial episodes in some other part of the world.

Fossil brachiopods are abundant and easy to find in Box Canyon, which cuts through Pennsylvanian shale south of Wellsville. (x 1)
—Drawings by Emily Silver

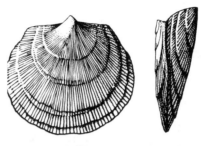

Just east of milepost 228 is a massive gray cliff of Mississippian Leadville limestone, probably once continuous with similar gray limestone found from Arizona to Alberta and Michigan. Farther west are brownish gray Devonian limestone (near the Chaffee County line) and Ordovician limestone, sandstone, and shale (across the river at milepost 227). Marine limestones form from untold billions of microscopic shells of marine organisms and are evidence that shallow seas teeming with life covered Colorado repeatedly during Paleozoic time, at least until the Ancestral Rockies formed.

Some of the hills near Salida, particularly soft, rubble-covered hills like the one with the **S**, are Tertiary intrusions and lava flows. Others are

Mt. Princeton and other Sawatch Range peaks top 14,000 feet. Mid-Tertiary sediments sloping from the mountains record regional uplift; they are capped with coarse Pleistocene gravel that reflects both uplift and glaciation of the mountain area. The light zone on Mt. Princeton's flank is an area leached by hot springs.
—R. E. Van Alstine photo, courtesy of U.S. Geological Survey

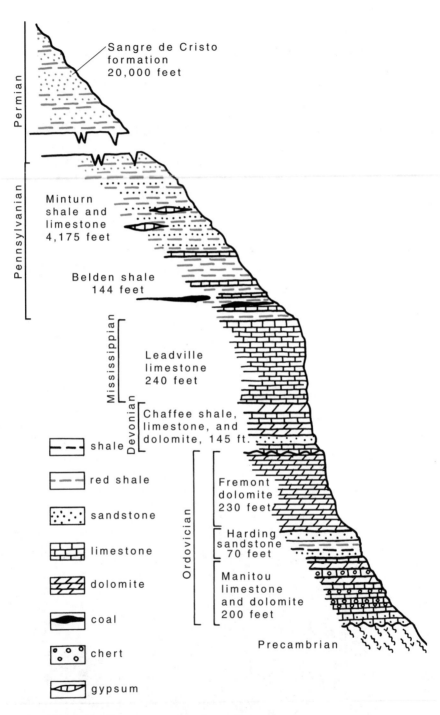

Rock layers exposed near Wellsville in Arkansas Canyon include sandstone-shale and sandstone-limestone cycles in the Minturn and Belden formations. Wavy lines are major unconformities representing long periods of erosion. The Fremont and Leadville limestones form prominent cliffs.

Precambrian metavolcanic rock—the remains of island arcs—not as intensely altered here as in some areas.

West of Salida, large pediments and alluvial fans slope from the Sawatch Range. This range contains many of Colorado's highest peaks. The three at its southern end, Mt. Shavano, Mt. Antero, and Mt. Princeton, are eroded from a single large Tertiary batholith. The Sawatch uplift includes the Arkansas Hills and the Mosquito Range east of the river, as well as the much higher Sawatch Range. The Arkansas River flows along a slim downfaulted block in the core of the uplift.

U.S. 50
Poncha Springs—Gunnison
62 miles (100 km)

West of Poncha Springs, U.S. 50 follows the South Fork of the Arkansas River, climbing gradually across deposits of Pleistocene gravel. Mt. Shavano and Mt. Antero, as well as Mt. Princeton farther north, all in the Sawatch Range, are parts of a Tertiary batholith more or less circular in outline and 15 to 20 miles across. The summit of Mt. Antero is a well-known gem locality, an area of granite and porphyry with open cavities that contain quartz, aquamarine, and other gem minerals.

West of Maysville, U.S. 50 reaches Precambrian rocks: schist, recognizable by its dark color and platy way of breaking, and light and dark gray gneiss. Around 1.8 billion years ago, these rocks were part of a volcanic island arc that included sedimentary and volcanic rocks. Colliding with the southeastern edge of the Wyoming Province about 1.7 billion years

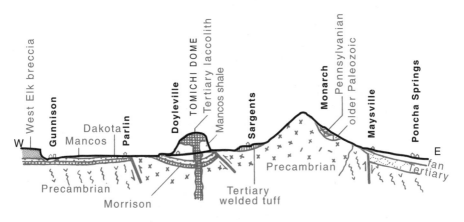

Section along U.S. 50 between Poncha Springs and Gunnison.

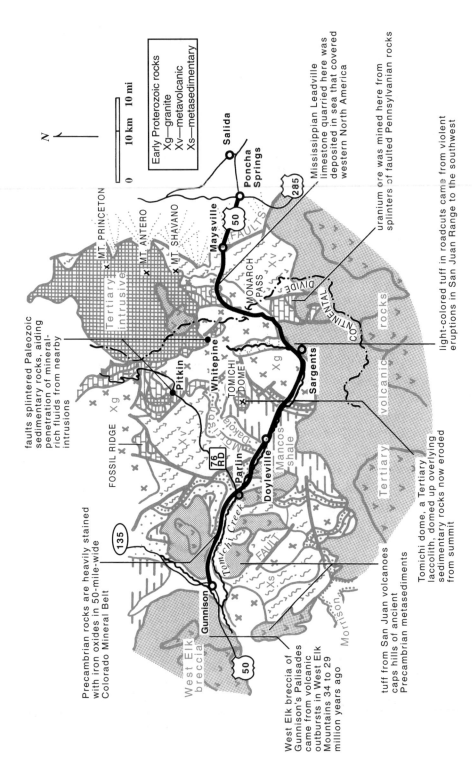

Early Proterozoic rocks
Xg—granite
Xv—metavolcanic
Xs—metasedimentary

N

0   10 km   10 mi

Salida

Poncha Springs

285

Maysville

50

MT. PRINCETON

MT. ANTERO

MT. SHAVANO

FAULTS

Tertiary Intrusive

MONARCH PASS

Xv

CONTINENTAL DIVIDE

Pitkin

Whitepine

TOMICHI DOME

Xg

Sargents

FOSSIL RIDGE

Xv

Dakota

Morrison

Mancos shale

Parlin

76 RD

Doyleville

FAULT

Xs

Tomichi Creek

135

Gunnison

West Elk breccia

50

Tertiary volcanic rocks

faults splintered Paleozoic sedimentary rocks, aiding penetration of mineral-rich fluids from nearby intrusions

Precambrian rocks are heavily stained with iron oxides in 50-mile-wide Colorado Mineral Belt

West Elk breccia of Gunnison's Palisades came from volcanic outbursts in West Elk Mountains 34 to 29 million years ago

tuff from San Juan volcanoes caps hills of ancient Precambrian metasediments

Tomichi dome, a Tertiary laccolith, domed up overlying sedimentary rocks now eroded from summit

light-colored tuff in roadcuts came from violent eruptions in San Juan Range to the southwest

uranium ore was mined here from splinters of faulted Pennsylvanian rocks

Mississippian Leadville limestone quarried here was deposited in sea that covered western North America

Geology along U.S. 50 between Poncha Springs and Gunnison.

ago, they piled up as a wide belt of new crust added to the young continent. Gray granite, intruded at about the same time, is exposed in the massive cliff above milepost 206.

Coarse gravel near the highway contains rounded boulders, indicating that it was stream-deposited, most of it by outwash streams from Ice Age glaciers.

Farther west the highway passes through a steeply tilted wedge of Paleozoic sedimentary strata. Dark red Ordovician limestone lies right on the granite, but the contact is hard to see. The erosion surface on the granite below this limestone represents well over a billion years, including the end of the Precambrian era and all of Cambrian time, when it was part of an island in the Cambrian sea.

Cliffs above the highway at Garfield are Paleozoic limestone. Just across the river they are smoothed and polished by movement along a fault. Steeply tilted gray rocks in the huge quarry near Monarch are Mississippian Leadville limestone, part of a sheet of similar limestone deposited in a shallow sea that covered much of the western half of the continent. Notice how the broad, steeply sloping surfaces of individual limestone beds make convenient quarrying planes. Limestone obtained here formerly went to the Colorado Fuel and Iron Company smelter at Pueblo, where it was used as a furnace flux to help float impurities from molten iron ore.

The Madonna Mine, above the quarry, once produced lead, zinc, and silver. Mine tours go about 1,500 feet into the mountainside. This area is within the Colorado Mineral Belt, a 50-mile-wide zone extending from Boulder County to southwestern Colorado. The mineral belt began to form during the Laramide Orogeny, with most mineralization taking place during the Tertiary volcanic episode. Magma and mineral-rich fluids rose along a major Precambrian shear zone, depositing ore minerals upon encountering and reacting with limestone.

Between the quarry and Monarch Pass, and for some distance down the western side of the pass, light gray Precambrian granite appears, with rusty iron oxide on fracture faces. Such iron-staining is common in the mineral belt.

Looking north from Monarch Pass, Mt. Aetna, the most conspicuous nearby peak, is part of the same Tertiary batholith as peaks farther north. High country well south of Monarch Pass is capped with Tertiary volcanic rocks of the San Juan volcanic field, mostly welded tuff that solidified from avalanches of volcanic ash so hot it welded itself together as it settled. Mt. Ouray, a truncated pyramid that rises above the volcanic rocks, marks the southern end of the Precambrian core of the Sawatch uplift.

Agate Creek canyon west of Monarch Pass was never glaciated. Note its V-shaped canyon profile and the deeply weathered rock that would have been the first thing scraped off by glaciers. At the foot of the pass, however, U.S. 50 crosses the moraine of a glacier that flowed down Tomichi Creek valley nearly to the little town of Sargents.

Along both sides of this part of the Sawatch Range, splinters and wedges of Paleozoic rock are caught between faults, as at the Monarch quarry. They usually contain Ordovician, Devonian, Mississippian, and Pennsylvanian sedimentary rocks. Many of them are mineral-enriched and were centers of mining activity in the 1870s to 1890s. A century later, uranium was mined from one of the rock splinters east of Sargents—originally a bustling railroad town supplying nearby mining communities.

Along Tomichi Creek west of Sargents, Precambrian rocks—granite and dark basalt metamorphosed into hornblende gneiss—are cut by many dikes, some of them well exposed in roadcuts. Dark dikes are similar in composition to basalt; light ones resemble granite in composition. Also visible in the highway cuts are xenoliths of surrounding gneiss broken off and floating in the granite. Intruded as a batholith 1.7 billion years ago, the granite apparently worked its way into the gneiss when the gneiss itself was slushy and almost melted.

Just west of milepost 182, Precambrian rocks of the Sawatch Range core were faulted over Cretaceous Dakota sandstone, which was bent up, dragged, and turned upside down in the process. The Dakota sandstone appears in the vertical cliff about a mile north of U.S. 50. Between the fault and Parlin the highway crosses a saucer-shaped basin of Mesozoic rocks, with a hogback of Dakota sandstone rimming a broad, sage-covered valley floored with Mancos shale.

One of the most striking landmarks in this area is round-topped Tomichi dome, a Tertiary laccolith formed when thick, gummy molten rock was forced up through the Dakota sandstone and into Mancos shale. There it spread sideways, doming up younger strata, most of which have since eroded away.

West of Tomichi dome, U.S. 50 enters an area that in late Paleozoic and early Mesozoic time was part of Uncompahgria, the western of two ranges of the Ancestral Rockies. The crest of the old highland was near the present town of Gunnison. Where Paleozoic rocks were eroded from the range, Jurassic strata lie directly on the Precambrian surface.

Among Mesozoic formations, three are quite distinctive and easily recognized. From top to bottom, they are:

- Mancos shale, slope-forming black shale that weathers light gray or yellow, deposited in the extensive Cretaceous sea. It is exposed in roadcuts and on slopes between Tomichi dome and Doyleville.

- Resistant tan Dakota sandstone, forming cliffs where it is horizontal, hogbacks where it is tilted, and high walls where it is vertical. Depos-

Outcrops of pink granite near Parlin weather along three intersecting sets of joints. The weathering rounds protruding corners and causes large curving rock flakes to peel off the boulders. This type of weathering is typical of granite and other even-grained rocks with intersecting joint sets. —M. H. Staatz photo, courtesy of U.S. Geological Survey

ited in early Cretaceous beaches and river channels, it appears at the fault near milepost 182 and again near milepost 176, and caps many of the hills north of U.S. 50.

- Morrison formation, colorful greenish and reddish Jurassic shale, a slope-former, deposited on an irregular surface of Precambrian granite. The formation, famous for its dinosaur fossils, is well exposed in roadcuts near mileposts 175 and 173.

West of Parlin are more Precambrian island arc rocks, far less metamorphosed than usual here. Greenish black basalt is the most common rock type. Many of the rocks are heavily iron-stained, evidence that they are within the Colorado Mineral Belt. In this area there are patches of Tertiary welded tuff, a fine tan or bluish gray rock speckled with black crystals of biotite or hornblende. Under a microscope, this rock shows many tiny, glassy shards like broken bits of volcanic froth.

West of milepost 163, Tertiary volcanic rocks become increasingly common, particularly light tawny ones containing many tuff fragments. These rocks came from volcanic centers in the West Elk Mountains, northwest of Gunnison, and are known as the West Elk breccia. (*Breccia* rhymes with *betcha* and is related to the English word "break.") It consists of broken fragments of lava mixed with volcanic ash, either ejected from volcanoes or washed here in volcanic mudflows. Breccia often weathers into strange pinnacles and cliffs like the Palisades along the river near Gunnison, where it lies on an Eocene erosion surface that cuts across Precambrian granite and Mesozoic rocks.

During the first years of its existence, Gunnison supplied mines on the western side of the Sawatch Range, in the West Elk Mountains to the north, in the Gunnison Gold Belt a few miles south and west of town, and in the San Juans to the south. From the 1960s to the 1980s, uranium ore from mines near Sargents was processed here. The mill has since been demolished and radioactive residue removed for safe storage.

U.S. 160
Walsenburg—Alamosa
73 miles (117 km)

At Walsenburg, sedimentary rocks that are flat-lying or gently warped on the plains begin to slant upward toward the mountains. The town itself is close to the first sharp rise, and U.S. 160 climbs through Cretaceous sandstone, older than the mountain uplift, onto Tertiary sedimentary rocks younger than the uplift. The broad sloping plain of Tertiary sediments, mostly sandstone and conglomerate derived from the mountains themselves, rises westward toward the Sangre de Cristo Range.

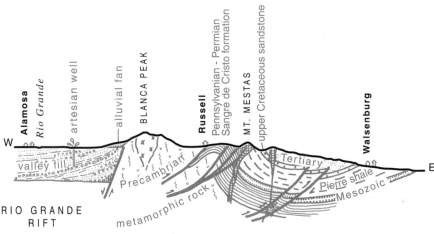

Section parallel to U.S. 160 across the Sangre de Cristo Range.

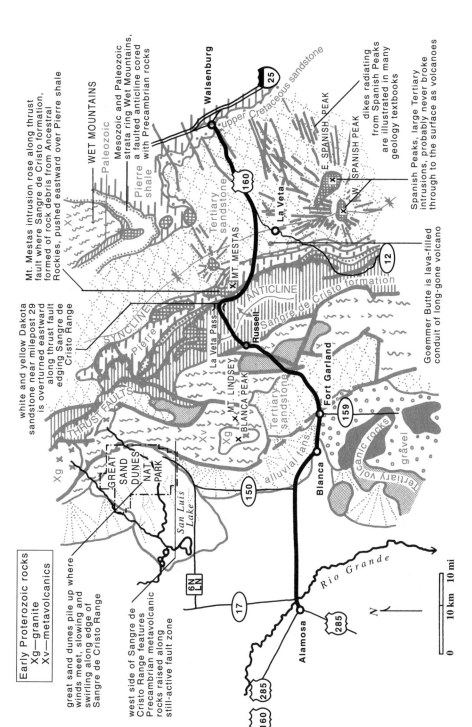

Geology along U.S. 160 between Walsenburg and Alamosa.

Early Proterozoic rocks
Xg—granite
Xv—metavolcanics

Mt. Mestas intrusion rose along thrust fault where Sangre de Cristo formation, formed of rock debris from Ancestral Rockies, pushed eastward over Pierre shale

WET MOUNTAINS

Paleozoic

Mesozoic and Paleozoic strata ring Wet Mountains, a faulted anticline cored with Precambrian rocks

Pierre shale

Xv

upper Cretaceous sandstone

dikes radiating from Spanish Peaks are illustrated in many geology textbooks

E. SPANISH PEAK

W. SPANISH PEAK

Spanish Peaks, large Tertiary intrusions, probably never broke through to the surface as volcanoes

La Veta

160

12

Walsenburg

25

Tertiary sandstone

x MT. MESTAS

ANTICLINE

Sangre de Cristo formation

Goemmer Butte is lava-filled conduit of long-gone volcano

white and yellow Dakota sandstone near milepost 29 is overturned eastward along thrust fault edging Sangre de Cristo Range

SYNCLINE

Pierre

La Veta Pass

Russell

x MT. LINDSEY
x x BLANCA PEAK

Xg

Xv

Tertiary sandstone

alluvial fans

Fort Garland

159

Tertiary volcanic rocks

gravel

THRUST FAULTS

Xg

GREAT
SAND
DUNES
NAT.
PARK

Blanca

150

great sand dunes pile up where winds meet, slowing and swirling along edge of Sangre de Cristo Range

San Luis Lake

west side of Sangre de Cristo Range features Precambrian metavolcanic rocks raised along still-active fault zone

6N
LN

17

285

160

285

Alamosa

Rio Grande

N

0   10 km   10 mi

Some distance north of U.S. 160, upturned Paleozoic and Mesozoic strata encircle the southern tip of the Wet Mountains. The narrow valley west of them is floored with Tertiary sediments renowned for fossilized skeletons of *Merychippus,* a small four-toed ancestor of the horse. Horses were unknown in America at the time of Columbus, but their ancient lineage began in the Western Hemisphere, spreading via the Bering Strait to Asia and Europe, where the lineage continued to develop while it died out in the Americas.

Spanish Peaks, visible to the south, are Tertiary intrusions that pushed their way up through Cretaceous and Tertiary sedimentary rocks after most of the rest of the Colorado Rockies were formed. They include a curious assemblage of different kinds of igneous rocks, from granite through very dark, basaltlike diabase. Dikes radiate outward for miles from the two intrusions.

U.S. 160 curves northward around Mt. Mestas, another Tertiary intrusion. Magma forming this mountain rose along one of the principle faults on the eastern side of the Sangre de Cristos, where red Permian rocks, visible west of the highway, are faulted eastward over younger Dakota sandstone. The intrusion is a white rock called microgranite that contains

Dikes radiating from the Spanish Peaks cut through surrounding Cretaceous and Tertiary strata on the eastern slope of the Sangre de Cristo Range. —Jack Rathbone photo

microscopic crystals of quartz, feldspar, and biotite—the same mineral trio that characterizes granite.

The fault west of Mt. Mestas is easy to find because a spring flows from it close to the road. Beyond and above the spring, dark red Pennsylvanian and Permian sandstone and shale of the Sangre de Cristo formation tint the slopes.

The Sangre de Cristo formation extends on across La Veta Pass. Though it is folded and faulted and may tilt in different directions, it can always be recognized by its reddish color. It originated as an alluvial apron west of Frontrangia, the main range of the Ancestral Rockies, and corresponds to the Fountain formation east of the Front Range and the Maroon formation around Aspen and Vail.

Just west of La Veta Pass the summits of Blanca Peak and Mt. Lindsey can be glimpsed. They are parts of a large fault block of Precambrian rock that juts westward from the Sangre de Cristo Range. Blanca Peak, with unusual black intrusive rocks, dates back 1.7 billion years, to the first orogeny known to have affected the Precambrian rocks. Mt. Lindsey wears a cap of Precambrian metavolcanic rocks thrust over a downfaulted block of Paleozoic sediments.

Descending the western side of La Veta Pass U.S. 160 crosses, from east to west:

- A broad swale of soft red shale and sandstone, the Sangre de Cristo formation.
- A thick sequence of fossil-bearing, gray Pennsylvanian shale, sandstone, and limestone, well exposed between mileposts 276 and 273. Small Tertiary intrusions penetrate these rocks.
- Precambrian granite, gneiss, and schist, some with yellowish green epidote showing on fracture surfaces. Dark green to black colors in many of these rocks come from iron and magnesium, showing that they originated as dark, basaltlike rock, probably lava flows, dikes, and sills mixed with sedimentary rocks of a volcanic arc.
- Lavender and purple volcanic rocks, west of milepost 268, related to the San Luis volcanic field at the southern end of the San Luis Valley.

Following major thrust faulting during the Laramide Orogeny, the San Luis Valley has had an unusual history. In mid-Tertiary time, much of the Southwest was uplifted and stretched. From central Colorado southward through New Mexico, the stretched crust broke along two large north-south faults, with the land between dropping downward in faulted blocks to form the Rio Grande Rift, of which the San Luis Valley is a part. Now covered by up to 13,000 feet of sediments and volcanic flows, many blocks of Precambrian rock beneath the valley are well below sea level. Triangular facets at the bases of mountain ridges, as well as a small fault scarp on alluvial fans in the northern part of the valley, show us that movement on the great rift faults is still going on.

Huge alluvial fans at the base of Blanca Peak coalesce with others along the western side of the Sangre de Cristo Range. They are probably similar to alluvial aprons that encircled the Ancestral Rockies in Pennsylvanian and Permian time, forming the Sangre de Cristo formation in this area and other coarse red rocks farther north and west. Remnants of higher terraces, formed during the Ice Age from glacial outwash, can be recognized by their even, sloping surfaces covered with rounded pebbles and cobbles.

The San Luis Valley is the only true desert in the Colorado Rockies; rainfall is less than 8 inches a year. It looks perfectly flat, but the southern half of the valley is drained by the Rio Grande through a steep-walled gorge that crosses the Colorado–New Mexico line. The north half of the valley is a closed basin; streams entering it from the mountains never reach the Rio Grande. Some of the valley sediments are lakebeds deposited during episodes of volcanic damming or periods of wetter climate.

When prevailing southwest winds sweep across the valley, they carry sand from its desert floor toward the Sangre de Cristos. Rising to cross low points in the range, the wind drops the sand in a sheltered corner just north of Blanca Peak, building some of America's tallest dunes—the Great Sand Dunes.

Some of the irrigation water in the San Luis Valley comes from deep artesian wells. Runoff from the mountains sinks into sand and gravel at the edges of the valley and flows slowly through porous, gravelly layers that lie between gently sloping, impervious beds of clay or volcanic rock. In wells that go down to the gravel aquifers, water rises and flows without pumping because the well head is lower than the point where the water became trapped in the gravel. Some deep wells in this valley tap as many as ten artesian aquifers. Though the wells flow on their own, most are pumped to increase flow. Valves on the wells are shut down to a trickle in winter to give underground reservoirs time to recharge. Snowmelt from surrounding mountains is also used for irrigation. Hot water wells, heated by Tertiary igneous rocks between the sedimentary layers, supply swimming pools in the Alamosa area.

Groundwater leaches salt from the subsoil here. To prevent salt from concentrating near the surface, farmers encircle fields with drainage ditches to lower the water table below the roots of their crops. They then water lightly with fresh water, leaving a gap of dry earth below the roots and above the water table. The story of agriculture in this valley is discussed more fully under **U.S. 285: Poncha Springs—New Mexico.**

Great Sand Dunes National Park and Preserve
46 miles (74 km) from U.S. 160
50 miles (80 km) from U.S. 285

The San Luis Valley is a desert; rainfall averages less than 8 inches a year. Prevailing winds blow from the southwest, gathering speed as they sweep across the valley toward three low passes in the Sangre de Cristo Range. In spring, particularly strong winds pick up sand from the valley floor and lift it or bounce it along until it reaches the protected pocket below the three passes. The southwest winds often meet a southeast wind blowing across the mountains, and where they swirl together they build up higher and higher dunes.

Comparison of old and new photographs shows little net movement of the main dune ridges, even over many decades. Most of the dune sand probably accumulated in Pleistocene time, when the Rio Grande looped farther into the valley, carrying great amounts of sediment eastward off the San Juan Mountains. Dune sand extends more than 100 feet below the surrounding surface, showing that the dunes have been here for some time.

Some dunes crest 700 feet above the valley floor. The dune sand is mostly fine, rounded grains of quartz and of volcanic rocks from the San Juans. This mixture gives the dunes a darker color than most dune and beach

Dunes on the far eastern side of the dune field are asymmetrical: their long windward slopes contrast with steep leeward slopes. Counterdunes edge the dune crests, a common sight at the Great Sand Dunes. —Felicie Williams photo

sand. Because dark surfaces soak up solar radiation, these dunes can be too hot for comfort. Wear thick-soled shoes if you climb on them, and avoid walking on the dunes at midday in summer.

When the wind blows strongly in one direction, sand drifts up the windward slopes of the dunes and then slides, rolls, and avalanches down the leeward slopes. If wind blows in one direction long enough, it will sculpt the dune until its windward side slopes gently and the leeward side is steep. When the wind changes, long ribbonlike counterdunes form along dune crests.

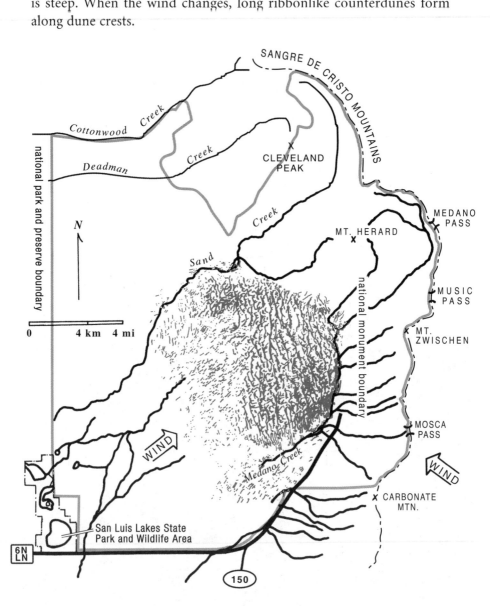

Great Sand Dunes National Park and Preserve. **X** *marks the highest dune.*

After rains, sweeping laminations characteristic of wind-deposited sand may appear on windward surfaces, laminations that make it possible to identify dune sand even when it becomes cemented into sandstone.

Shaped mostly by winds that change from southwest to southeast and back again, most of the dunes at Great Sand Dunes National Park and Preserve form north-south ridges of reverse dunes. Transverse dunes rim the far eastern side of the dune field, the side most commonly visited, where their migration is checked by flowing streams. A few crescent-shaped barchan dunes show up around the fringes of the main dune mass, where sand is less abundant. Several star dunes lie in parts of the dune field where wind direction varies the most.

The interplay between the waters of Medano Creek and the dunes is fascinating. At times, dune ridges encroach on the stream, which rarely flows all year. During spring runoff and after heavy rains, the stream carves back the dunes, and large masses of dry sand cascade down steep lee slopes and off low cliffs undercut by flowing water. The creek flows in the braided pattern typical of overloaded streams. Much of its water actually flows *through* rather than around the dunes, reappearing in marshy springs east of the dunes.

Medano Creek, which flows only in spring or after heavy summer storms, helps control the shape of the dune field by washing sand back from its eastern side. —Halka Chronic photo

WIND

Wind gradually jumps sand grains up the windward slope of a growing dune. As sand settles on the leeward slope where wind velocity lessens, the leeward slope steepens and sand slumps downslope in sand avalanches.

WIND

WIND   REVERSED WIND

Transverse dunes, with long gentle windward slopes, form when sand is abundant, wind is from one direction, and there are few plants. These dunes are found on the dune field's eastern edge.

Reverse dunes occur where wind reverses direction frequently. Dune slopes are fairly symmetrical. Small **counterdunes** form along dune crests, and the dunes don't migrate. Such dunes are the most common kind at Great Sand Dunes National Monument.

WIND

WIND

Longitudinal dunes, common west of the main dune field, parallel the wind direction. They build when and where sand is scarce. Longitudinal dunes many tens of miles long occur in great sand deserts like the Sahara.

Barchan dunes result when wind blows sparse sand in one direction across a flat surface. A few of these fringe the main dune field.

WIND

WIND

WIND

WIND

WIND   WIND

Star dunes form when wind shifts through several directions, as in the northern area of the dune field where wind eddies around nearby mountains.

Parabolic dunes and **blowouts** develop when vegetation anchors a dune's arms while the wind blows away sand at its center.

Dunes vary in shape and size, depending on the direction and strength of the wind and the amount of available sand.

Examine some of the stream and dune surfaces, for they have tales to tell. Stream-formed ripple marks differ from wind-formed ripples in their shape and the size of their sand grains. Shrinkage cracks form in drying mud along the streambed. On sandbars and slopes near the stream, tracks of dune- and forest-dwelling animals are abundant, particularly in the early morning. Beetle and cricket tracks are perhaps the most common, with slime trails of snails and conical holes of ant lions fairly frequent also. Watch for the S-shaped groove made by a lizard's tail. Birds, deer, coyotes, rabbits, and humans—barefoot and otherwise—all leave characteristic trails. Grasses blown by the wind inscribe graceful circles in the sand.

Many such patterns are preserved as shifting sand drifts over them. Tracks made by long-extinct animals and circles inscribed by long-gone grass are present in sandstones deposited millions of years ago in ancient dunes and near ancient rivers. Ripple marks and mudcracks, too, are often preserved in sedimentary rocks. Perhaps some paleontologist of the distant future will puzzle over the crooked trail made by a crushed paper drinking cup blown from the national monument picnic area!

In 2000 Congress voted to enlarge the national monument and change it to a national park. The park designation became official in 2004. Most of the added area was part of an old Spanish land grant that had remained in private hands and was fairly pristine. East of the park, land up to the Continental Divide has been designated Great Sand Dunes National Preserve. Geology of the preserve area is best described in the sections on U.S. 285 and U.S. 160 north and east of Alamosa.

U.S. 285
Denver—Fairplay
85 miles (137 km)

As you drive toward the mountains on U.S. 285, the Tertiary pediment shows distinctly—an undulating surface about halfway up the Front Range, 8,000 to 10,000 feet in elevation. When regional uplift gave mountain streams new energy, they cut many deep canyons into the pediment.

Bear Creek Lake, about a mile west of Kipling Street, is purposely kept nearly empty, ready to catch flash floods that could endanger Lakewood and Denver.

Between Denver and the mountains, U.S. 285 crosses soft Tertiary and Cretaceous rocks rarely exposed in the urban area. Outcrops of Cretaceous and Jurassic sediments begin at the Dakota Hogback, upturned layers of

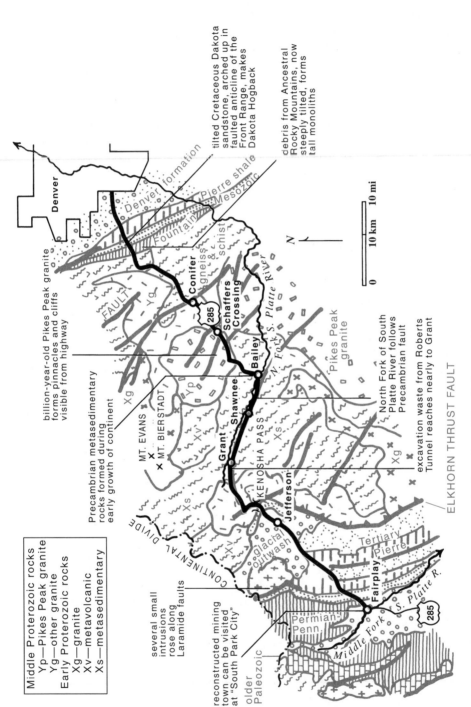

Legend:

Middle Proterozoic rocks
Yp—Pikes Peak granite
Yg—other granite
Early Proterozoic rocks
Xg—granite
Xv—metavolcanic
Xs—metasedimentary

billion-year-old Pikes Peak granite forms pinnacles and cliffs visible from highway

Precambrian metasedimentary rocks formed during early growth of continent

tilted Cretaceous Dakota sandstone, arched up in faulted anticline of the Front Range, makes Dakota Hogback

debris from Ancestral Rocky Mountains, now steeply tilted, forms tall monoliths

several small intrusions rose along Laramide faults

reconstructed mining town can be visited at "South Park City"

North Fork of South Platte River follows Precambrian fault

excavation waste from Roberts Tunnel reaches nearly to Grant

ELKHORN THRUST FAULT

Geology along U.S. 285 between Denver and Fairplay.

resistant Dakota sandstone originally deposited near a Cretaceous shore. Petrified ripple marks are visible in the highway cut; dinosaur tracks have been found here as well.

Between the hogback and the mountain front, the highway crosses successively older strata: Jurassic purple and green shale of the Morrison formation, Triassic and Permian red sandstone and limestone, then Pennsylvanian sandstone that stands up as tall pink monoliths. Once flat-lying, all these rocks were tilted by mountain uplift. West of them, separated by more than a billion years of time, are Precambrian rocks. Older Paleozoic strata, though deposited here, were eroded away when this area was part of the Ancestral Rockies.

Precambrian gneiss and schist form much of the mountain core along this route. Their radioactive clocks tell us they were metamorphosed 1.8 to 1.7 billion years ago. They are the oldest known rocks in most of Colorado,

Ripple marks from a Cretaceous shore corrugate a surface of the Dakota sandstone. Most rock-hammer handles are about a foot long, so geologists use them in photographs to indicate scale. —W. R. Hansen photo, courtesy of U.S. Geological Survey

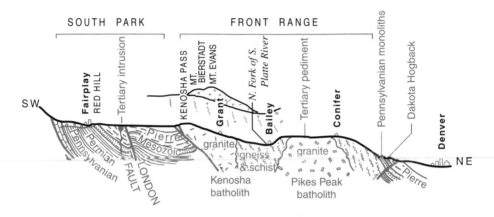

Section along U.S. 285 between Denver and Fairplay.

the roots of ancient mountains worn away by the end of Precambrian time. They vary in color from light gray to almost black depending on the proportion of black minerals (hornblende and biotite) and light minerals (quartz and feldspar). Very dark parts of them may have been basalt flows, dikes, and sills; light-colored portions were either light-colored volcanic rocks or sediments derived from them. Here and there, pegmatite dikes contain large crystals of biotite, quartz, and feldspar.

Careful study of faults, dikes, and the grain or fabric, of the rocks shows us that the trend of Precambrian mountains was northeasterly. But most faults here trend northwest and developed in Pennsylvanian time, when the Ancestral Rockies formed. Many faults were reactivated during the Laramide Orogeny when the Front Range took shape, even though they don't parallel the north-south trend of the Front Range.

Between Conifer and Bailey, U.S. 285 crosses the Tertiary pediment, with outcrops of granite and metamorphic rocks. The highway follows the bulbous edge of the Pikes Peak batholith, an immense mass of coarse pink granite that pushed and melted its way into the metamorphic rocks about a billion years ago. It is easy to recognize because it is lighter in color and more uniform in texture than the metamorphic rocks and usually has large, coarse crystals. In places it forms knobs, massive cliffs, and pinnacles of monumental rounded blocks. With long slow weathering, it decomposes into pink or salmon-colored sand.

U.S. 285 drops into the canyon of the North Fork of the South Platte River at Bailey, encountering good exposures of gneiss and schist. The river and highway follow a fault zone, where the rocks are broken and weak, and pass below the steep south slope of Mt. Evans (14,264 feet) and Mt. Bierstadt (14,060 feet). The summits of both mountains are composed of granite of the Routt plutonic suite, intruded around 1.7 billion years ago. Some of the granite appears near the highway west of Shawnee.

Flat-topped terraces along the river date from Pleistocene time, when the river carried much more water, sand, and rock than it does now. The highway crosses a hilly glacial moraine near milepost 213.

Just west of Grant, U.S. 285 passes the portal of 23-mile-long Harold D. Roberts Tunnel, which brings water from Dillon Reservoir eastward under the Continental Divide to the North Fork of the South Platte River for Denver's use. Colorado's eastern slope population draws increasingly on western slope water.

Several miles west of the tunnel portal the highway crosses into granite of the Kenosha batholith, part of the 1.4-billion-year-old Berthoud plutonic suite. The contact is irregular, and isolated granite fingers seem to float among the flow lines of the metamorphic rock. The granite is quite badly fractured because several large faults come together here.

There are many beaver dams in this valley. Active little geologic agents, beavers create ponds that in time fill in and become meadows; their dams decrease erosion and lessen the severity of floods.

Like many Colorado passes, Kenosha Pass developed in an area weakened by faults. Most faults in this area trend northwest-southeast. Hummocky areas near the pass are glacial ground moraines.

The overlook just beyond the pass at milepost 203 offers a splendid view of South Park, one of three large intermontane valleys in Colorado. The valley lies in a faulted syncline between the Front Range and the Sawatch uplift. Its eastern side is edged by the Elkhorn thrust fault, where Front Range Precambrian rocks pushed up and westward over Cretaceous and early Tertiary rocks.

On the western side of South Park, Paleozoic formations rise to the crest of the Mosquito Range. Paleozoic and Mesozoic strata underlie much of South Park but are mostly concealed by a veneer of glacial outwash gravel. The sedimentary rocks dip east, so U.S. 285 crosses older and older rocks as it travels westward, in much the same pattern as on the eastern side of the Front Range. Descending from Kenosha Pass, the highway crosses the following features:

- a wide valley underlain by Cretaceous Pierre shale;
- at milepost 190 a large ridge-forming igneous dike intruded during the Laramide Orogeny;
- Red Hill, a long hogback of tilted Cretaceous Dakota sandstone capping red slopes of Permian and Jurassic shale;

• another valley, this one underlain by Permian and Pennsylvanian redbeds similar to the Fountain formation near Denver. Covered in the valley by glacial deposits, these rocks extend up onto the shoulder of the Mosquito Range.

Fairplay, established in 1859, was a lively and often unruly gold camp. At first, gold was panned from river gravel or separated in crude homemade rockers. Then for many years gravel was washed with high-pressure hoses, the streams of rock and sand run through sluices where gold was caught in corrugated riffles lined with gunnysacking. In 1922 a gold dredge was assembled on the Middle Fork of the South Platte to process river gravel; a similar dredge mined gold south of Fairplay until 1952, leaving giant arcuate heaps of dredge tailings that can still be seen near Fairplay.

By following gold-bearing gravel up the valleys, miners found more riches in the solid rock among the peaks. Mineral veins there were similar geologically to those at Leadville, and it's no wonder: Leadville's mines are only 16 miles to the west as the crow flies, on the western slope of the Mosquito Range.

The South Park Ranger District office distributes brochures for several self-guided driving tours of mining areas near Fairplay. South Park City, an outdoor museum, contains many original buildings and artifacts dating from gold-camp days.

A gold dredge, built in 1941, operated just south of Fairplay until 1952. It dug gravel 70 feet below water level, collapsing the bank 35 feet above water level, and thus churned through 105 feet of gravel as it slowly advanced, dumping ridges of tailings behind. The gravel was sorted and sluiced right on the dredge, which in 11 years recovered more than 3½ tons of gold from about 33 million cubic yards of gravel. —L. C. Huff photo, courtesy of U.S. Geological Survey

West of Fairplay in the Mosquito Range, glacial erosion has exposed Paleozoic sedimentary rocks in Horseshoe Cirque (left of center). A glacial lake lies in the cirque, below rock glaciers of fallen rock lubricated with ice. Mt. Sheridan and Mt. Sherman to the north (right) are peppered with old mines, from which ore was transported in ore buckets strung on a cable lift. —T. S. Lovering photo, courtesy of U.S. Geological Survey

U.S. 285
Fairplay—Poncha Springs
57 miles (92 km)

Where U.S. 285 crosses the Middle Fork of the South Platte near Fairplay, bouldery piles of tailings from gold dredging edge the stream. A short distance south of Fairplay, a large gold dredge, inactive after 1952, floated on its self-made pond until the mid 1980s, when it was dismantled and shipped to Colombia, South America, to mine tin.

The highway rises almost immediately onto a terrace of stream-deposited glacial gravel, with occasional exposures of red Permian Maroon formation. For an interesting side trip to older Paleozoic rocks of the Mosquito Range, drive up Fourmile Creek to Horseshoe Cirque. Self-guided tour brochures are available at Fairplay Ranger Station. Above the old mining town of Leavick, glacial erosion has exposed sedimentary layers that range in age from Cambrian to Pennsylvanian.

Buffalo Peaks, the dark double-summited mountain southwest of Fairplay, is a remnant of many thick flows of lava and layers of volcanic ash that once filled a Tertiary valley. The flows, composed of dark fine-grained andesite, are quite resistant, but soft valley walls that contained them have

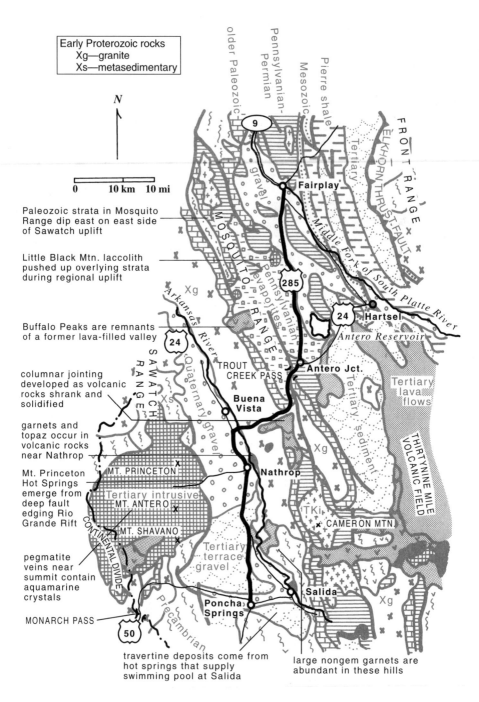

Geology along U.S. 285 between Fairplay and Poncha Springs.

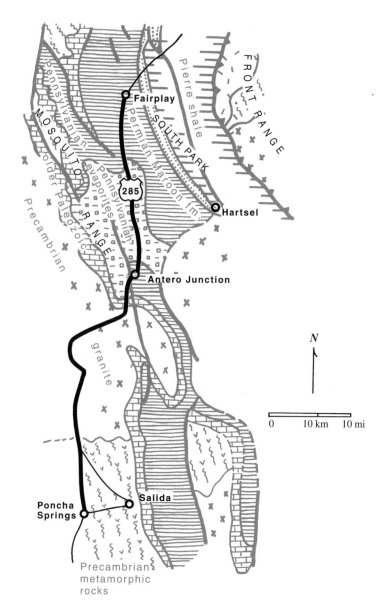

The geologic map of this area (opposite) looks hopelessly complex. But if we omit Tertiary and Quaternary deposits (this page), a simpler pattern emerges: north-south bands of Paleozoic and Mesozoic strata caught between two uplifted blocks of Precambrian rocks. The youngest Paleozoic strata are on the east side of South Park and the oldest are on the west because this is the eastern limb of a large faulted anticline.

worn away, so that what used to be a valley is now a mountain. Remnants of related flows cap several small buttes nearby.

Little Black Mountain, the round-topped mountain between Buffalo Peaks and milepost 178, is a small laccolith in which magma domed up overlying strata but did not escape to the surface. The igneous rocks penetrated the crust during mid-Tertiary regional uplift, when the crust was stretched and broken.

Pikes Peak is in view to the southeast beyond humpy hills of the Thirtynine Mile volcanic field. A hogback ridge of Paleozoic sedimentary rocks that edges South Park on the west appears straight ahead, with distant summits of the Sangre de Cristo Range on the southern skyline.

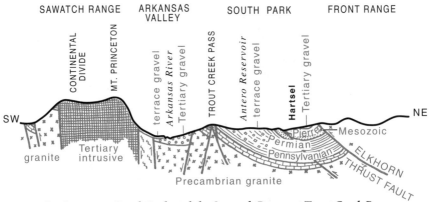

Section across South Park and the Sawatch Range at Trout Creek Pass.

As U.S. 285 turns toward Trout Creek Pass south of Antero Junction, Pennsylvanian black shale, evaporites, and thin gray limestone appear in roadcuts. These sediments accumulated in a basin between the two ranges of the Ancestral Rockies.

Trout Creek Pass is right on a large fault, one of several that run along the Mosquito Range. Strata west of the pass dropped down relative to those east of it. Between mileposts 223 and 224 the highway crosses a small syncline of Paleozoic rocks.

The strata here are:

- Pennsylvanian black marine shales and thin limestones, with some coaly layers and fossil plants;
- Mississippian Leadville limestone, gray and massive, forming skyline ridges above both sides of the valley;
- Devonian Chaffee formation, light brown limestone and fine white sandstone poorly exposed in slopes and ledges;

- Ordovician cliff-slope-cliff sequence: two massive limestones, the Manitou and Fremont formations, sandwiching the Harding sandstone. These rocks are visible at the narrows just above the bridge over Trout Creek.

Fossils are fairly common in Paleozoic limestone west of Trout Creek Pass. Horn corals like this one are easy to find in the Fremont limestone, the upper of the Ordovician cliffs. (× 1)
—Drawing by Emily Silver

U.S. 285 descends toward the Arkansas Valley through rugged country studded with rounded granite boulders. Intruded during the Early Proterozoic Orogeny, the 1.7-billion-year-old rock is cut by white pegmatite veins containing many rare and unusual minerals such as beryl, tantalite, and xenotime.

Weathering of granite follows a typical pattern, with mica and feldspar grains decomposing into clay minerals and, as they expand, popping off quartz grains and sometimes whole curving sheets of the rock surface. Sand around the base of the boulders is feldspar and quartz grus made up of fallen crystals of the parent rock.

The granite is part of the core of the Sawatch uplift, a broad faulted anticline. The Mosquito Range is the eastern limb of the anticline, the Sawatch Range the western limb. Slicing between them, the Rio Grande Rift forms a fault valley occupied by the Arkansas River.

Several high peaks of the southern Sawatch Range—Fourteeners all— are visible to the west as you descend into the Arkansas Valley. Though the northern part of this range is composed of Precambrian rock, Mt. Princeton, directly in line with the highway, and Mt. Antero and Mt. Shavano farther south are parts of a batholith intruded only 36 million years ago, well after the Rockies rose. The intrusion is the solidified magma chamber from which a particularly large eruption exploded; its volcanic ash has been traced across the Front Range and onto the Great Plains south of Denver. The volcano, if it survived the eruption, has since eroded away.

Near the summit of Mt. Antero is a remarkable area where open cavities in granite and porphyry are lined with crystals of gem-quality aquamarine, phenacite, smoky quartz, and fluorite. The crystals grew from superheated fluids escaping from cooling magma near the top of the batholith. No permission is needed to search for them; the 14,000-foot elevation, difficult access, and afternoon thunderstorms are obstacles enough!

Glacier-carved cirques appear at high elevations on all the large peaks of the Sawatch Range. Glaciers left their signatures between the peaks as well: U-shaped valleys and moraines down to 8,000 feet, the lower limit of glaciation in Colorado. The Arkansas Valley, just below 7,800 feet elevation, was never glaciated. Along the base of the Sawatch Range, coalesced alluvial fans contain sorted and stratified glacial outwash, products of streams draining the glaciers. No such large fans developed along the eastern side of the valley, as this part of the Mosquito Range was not glaciated.

The eroded white patch on the southeastern flank of Mt. Princeton, visible between mileposts 143 and 141, contains kaolinite, a soft white clay mineral formed as hot water rising along faults leached feldspar minerals of the batholith. The powdery white rock gave Chalk Creek its name, though kaolinite's composition is quite different from that of real chalk. Hot springs, fed by rain and snowmelt that sink down a fault in the batholith, rise to the surface near the mouth of Chalk Creek Canyon, where an east-west fault intersects the north-south one at the edge of the valley. The faults are hidden by Tertiary and younger gravel, but we know they exist because of horizontal offset of the batholith along the straight valley of Chalk Creek, and the huge vertical drop of the Arkansas Valley. At Mt. Princeton Hot Springs a spa and pool invite a geologic swim! Thermal waters also heat nearby greenhouses.

The Arkansas River follows the long, straight valley that is the northern segment of the Rio Grande Rift. The river once continued southward to join the Rio Grande; an ancient channel, now filled with volcanic rocks, underlies Poncha Pass south of Poncha Springs.

More hot springs emerge fairly high on the slope south of Poncha Springs. Their flow is only about 15 gallons per minute, but they are quite hot, 150 degrees Fahrenheit. Most of the water flows through an insulated pipe to the big indoor swimming pool at Salida.

U.S. 285
Poncha Springs—New Mexico
126 miles (203 km)

South of Poncha Springs, U.S. 285 crosses Poncha Pass, the saddle between the Sawatch and Sangre de Cristo Ranges. For much of the way over Poncha Pass, Precambrian volcanic island arc rocks edge the highway. Around 1.8 billion years ago, these were sediments and light-colored volcanic rocks intermixed with dark basalt flows and penetrated by dikes. During mountain building 1.7 billion years ago, they were metamorphosed into hornblende gneiss, schist, and striking black-and-white banded gneiss.

Much of the Precambrian rock has more recently been altered hydrothermally as percolating hot water changed feldspar minerals into kaolinite, a white clay mineral, and produced bright green coatings of epidote on fracture surfaces.

Patches of purplish Tertiary volcanic rocks begin to show up by the road at milepost 122. Farther southwest, similar dull red or purple ash sheets and flows have piled up thousands of feet thick to form the San Juan Mountains, by far the largest volcanic region in Colorado. U.S. 285 skirts the eastern side of these mountains between Saguache and Alamosa, slicing through a flow that reaches out into the San Luis Valley.

Poncha Pass marks the divide between Arkansas River drainage to the north and the Rio Grande watershed to the south. However, in Miocene and Pliocene time water from what is now the Arkansas Valley flowed straight south through a channel just west of Poncha Pass, into the San Luis Valley and the Rio Grande. During Pliocene or early Pleistocene time, erosion at the headwaters of an east-flowing river—the ancestral Arkansas—broke through into the Rio Grande Rift valley, offering a lower drainage route, so the northern part of the Rio Grande changed its course, probably with the added inducement of lava flows near Poncha Pass. In geologic lingo, the Arkansas River "pirated" the waters of the upper Rio Grande.

The San Luis Valley is bordered on the east by the Sangre de Cristo Range and on the west by the San Juan Mountains. It extends from Poncha Pass almost to the New Mexico border. As large as Connecticut, and far deeper geologically than it looks, it is a true rift in the continent, bordered by deep faults that reach down through the crust to Earth's mantle. Rift faulting began with mid-Tertiary regional uplift and continues right up to the present. The valley's bedrock floor, the solid rock beneath layer after layer of gravel, sand, clay, lava, and volcanic ash, is 5,000 to 13,000 feet below the present surface—much of it well below sea level.

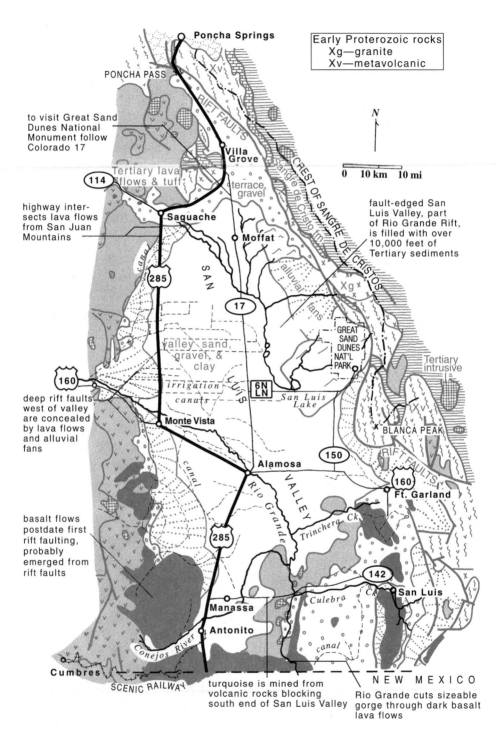

Early Proterozoic rocks
Xg—granite
Xv—metavolcanic

Poncha Springs

PONCHA PASS

to visit Great Sand
Dunes National
Monument follow
Colorado 17

Villa
Grove

RIFT FAULTS

CREST OF SANGRE DE CRISTOS

Sangre de Cristo m.

Xv

Tertiary lava
flows & tuff

114

terrace
gravel

highway inter-
sects lava flows
from San Juan
Mountains

Saguache

Moffat

fault-edged San
Luis Valley, part
of Rio Grande Rift,
is filled with over
10,000 feet of
Tertiary sediments

SAN

285

17

alluvial fans

Xg

valley sand,
gravel, &
clay

GREAT
SAND
DUNES
NAT'L
PARK

Tertiary
intrusive

160

irrigation
canals

LUIS

6N
LN

San Luis
Lake

Xv

deep rift faults
west of valley
are concealed
by lava flows
and alluvial
fans

Monte Vista

BLANCA PEAK

RIFT FAULTS

Alamosa

150

VALLEY

160

Ft. Garland

basalt flows
postdate first
rift faulting,
probably
emerged from
rift faults

canal

Rio Grande

Trinchera Ck.

285

142

Culebra

Ck

San Luis

Manassa

Antonito

Conejos River

canal

NEW MEXICO

Cumbres

SCENIC RAILWAY

turquoise is mined from
volcanic rocks blocking
south end of San Luis Valley

Rio Grande cuts sizeable
gorge through dark basalt
lava flows

N

0   10 km   10 mi

Geology along U.S. 285 between Poncha Springs and New Mexico.

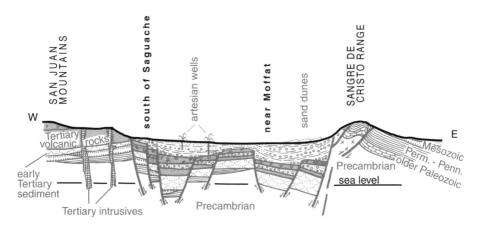

Section across northern part of San Luis Valley.

The northern half of the valley now has no outlet. Streams entering it disappear into porous sands of alluvial fans and valley floor. If the climate were less desertlike, as it has been in the past, a lake would develop here.

Since Pleistocene time, prevailing winds from the west and southwest have blown sand from the valley floor to the east side of the valley, where it has accumulated between Blanca Peak and the northern part of the Sangre de Cristo Range. The resulting dunes rise 700 feet high.

Separated from the northern closed basin by an almost imperceptible rise near Monte Vista and Alamosa, the southern part of the San Luis Valley drains into the Rio Grande. This river and its tributaries rise high in the San Juan Mountains, where heavy winter snowpack and seasonal summer rains provide plenty of water for eroding the mountains and carrying debris into the valley. The deepest deposits in the valley, known from water well records, are from Paleocene time, showing that the valley became established soon after the Rockies began to rise. Deposits continued through mid-Tertiary regional uplift and into Quaternary time. Layers of volcanic ash and lava from the San Juan Mountains alternate with layers of gravel, sand, and clay.

Recent studies have added to our understanding of how the Rio Grande Rift, of which the San Luis Valley is just a part, came into being. During regional uplift in mid-Tertiary time, Earth's curving crust was stretched to the breaking point. Under tension, it cracked along a pair of especially deep faults that stretch from mid-Colorado southward through New Mexico, defining the long, narrow wedge of the rift. Helped by clockwise twisting of the Colorado Plateau of Arizona, Utah, and western Colorado, the edges of the rift pulled apart, allowing the slender sliver between them to drop many thousands of feet—from perhaps 9,000 feet above sea level to well below sea level.

On a tectonic scale, the underlying cause of these changes is not well understood. Before Miocene time, the North American plate, drifting westward, overrode much of the Pacific plate. As it reached the Pacific mid-ocean ridge, huge new forces must have developed, forces great enough to reach the continent's interior, lifting it up as much as 5,000 feet, stretching it, and breaking it.

Though movement along the San Andreas fault system in California may now have begun to relieve the pressure, sporadic movement on the faults that edge the Rio Grande Rift continues to this day. The steep west face and high jagged crest of the Sangre de Cristo Range are the eroded scarp of the large fault zone bordering the east side of the San Luis Valley. Notice the triangular facets at the bases of mountain ridges, fault facets touched only lightly by erosion. Within the last few thousand years, several smaller fault scarps have developed in alluvial fans along the base of the range. Several hot springs bring thermal water to the surface along the faults. Mineral Hot Springs, 6 miles south of Villa Grove on Colorado 17, is open to the public and so is Hooper Pool, also known as the Sand Dunes Pool, west of Great Sand Dunes National Park and Preserve.

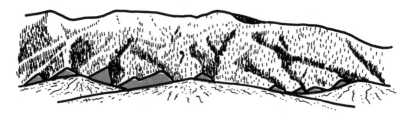

Northeast of Villa Grove, triangular facets across the toes of mountain ridges have scarcely eroded since they were formed by movement on faults at the edge of the valley.

Groundwater geology of the San Luis Valley is particularly interesting. Canals built around 1880 once brought irrigation water from Rio Grande tributaries into the nondraining northern half of the valley. Farming was good for a while, but eventually the water table, the level below which the sediments are saturated by groundwater, rose nearly to the surface, and farmland became waterlogged and salty. Farmers moved south to the drained part of the valley near Monte Vista and Alamosa. Unfortunately, starting in the 1930s, the region suffered from a long drought, with runoff below normal for 20 to 30 years. Desperate farmers turned to subsurface water, drilling wells deeper and deeper into the valley floor. This time they were luckier. Because of the trough shape of the valley and the alternating layers of porous gravel and impervious clay, ash, and lava that fill it, their wells were artesian—water rose and flowed from them without pumping. More than 7,000 flowing wells now supplement the Rio Grande irrigation water.

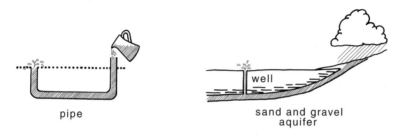

In an artesian system, the point of entry is higher than the well head, so well water rises without pumping. Impervious shale or clay layers prevent the water from rising except in wells that penetrate those layers.

But all that irrigation raised the water table. Subsoil in the San Luis Valley is salty, and groundwater brought salt to the surface, leading to disaster as far as farming was concerned. To counteract this problem, farmers now encircle their fields with drainage ditches to keep the water table below the roots of their crops. Then they flush the topsoil with fresh water from the mountains or from their deep artesian wells. After letting the soil dry thoroughly—not difficult in the San Luis Valley's dry climate—they water lightly from the top down, keeping the topsoil wet enough to satisfy their crops while leaving a gap of dry earth below the roots and above the water table.

Blanca Peak bulges westward from the Sangre de Cristo Range where the great fault system that edges the range fans out, its exact location hidden under the alluvial fans.

Near Monte Vista, U.S. 285 crosses the Rio Grande, less grand now than it used to be because water is diverted from it all along its course from the high peaks of the San Juans to the Gulf of Mexico. Turning south near Alamosa, the river plunges into a narrow, black-walled canyon cut through late Tertiary basalt. Looking east from milepost 12 you can see volcanic hills near Manassa, some of them capped by remnants of flat-lying lava flows. There, turquoise has been mined, a product of hydrothermal alteration of copper minerals in the volcanic rock.

From Antonito a narrow-gauge railroad winds its way up Pine Canyon and across Cumbres Pass to Chama, New Mexico. Rocks along the railroad range in age and composition from Precambrian intrusive and metamorphic rocks to Pleistocene glacial deposits. Guidebooks available in the depot in Antonito outline the geology along the rail trip, which runs only in summer.

The dome-shaped mountain to the south, San Antonio Peak, is one of a cluster of small volcanoes that, with their related lava flows, dam the San Luis Valley, preventing escape of groundwater.

U.S. 287
Wyoming—Loveland
54 miles (87 km)

One of the most interesting routes by which to enter Colorado is U.S. 287, which crosses the Wyoming line among hills of pink granite, most of it surfaced with gray lichens. Part of the core of the Front Range, the granite is one lobe of the Sherman batholith, part of the Berthoud plutonic suite. Intruded during a Precambrian mountain-building episode, the granite is around 1.4 billion years old. And what a history it has had!

- worn down to sea level during the long erosion interval at the end of Precambrian time;
- covered with many layers of early Paleozoic marine sediments;
- intruded by kimberlite pipes, volcanic conduits that preserved blocks of Ordovician and Silurian rocks;
- uplifted during Pennsylvanian time as part of the Ancestral Rockies;
- stripped and laid bare once more by erosion;
- covered again, with more than 15,000 feet of Pennsylvanian, Permian, and Mesozoic marine and continental sediments;
- lifted again during Laramide mountain-building as part of the Front Range;
- cleaned off anew by another cycle of erosion;
- lifted about 5,000 more feet to its present elevation.

The granite is composed of quartz, mica, and pink feldspar grains of almost uniform size. Everywhere it is jointed in a more or less cubic pattern, each set of joints reflecting a different direction of stress to which the rock has at some time been exposed. It weathers into titanic boulders and decomposes into coarse pink sand. It extends into the Sherman Mountains, part of the Laramie Range, in Wyoming, as well as south from the state line along the highway for about 10 miles.

Two miles south of Virginia Dale, U.S. 287 rises to a relatively smooth, bare granite surface that is probably a pediment smoothed during erosion

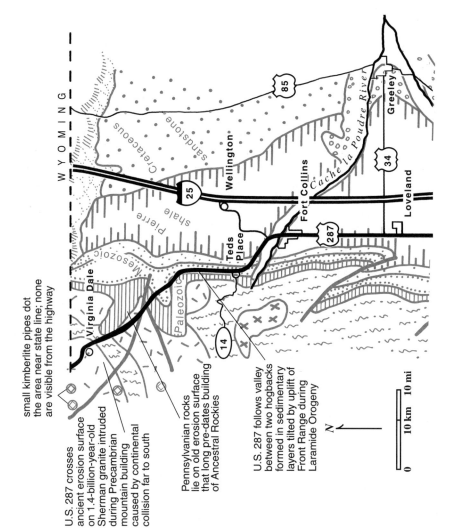

small kimberlite pipes dot the area near state line; none are visible from the highway

U.S. 287 crosses ancient erosion surface on 1.4-billion-year-old Sherman granite intruded during Precambrian mountain building caused by continental collision far to south

Pennsylvanian rocks lie on old erosion surface that long pre-dates building of Ancestral Rockies

U.S. 287 follows valley between two hogbacks formed in sedimentary layers tilted by uplift of Front Range during Laramide Orogeny

WYOMING

Virginia Dale

Mesozoic

Paleozoic

Teds Place

14

287

25

Wellington

85

Pierre shale

Cretaceous sandstone

Fort Collins

Cache la Poudre River

34

Loveland

Greeley

N

0   10 km   10 mi

Geology along U.S. 287 between Wyoming and Loveland.

of the Ancestral Rockies—a 270-million-year-old roadbed! Nearby, deep red Pennsylvanian sandstone and shale of the Fountain formation lie directly on the granite surface. The contact seems sharp when seen from a distance, but the surface of the granite beneath the Pennsylvanian sediments is deeply weathered.

We know that older Paleozoic rocks—Cambrian through Devonian—once covered the granite here because fragments of them were preserved when they were carried downward into numerous small kimberlite pipes that dot this area. The pipes are former volcanic conduits that pierced the Sherman granite and overlying Paleozoic sediments in Devonian time. Though they can't be seen from the highway, the cluster of pipes, known as the State Line kimberlites, are worth discussing.

Kimberlite is an unusual igneous rock formed from gaseous magma that escapes rapidly from deep in Earth's mantle through narrow cracks in the crust. When it's within a mile or two of the surface, it blasts out the rock above it, forming a steep cone-shaped pipe. Much of the blasted-out material falls back down the pipe, becoming embedded in the cooled kimberlite magma.

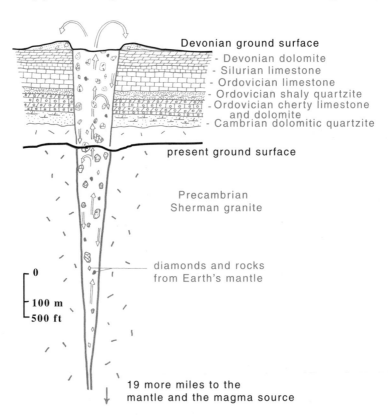

Section of a kimberlite pipe through the Sherman granite as it would have appeared in late Devonian time. Since then, erosion has stripped away overlying sedimentary rocks.

In addition to preserving samples of overlying rock, kimberlite pipes contain rarely seen minerals formed deep within the upper mantle—including diamonds. Since diamonds are mined from similar pipes near Kimberley, South Africa, discovery of the pipes in Colorado and Wyoming set off a small diamond rush, and gem-quality stones were found in 1975. The largest unearthed by 2000 was a 28.3-carat stone that, after cut to a 16.9-carat gem, sold for around $300,000. The gems, however, are rare, at most around one carat per ton of rock.

Ironically, in 1871 a team of swindlers scattered imported diamonds in an area about 170 miles west of here to pry money from investors. Cagey about the location, they sometimes boarded the train in Laramie. Had they picked a site closer to Laramie, they might have been selling real diamond mines.

All along this side of the Front Range, Pennsylvanian and younger rocks bend up against the mountains. Where U.S. 287 crossed the granite of the Sherman batholith near the Wyoming border, deep red Pennsylvanian deposits lie directly on the granite surface; they represent sand and silt that accumulated along the northeastern edge of Frontrangia as the Ancestral

Uncut diamonds from the State Line kimberlites have the natural shape of diamond crystals: two pyramids, base to base. The large stone in the center weighs 14.2 carats and is half an inch across. —Photo courtesy of Howard Coopersmith, Kelsey Lake Mine

Rockies rose and eroded. Farther south, resistant Lyons sandstone is bent up into cuestas and hogbacks east of the mountains. Deposited as dune sand, the pinkish rock, often discolored by lichens, parallels U.S. 287 for several miles. Passing through it, the highway is confronted with another prominent ridge, one known to geologists and nongeologists alike as the Dakota Hogback because its backbone, a resistant layer that lies between softer strata, is the Dakota sandstone.

For 10 miles the highway runs between the Lyons sandstone hogback and the Dakota Hogback along a valley eroded in soft Triassic and Jurassic shale and mudstone. In places, Triassic strata, deposited in shallow seas and salty lakes, contain gypsum; gypsum and limestone—the ingredients of cement—are mined near Livermore. U.S. 287 goes through the Dakota Hogback just south of Teds Place, where the Cache la Poudre River has carved a ready-made break. The highway emerges among younger Cretaceous rocks, the Niobrara limestone and Pierre shale, near Laporte. However, good exposures of these formations are rare.

Quaternary stream deposits fill river valleys here, bordered in places by Pleistocene terraces. Parts of Fort Collins and Loveland, as well as towns

Tilted Paleozoic and Mesozoic strata in hogbacks fringing the mountains are all that remains of a blanket of sedimentary rock that extended far to the west before the Front Range rose. —Jack Rathbone photo

farther south, lie on the terraces, which were established as river flood-plains by heavily loaded Ice Age streams. Less heavily burdened streams channeled through the terraces after the glaciers disappeared.

Cache la Poudre River and Big Thompson River at Loveland, as well as St. Vrain Creek and Boulder Creek farther south, are tributaries of the South Platte River. They helped excavate the Colorado Piedmont, the wide valley that separates the Tertiary pediment, halfway up the mountainside, from its original connection with the Great Plains to the east.

Reservoirs near the mountain front store spring runoff from the mountains for farms and towns in the piedmont area, including some western slope water piped in tunnels through the mountains. The Dakota Hogback forms a natural eastern wall for Horsetooth and Carter Reservoirs near Fort Collins. Soil derived from the Pierre shale is fertile and farmed; less fertile gravel terraces are suitable for grazing but increasingly used for homesites.

The high part of the Front Range, west of Fort Collins and Loveland, lies along the Continental Divide, which swings farther east here than at any other place in the United States. The prominent peak to the southwest, Longs Peak, is a Fourteener in good standing at 14,255 feet in elevation. Its flat summit, about the size of a city block, may be a remnant of the surface eroded at the end of Precambrian time. Or it may date back to erosion of the Ancestral Rockies, in which case it corresponds to the Sherman granite surface near the Wyoming line. The same surface lies 13,000 feet underground near Denver, giving us a measure of vertical displacement along the faults that edge the Front Range—27,000 feet, about 5 miles.

Colorado 9
Kremmling—Fairplay
75 miles (121 km)

Between Kremmling and Frisco, Colorado 9 follows a long valley eroded by the Blue River in soft Pierre shale, which overlies older Mesozoic limestone, sandstone, and shale. These strata rise westward, pushed up with the Gore Range during the Laramide Orogeny.

To the east, the Williams Fork Mountains mark the true western edge of the Front Range fault block. Near their crest, along the Williams Range thrust fault, Precambrian gneiss and granite have pushed westward across the Pierre shale and other Mesozoic rocks. Though forested slopes hide the fault from view, it has been traced southward for about 40 miles. Near the highway, Cretaceous sediments form sharp ridges where they crumpled as the immense fault block rode westward over them.

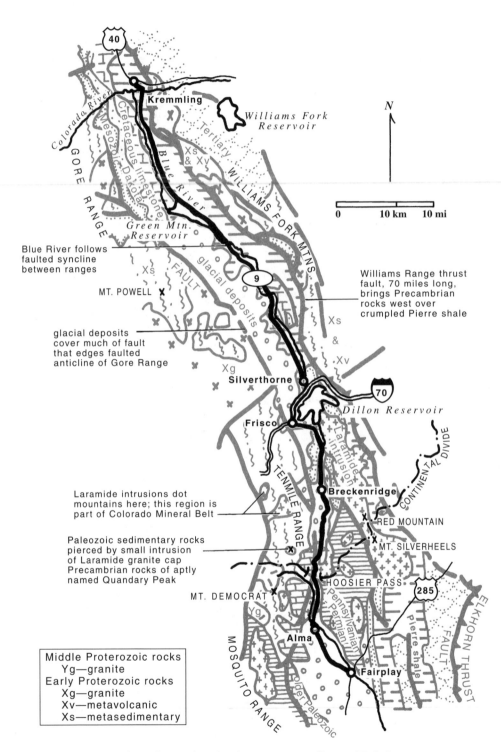

Geology along Colorado 9 between Kremmling and Fairplay.

A gash in the hills west of Kremmling marks the deep canyon carved by the Colorado River through the Gore Range. Unpaved 1-Road, leaving Colorado 9 just south of the Colorado River bridge, follows the canyon to State Bridge and Colorado 131, with close-up views of the range's Precambrian core as well as younger sedimentary rocks. —Felicie Williams photo

Green Mountain Reservoir is part of the Big Thompson Project, a series of reservoirs, tunnels, and canals that route western slope water under the Continental Divide to farms and cities east of the mountains.

Steep U-shaped glaciated canyons cut into the eastern slope of the Gore Range, and glacial terraces flank the range. Neither terraces nor other glacial debris exist on the other side of the Blue River valley, however, where lower summits and gentler, southwest-facing slopes were not as conducive to glaciation.

For most of the distance between Kremmling and Dillon, Colorado 9 rides on Pierre shale, a relatively thin-bedded, fragile rock famously prone to sliding. Retaining walls help to protect the highway from all-too-common slides.

The highway skirts the shore of Dillon Reservoir, part of Denver's water system. A tunnel carries water through the Front Range to the drainage of the South Platte River, where it is diverted into reservoirs for the city.

Southwest of Frisco a bold cliff marks one of the major Laramide faults involved in uplift of the Tenmile Range. Highly fractured Precambrian gneiss

Green Mountain projects from the western side of the valley, its shale and lime-stone intruded and strengthened by a Tertiary laccolith and several sills. The laccolith provides a convenient foundation for Green Mountain Reservoir dam, and impervious Pierre shale waterproofs the reservoir. —Felicie Williams photo

and schist—ancient sedimentary rocks metamorphosed 1.7 billion years ago—core the huge uplifted block of the range, which extends from Frisco to the Continental Divide at Hoosier Pass.

The highway ascends the valley of the Blue River across coarse ground moraine left behind by a receding glacier. The completely unsorted mix of large and small angular rock fragments, interspersed with sand and clay, is characteristic of glacial moraines.

Glaciers flowing down this valley and its side canyons scoured the mountain walls, exposing and grinding up gold-bearing veins. Swift mountain streams, able to transport heavy loads, carried the glacial debris downhill, rounding its boulders and cobbles, sorting it by size, and dropping it, gold and all, where they slowed down in the gentler valley below. Nineteenth-century prospectors discovered these stream-formed placer deposits and with simple tools like shovels, goldpans, rockers, and sluices recovered what gold they could from near the top of the deposits.

But gold is six times as heavy as most rocks and tends to settle in deeper parts of stream deposits. Near Breckenridge the valley has been worked over by a large gold dredge able to scoop up deep coarse gravel mechanically and to separate the gold from it right on the dredge, leaving the leftover gravel tailings in bouldery ridges. Modern gravel quarriers are now recycling the spent gravels, creating new ridges that can be distinguished from the old by the absence of trees.

Like most ski towns in Colorado, Breckenridge started as a mining town. Placer mines gave way to underground mines in mountains east and west of town. Breckenridge lies within the Colorado Mineral Belt, a northeast-trending band of mineralized rock that hosts most of Colorado's mines. Many small late Cretaceous and Tertiary intrusions rose along this trend, actually a network of Precambrian faults. Hot, mineral-rich fluids from the intrusions penetrated surrounding rocks, depositing metallic minerals along faults and joints. Most of the mine dumps that once peppered slopes around Breckenridge are now smoothed over into ski runs and homesites.

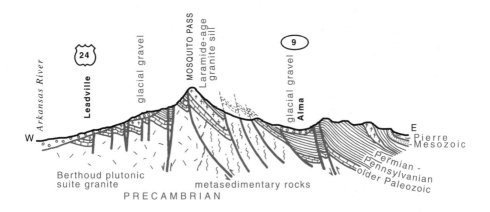

Section across Colorado 9 at Alma.

The Blue River valley south of Breckenridge is floored with glacial deposits almost all the way to Hoosier Pass. Glaciated canyons cut deeply westward into the Tenmile Range's pegmatite-laced Precambrian metamorphic rocks. As the highway climbs more steeply, some of the highest peaks of the range come into view, among them Quandary Peak, capped with red-tinted Paleozoic sedimentary rocks, tiny remnants of the strata that once arched across the range.

East of the highway, forest-covered redbeds of the Minturn and Maroon formations define the former extent of Frontrangia; these sediments washed off the western flank of that range in Pennsylvanian and Permian time.

Hoosier Pass marks the Continental Divide. The Blue River and its tributaries drain north into the Colorado River and ultimately the Pacific; south of the divide the Middle Fork of the South Platte drains ultimately into the Mississippi and the Gulf of Mexico. The divide also separates the Tenmile Range from the Mosquito Range to the south.

Two high peaks south of Hoosier Pass, pointed Mt. Lincoln (14,286 feet) and rounder Mt. Bross (14,172 feet), are part of the Mosquito Range. Early Paleozoic limestone and sandstone, deposited long before uplift of the Ancestral Rockies, tilt up against the east slope of these mountains. —Felicie Williams photo

Brilliant ruby-colored rhodochrosite crystals occur in a vein in the Sweet Home Mine, high in the peaks west of Alma. Composed of manganese carbonate, rhodochrosite is not uncommon in lead mines, though excellent crystals of it are rare. This 6-inch beauty, the Alma King, is at the Denver Museum of Nature and Science. —Felicie Williams photo

Peaks south of Hoosier Pass are cored with Precambrian igneous and metamorphic rocks—a continuation of the fault block of the Tenmile Range. The ancient rocks, laced with light-colored pegmatite veins, can be seen well from the pullout just south of the pass, as can the Paleozoic sedimentary rocks tilted up against them.

Silver, gold, lead, and zinc have been mined south of Hoosier Pass, where they occur in faulted Paleozoic limestone and Laramide sills. Mineralization is due to mid-Tertiary igneous activity and is similar geologically to that at Leadville, across the Mosquito Range from Alma.

As at Breckenridge, gold washed into streambeds from glacial moraines first drew miners to the Alma and Fairplay areas. Placer mining of stream deposits, moraines, and even some areas of soil high in the peaks has occurred in fits and starts since 1859, varying with the price of gold. Mining claims are private property, so ask for permission if you want to pan for gold.

Fairplay, "the biggest little city in South Park," lies at the edge of the wide fault-rimmed South Park syncline between the Front Range and the Mosquito Range. See road guides for U.S. 285 for descriptions of the town and its surroundings.

Mounds of coarse placer-mined gravel fill the valley near Alma. —Felicie Williams photo

Colorado 14
Teds Place—Gould
66 miles (106 km)

Teds Place lies in a long north-south valley between the Dakota Hogback, the ridge formed by the Dakota sandstone, and Precambrian rocks of the Front Range. Though the rock sequence is offset by faults, the main units can be identified. From east to west, they are:

- Dakota sandstone of the Dakota Hogback;
- greenish Morrison shale on the western slope of the hogback;
- soft Triassic shales of the valley floor, mostly concealed by stream deposits;
- a low ridge of pink Lyons sandstone, often quarried for building stone;
- a resistant ridge of Fountain formation;
- metamorphic rocks of the mountain front.

Rocks that wall Poudre Canyon, 1.8-billion-year-old sediments and volcanic flows of an island arc, were steeply folded and highly altered by the immense heat and pressure of the Early Proterozoic Orogeny 1.7 billion

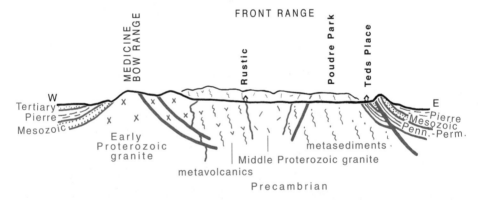

Section along Colorado 14 between Teds Place and North Park.

════ IN CASE OF FLOOD, CLIMB TO SAFETY ════

This is no idle warning. A sudden rainstorm up Poudre Canyon can cause the stream to rise over 10 feet in a matter of minutes. The water's force can roll giant boulders and swallow cars. A stream on the rampage moves more sediment downstream in a day than is moved in a hundred years of nonflood conditions. Raging waters move faster than you can drive down the winding canyon road, which is why signs warn you to leave your car and climb!

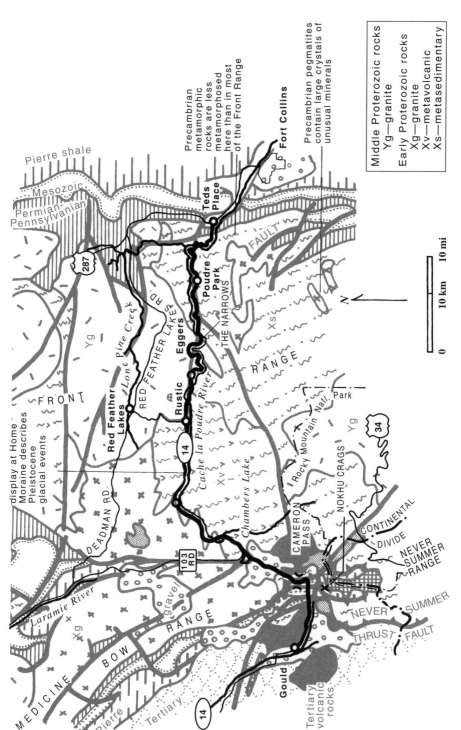

Geology along Colorado 14 between Teds Place and Gould.

Middle Proterozoic rocks
Yg—granite
Early Proterozoic rocks
Xg—granite
Xv—metavolcanic
Xs—metasedimentary

Precambrian metamorphic rocks are less metamorphosed here than in most of the Front Range

Precambrian pegmatites contain large crystals of unusual minerals

Fort Collins

Teds Place

FAULT

Poudre Park

THE NARROWS

Xs

RANGE

Eggers

Rustic

RED FEATHER LAKES RD.

Lone Pine Creek

14

Cache la Poudre River

Rocky Mountain Natl. Park

Chambers Lake

34

Red Feather Lakes

FRONT

Yg

287

display at Home-Moraine describes Pleistocene glacial events

DEADMAN RD.

103 RD

Xv

CAMERON PASS

NOKHU CRAGS

Yg

CONTINENTAL DIVIDE

NEVER SUMMER RANGE

Laramie River

gravel

Xg

MEDICINE BOW RANGE

Pierre

Tertiary

14

Gould

Tertiary volcanic rocks

NEVER SUMMER THRUST FAULT

Pierre shale

Mesozoic

Permian-Pennsylvanian

N

0   10 km   10 mi

Vertical faces on Precambrian metamorphic rocks in Poudre Canyon were once the surfaces of flat-lying sedimentary layers; dark and light layering reflects original shale and sandstone. In the first few miles of the canyon, these rocks are less intensely metamorphosed than elsewhere in Colorado. —Felicie Williams photo

years ago. About 1.4 billion years ago, the ancient metamorphic rocks were intruded by granite of the Berthoud plutonic suite, some of which appears on the north wall of the canyon between the Narrows and Rustic. Schist and gneiss east of the Narrows was originally sedimentary rock; compare it with light- and dark-banded gneiss west of Rustic, derived from volcanic rocks.

Large-crystalled pegmatite veins that pattern these ancient rocks developed from crack-filling fluids during the Early Proterozoic Orogeny. Since minerals crystallize at known pressures and temperatures, geologists have learned from pegmatite crystals that these rocks were buried 7 to 11 miles deep and heated to between 500 and 700 degrees Celsius when they formed. These are among Colorado's oldest rocks, the deep roots of an ancient mountain range.

A display at Home Moraine describes glaciers that scoured the upper Poudre Valley in Pleistocene time. Here, the narrow V-shaped canyon gives way to a wider U-shaped valley. Moraines contain unsorted gravel, sand, and clay, with angular cobbles and boulders that often display scratches from grinding against other rocks. Flat-topped lateral moraines that edge the valley accumulated next to the glacier flanks.

Because erosion cuts more easily through highly fractured schist and gneiss than through massive granite, the valley turns toward the southwest, staying south of the contact between 1.7-billion-year-old Precambrian metamorphic rocks of the Mummy Range to the south and 1.4-billion-year-old granite of the Laramie Range to the north. The glacier straightened the stream channel down which it came; compare this valley with the stream-cut canyon below Home Moraine.

Mid-Tertiary volcanic rocks near Chambers Lake, in the valley south of it, and on ridges to the southeast include flows of basalt, andesite, and rhyolite from volcanoes southeast of Cameron Pass. Along the highway

At Poudre Falls, zoned pegmatite veins stand out sharply against dark amphibolite. Zoning—banding parallel to the edges of veins—is common in pegmatites, where each kind of mineral grows at its own pace, spreading slowly inward from the wall as temperature, pressure, and chemical conditions become suitable. —Felicie Williams photo

many of these rocks are hidden by thick forests. A terminal moraine dams Chambers Lake.

At Cameron Pass, Colorado 14 crosses the Never Summer thrust fault, where the Front Range fault block slid westward during the Laramide Orogeny. Rocks close to the fault are ground up and soft, setting the stage for the erosion of the pass. South of the pass, tilted Paleozoic and Mesozoic sedimentary rocks show up well in a large, colorful highway cut; some are actually overturned.

Soon after crossing Cameron Pass, Colorado 14 climbs onto a lateral moraine. Watch for the glacier's terminal moraine between mileposts 61 and 60. Gould lies in North Park just below the lowest moraine. A large intermontane valley, North Park was never glaciated, though its floor is veneered with coarse glacial outwash gravel carried in by streams.

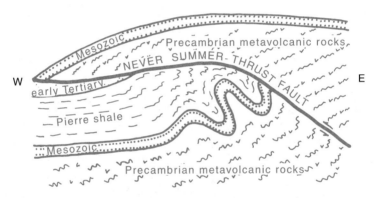

During the Laramide Orogeny, Precambrian rocks were pushed westward along the Never Summer thrust fault as much as 10 miles over Mesozoic rocks. Such thrusting seems hard to believe, but similar overthrusts occur in many western ranges.

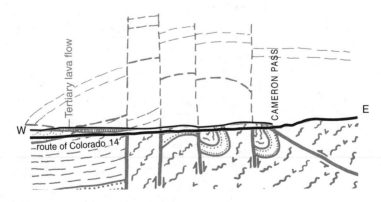

Tertiary faulting and volcanism superimposed on the thrusting further complicated the picture, as did subsequent erosion.

The Pierre shale of Nokhu Crags, south of Cameron Pass, was bent nearly vertical as Precambrian rocks thrust westward over it along the Never Summer thrust fault. Later, the tilted beds were faulted upward in a block, so now they loom over the overthrust Precambrian rocks to the west. Mt. Richthofen, behind the crags, is part of a mid-Tertiary intrusion. —Felicie Williams photo

Colorado 14
Gould—U.S. 40
57 miles (92 km)

The high, wide valley of North Park is a syncline between faulted anticlines of the Front Range and the Park Range, with the volcanic Rabbit Ears Range to the southwest. A major thrust fault, which can be traced for about 40 miles, borders the north end of North Park. There, Precambrian rocks above the fault slid 4 to 12 miles southwest over younger rocks.

Thick Tertiary sediments blanket the valley floor, some of the vast amounts of gravel, sand, and silt washed off the Rockies in Eocene time. Carved into terraces by glacial meltwater, they are covered in places by patches of glacial outwash. Gentle folds and many small faults in the rocks of the park's floor have been studied thoroughly because they form

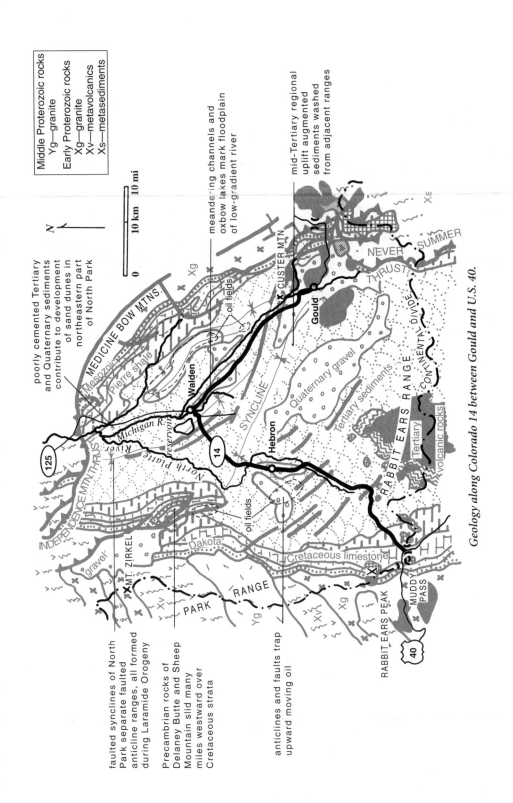

Legend:

Middle Proterozoic rocks
Yg—granite
Early Proterozoic rocks
Xg—granite
Xv—metavolcanics
Xs—metasediments

N

0   10 km   10 mi

faulted synclines of North
Park separate faulted
anticline ranges, all formed
during Laramide Orogeny

Precambrian rocks of
Delaney Butte and Sheep
Mountain slid many
miles westward over
Cretaceous strata

anticlines and faults trap
upward moving oil

poorly cemented Tertiary
and Quaternary sediments
contribute to development
of sand dunes in
northeastern part
of North Park

meandering channels and
oxbow lakes mark floodplain
of low-gradient river

mid-Tertiary regional
uplift augmented
sediments washed
from adjacent ranges

MEDICINE BOW MTNS.

Mesozoic

Pierre shale

oil fields

Xg

X CUSTER MTN.

Gould

Walden

Michigan R.

River

North Platte

reservoir

14

Hebron

SYNCLINE

Quaternary gravel

Tertiary sediments

RABBIT EARS RANGE

CONTINENTAL DIVIDE

NEVER

SUMMER

Xs

THRUST

Tertiary

volcanic rocks

125

INDEPENDENCE MTN THRUST

gravel

X MT. ZIRKEL

Dakota

oil fields

Cretaceous limestone

PARK   RANGE

Yg

Xv

Xg

Xv

X

RABBIT EARS PEAK

MUDDY
PASS

40

Geology along Colorado 14 between Gould and U.S. 40.

structural traps that have collected oil in the underlying Mesozoic strata. Coal has been mined from Tertiary rocks in North Park as well.

Since Pleistocene time, streams have channeled the former surface of North Park, leaving flat-topped terraces veneered with glacial outwash gravel. Near milepost 37, where the road crosses one of these terraces, an overlook gives a bird's-eye view of the valley below, with its meandering stream and crescents of abandoned channels.

Southwest of Walden, low hills range across the valley like waves in a sea, rippled by many small faults and folds in the Tertiary sediments. Watch in these hills for knobby rust-colored limonite concretions, some up to 2 feet across, formed as groundwater percolating through the sediments deposited

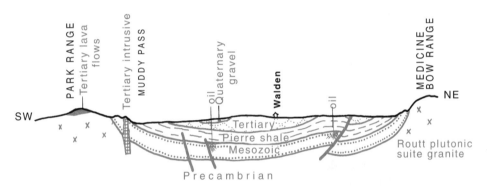

Northeast-southwest section across North Park through Walden.

The lower edge of Custer Mountain, east of Colorado 14 at milepost 51, is a long thin dike of Tertiary intrusive rock. Trees grow well on Precambrian rocks on the upper part of the mountain but not on the dike itself. —Felicie Williams photo

iron oxide around bits of plant material or grains of iron-rich minerals. Through many years, the concretions grew as more layers of limonite were added. As the sediments wear away, the strong, dense concretions remain at the surface.

From about 10 miles southwest of Walden, Delaney Butte and Sheep Mountain can be seen jutting up in front of the Park Range. Each is a sliver of Precambrian and Mesozoic rocks thrust westward over Pierre shale. Movement on these thrust faults may equal or exceed that at the northern end of North Park.

The Rabbit Ears, atop the Park Range to the west, are hard volcanic breccia with volcanic rock fragments embedded in lava, probably formed near or even within a volcanic vent. Except for these much younger volcanic rocks, this part of the Park Range is cored with Precambrian metamorphic rocks, originally island arc deposits. They were metamorphosed and intruded by granite some 1.7 billion years ago as they pressed against the edge of the Wyoming Province. Lower slopes of the range are uptilted Triassic, Jurassic, and Cretaceous sedimentary rocks.

Tertiary sedimentary rocks are exposed in a 15-foot roadcut near milepost 8. Outcrops of these rocks are rare because they erode rapidly into rounded hills. Note the typical horizontal beds and the crossbedding in river-deposited sand. —Felicie Williams photo

The peaks of the Rabbit Ears Range, the mountainous area southeast of the highway at the south end of North Park, are a scattering of small mid-Tertiary intrusions surrounded by slightly younger lava flows. The northern and western slopes of the range are immersed in immense sloping aprons of landslides, thanks to weak Tertiary sediments that underlie the lava flows. High peaks of the range were glaciated, and glacial steepening of valley walls may have contributed to the abundance of the landslides.

Colorado 82
Glenwood Springs—Aspen
42 miles (68 km)

**with side trips to Marble via Colorado 133
and Maroon Bells via Forest Route 125**

Glenwood Springs squeezes into the narrow canyon of the Colorado River and the lower valley of Roaring Fork River with little room to spare. Steep canyon walls bring geologic risks: mudslides and earthflows have repeatedly destroyed buildings and endangered human lives.

Southeast of Glenwood Springs, up the valley of the Roaring Fork River, we find evidence of three major events in Colorado's tectonic history:

- Distinctive dark red rocks of the Maroon formation, consisting of sand, clay, and gravel washed eastward from Uncompahgria, are evidence of uplift of the Ancestral Rockies in Pennsylvanian time, probably as a result of the collision between North America and Africa.

- The Maroon formation and Mesozoic sedimentary rocks dip steeply southwest off the Elk Mountains, an anticline dating back to the Laramide Orogeny, the result of North America's collision with a subcontinent on the Pacific plate. The tilted Mesozoic strata appear in the Grand Hogback, west of the highway.

- The Elk Mountains are capped with volcanic rocks that erupted along faults created during mid-Tertiary regional uplift, when all of Colorado rose to its present elevation.

Glaciation in Colorado's mountains was an important event, too, creating much of this area's attractive scenery. The floor of Roaring Fork Valley contains thick deposits of river gravel, much of it glacial outwash; the upper parts of this valley and its tributaries once contained sizable glaciers.

Each horizontal terrace along the valley represents a period of stability in the history of the river—a balance of gradient, water quantity, stream load, and gravel and sand supply. Farther toward Aspen, you will see as

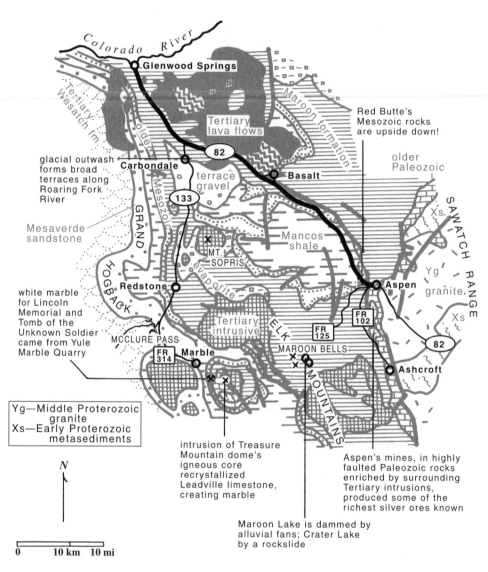

Geology along Colorado 82 between Glenwood Springs and Aspen.

Map labels:

Colorado River

Glenwood Springs

Tertiary lava flows

Maroon formation

Red Butte's Mesozoic rocks are upside down!

older Paleozoic

glacial outwash forms broad terraces along Roaring Fork River

Carbondale

82

terrace gravel

Basalt

133

Mesaverde sandstone

Mancos shale

Mesozoic

GRAND HOGBACK

MT. SOPRIS

evaporite

Xs.

SAWATCH RANGE

Yg granite

white marble for Lincoln Memorial and Tomb of the Unknown Soldier came from Yule Marble Quarry

Redstone

Tertiary intrusive

ELK MOUNTAINS

Aspen

FR 102

FR 125

Xs

MCCLURE PASS

FR 314 Marble

MAROON BELLS

82

Ashcroft

Yg—Middle Proterozoic granite
Xs—Early Proterozoic metasediments

intrusion of Treasure Mountain dome's igneous core recrystallized Leadville limestone, creating marble

Aspen's mines, in highly faulted Paleozoic rocks enriched by surrounding Tertiary intrusions, produced some of the richest silver ores known

Maroon Lake is dammed by alluvial fans; Crater Lake by a rockslide

N

0   10 km   10 mi

many as three terrace levels, representing three glacial episodes in Colorado; the highest terrace is the oldest. The terraces contain rounded cobbles of many colors: red rocks derived from the Maroon formation, black basalt from volcanic hills to the north, and pale gray granite from the heart of the Sawatch Range.

Mt. Sopris, south of the highway, is a large Tertiary stock, an intrusion of crystalline igneous rock called quartz monzonite, containing two types of feldspar, a little quartz, and a few dark minerals. Several well-developed

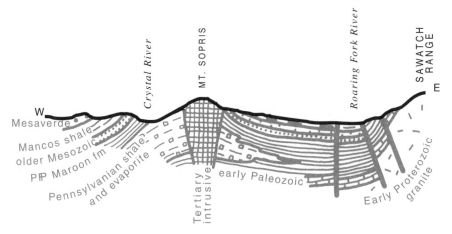

Section south of Colorado 82 through Mt. Sopris and Aspen.

MARBLE

For a scenic and interesting side trip, turn south on Colorado 133 through Carbondale. The road follows the canyon of the Crystal River between red slopes of Maroon sandstone and shale. Bear left at the McClure Pass junction and continue upriver into a glaciated valley carved mostly in soft, drab Mancos shale. Valley slopes, oversteepened by glaciation, are prone to turn into mudflows during heavy rains. The town of Marble was once destroyed by such a mudflow but sprang up again just west of its original site.

The Yule Marble Quarry, source of the exquisite white marble of the Lincoln Memorial, the Tomb of the Unknown Soldier, and more than sixty other public buildings across the country, is on the flanks of Whitehouse Mountain about 2 miles southeast of Marble. Snowy white and beautifully free from flaws, the marble was formed by metamorphism of Leadville limestone in an area domed up and heated by a Tertiary igneous intrusion. Marble was quarried between 1883 and 1941, when a flood shut down the mines. The quarries reopened in the late 1980s. Snow white sculptures made from this marble will catch your eye in many western Colorado towns.

rock glaciers with large angular boulders creep down the cirque valleys below the summit. Like their less rocky cousins, slow-moving rock glaciers reshape their valleys by grinding rock with rock and build up morainelike ridges at their lower ends. They develop in high, steep, glaciated areas where large quantities of jointed rock pried apart by frost become imbedded in ice.

The town of Basalt on Colorado 82 is surrounded by basalt boulders fallen from a lava-capped mesa to the north. The grayish black lava, some of it honeycombed with vesicles (small bubble holes) was so fluid that it spread into broad, almost horizontal sheets. It probably welled up from fissures and cracks created by mid-Tertiary crustal stretching, quietly flowing out onto a nearly level surface. The flow was once continuous with the lava cap west of Roaring Fork Valley.

Near Basalt, Triassic redbeds appear on hills on either side of the road, which runs eastward along a gentle syncline between the Elk Mountains and the Sawatch Range. The red rocks are more or less continuous with the Pennsylvanian-Permian redbeds of the Maroon formation, and show that in Triassic time Uncompahgria remained high enough to provide a source of sediments. Above the Triassic redbeds, Jurassic Morrison shale forms

Mt. Sopris is a stock, a large intrusion of igneous rock. Streams heavily loaded with glacial debris formed the foreground terraces. —Jack Rathbone photo

Photographed from above, a rock glacier carves deeply into the flank of Mt. Sopris. Its toe, reaching into the present-day forest, shows that the mass of rocks, lubricated with ice, still creeps slowly downslope. —Bob Beehler photo, courtesy of Peter Birkeland

mounded hills. It was deposited in swampy deltas in a wide region devoid of mountains, for the Ancestral Rocky Mountains were worn away by Jurassic time. With the many faults between the two mountain ranges, sedimentary rocks may be strangely tilted this way and that.

Near milepost 30 watch for a small anticline caused by flowing gypsum in Pennsylvanian rocks. Gypsum tends to flow under pressure like Silly Putty, only more slowly, and here, below the present surface, it has bulged into a pluglike mass. Because the gypsum is more buoyant than overlying rocks, the plug pushes upward, forming the little anticline.

Mesozoic sedimentary layers are upside down on Red Butte, to the left at milepost 40. When the edge of the Sawatch Range pushed upward and westward over this area during the Laramide Orogeny, these rocks were overturned by movement along the fault. The overlying edge of the range has since eroded away. Now, red Triassic strata at the top of the butte overlie younger pink Jurassic sandstone and varicolored Morrison shale, which in turn overlie still younger Cretaceous sandstone and shale. "Youngest on top" doesn't hold here!

At Aspen, older Paleozoic rocks are folded and faulted on the flank of the Sawatch Range. The most prominent Paleozoic formation is the

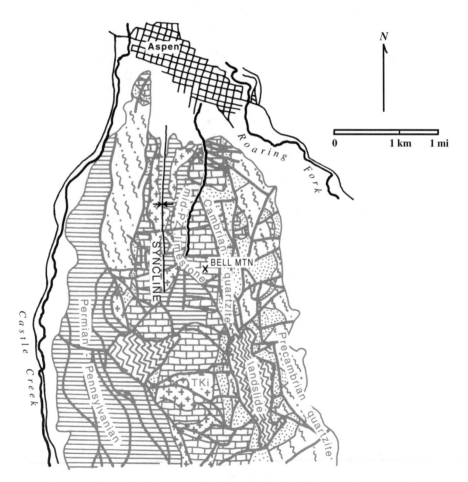

One can't help but be impressed by the complex faulting on the mountain slope south of Aspen. Several hundred faults have been mapped in the painstaking search for valuable minerals. All these structural complexities, now beneath Aspen's ski lifts, have been smoothed over and planted with grass.

massive gray Leadville limestone, a Mississippian rock that forms the slopes of West Aspen Mountain, straight ahead at milepost 42. Numerous faults transect the slopes under the ski lifts, slopes that in Aspen's mining heyday (1879–1893) were dotted with mines. Stair-stepping faults brought mineral-rich Paleozoic rocks, particularly Cambrian, Ordovician, and Mississippian limestones, to the surface again and again. Gold, silver, lead, and zinc were deposited during the Tertiary volcanic episode, when concentrated solutions rich in minerals penetrated thousands of new-formed faults and joints.

MAROON BELLS

Forest Route 125 to Maroon Bells follows the canyon of Maroon Creek. During the summer, access is by shuttle from Aspen.

High red walls of the Maroon formation rise above the aspen and spruce of the valley floor. Composed of sand and mud washed off Uncompahgria in Pennsylvanian and Permian time, this formation is in many places altered by heat and fluids from surrounding Tertiary intrusions, which caused a graying of the otherwise dark red rocks.

A succession of geologic events created the inspiring scenery at Maroon Bells: rise of the Ancestral Rockies in Pennsylvanian time, deposition of dark red sediments derived from them, long burial and cementing of these sediments into rock, Laramide uplift and then mid-Tertiary regional uplift, hardening by volcanic fluids, and glaciation during Pleistocene time. —Jack Rathbone photo

Between Maroon Creek and Ashcroft in the Elk Mountains, the valley of Conundrum Creek shows the U-shaped profile that is the signature of glacial erosion. High cirques carved by small tributary glaciers appear on the right. This valley is accessible only by trail. —B. H. Bryant photo, courtesy of U.S. Geological Survey

Rock glaciers in the Elk Mountains, like this one on Pyramid Peak, are fed with rocks loosened by freezing and thawing. The mass of broken rock imbedded in ice still creeps slowly down the valley. —B. H. Bryant photo, courtesy of U.S. Geological Survey

The lowest of several glacial moraines crosses the mouth of Maroon Creek less than a mile from Aspen. The valley's U-shaped profile is visible from there. Hanging valleys and high waterfalls ornament it farther up, marking sites where small tributary glaciers entered the main valley glacier. Maroon Lake is dammed not by moraines, as one might expect, but by large alluvial fans.

A 2-mile trail leads to Crater Lake in the glaciated valley of West Maroon Creek. Aspen, spruce, and fir rise against the splendid reddish gray cliffs of Maroon Bells, whose peaks top 14,000 feet. Geologic features abound. Note immense rock glaciers far up the valley and talus cones below avalanche tracks on the steep slopes of Maroon Bells.

As in most mining areas the Aspen region is a geologic whodunit: Are faults unusually abundant in this area, their very abundance accounting for the mineral enrichment? Or do there simply *seem* to be more faults here where the area has been gone over inch by inch and foot by foot, aboveground and below, in the search for elusive riches?

On Smuggler Mountain east of the valley, houses and parks rest on waste piles from lead and silver mines—the lead, in particular, a danger to human health. Laced with these and other metallic elements, the dump material can't be removed without moving the houses. Instead, the ground has been covered in many places with a thick layer of uncontaminated soil. Luckily, the Roaring Fork River remains free of contamination.

Aspen is above the lower limit of glaciation: the oldest moraines extend 2 to 3 miles below the town. Many glacial features can be seen by driving up the Roaring Fork toward Independence Pass, or up Castle Creek to Ashcroft.

Colorado 82
Aspen—U.S. 24 via Independence Pass
44 miles (71 km)

At the bridge just southeast of downtown Aspen, Cambrian sandstone forms a ridge that runs uphill to the south. The contact between it and underlying Precambrian granite follows the eastern side of the ridge. East of the contact, beyond a moraine, is the Precambrian core of the Sawatch Range—1.4-billion-year-old granite of the Berthoud plutonic suite.

Colorado 82 heads south up the valley of the Roaring Fork River, which for 5 miles courses lazily across a wide, flat expanse of glacial outwash. Moraines close the upper end of the valley and are plastered along steep slopes near the road. The glacier that occupied this great trough must in places have been thousands of feet thick because it shaped the angular shoulder far up on the valley wall. Old lateral moraines, as well as rock from the oversteepened canyon walls, have tumbled toward the valley floor, partly obscuring its U-shaped profile.

Leaving the open valley, the highway climbs the flank of Smuggler Mountain above a series of moraines separated by level meadows, then clings to a roadway blasted in the light gray granite of the canyon wall. Far below, the stream leaps madly down a boulder-clogged gorge. In places the canyon walls are smoothed and polished by the glaciers. The narrow, steep part of the gorge must have been the site of a magnificent icefall, with tumbled blocks of glacial ice broken by deep crevasses.

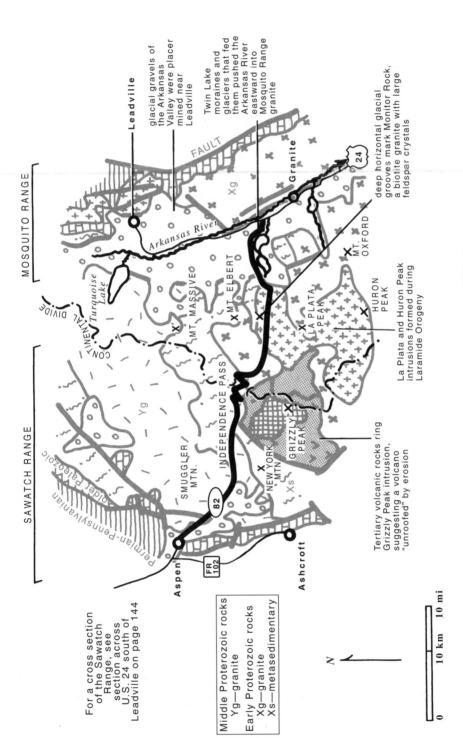

Geology along Colorado 82 between Aspen and U.S. 24.

MOSQUITO RANGE

SAWATCH RANGE

CONTINENTAL DIVIDE

Leadville

glacial gravels of the Arkansas Valley were placer mined near Leadville

Twin Lake moraines and glaciers that fed them pushed the Arkansas River eastward into Mosquito Range granite

deep horizontal glacial grooves mark Monitor Rock, a biotite granite with large feldspar crystals

FAULT

Xg

Granite

Arkansas River

Turquoise Lake

MT. MASSIVE

X MT. ELBERT

MT. OXFORD

LA PLATA PEAK

HURON PEAK

La Plata and Huron Peak intrusions formed during Laramide Orogeny

Yg

INDEPENDENCE PASS

SMUGGLER MTN.

NEW YORK MTN.

Xs

GRIZZLY PEAK

Tertiary volcanic rocks ring Grizzly Peak intrusion, suggesting a volcano "unroofed" by erosion

Permian-Pennsylvanian

older Paleozoic

82

FR 102

Aspen

Ashcroft

24

For a cross section of the Sawatch Range, see section across U.S. 24 south of Leadville on page 144

Middle Proterozoic rocks
Yg—granite

Early Proterozoic rocks
Xg—granite
Xs—metasedimentary

N

0   10 km   10 mi

Close to timberline, near the ghost town of Lincoln Gulch, many more glacial features are visible. The cirques behind Lincoln Gulch are carved in Precambrian granite and schist of New York Mountain. The last terminal moraines of melting glaciers lie right inside the cirques. These rocky lips emphasize the basinlike nature of the cirques. Hummocky ground moraines, glacial debris left as the last of the glaciers melted, coat less rugged parts of the upper Roaring Fork Valley.

Ancient metamorphic rocks exposed in the heart of the Sawatch Range are patterned with fascinating wavy bands that tell us the rock must have been nearly molten, perhaps thick and porridgelike, when it was crumpled into mountain roots 1.75 billion years ago. The banding and the chemical composition of these rocks suggest they were derived from sedimentary layers originally deposited on the ocean floor. As oceanic crust was drawn beneath the Wyoming Province, the sediments piled up into mountains against the continent's edge. Deeply buried, the sediments were gradually metamorphosed into gneiss.

In some areas, metavolcanic rocks are mixed with the metasediments. Pressures and temperatures that could so alter solid rock are attained only deep below the surface; the metasediments may have been buried by as much as 10 miles of overlying rocks, since eroded away. As the subducted crust melted, batholiths pushed and melted their way upward into the metamorphic rocks.

The last stretch of road west of Independence Pass is a narrow ledge cut into tightly folded, fractured, decomposing brown metamorphic rocks—not the most stable of roadbeds. Be sure to look back (if you are not the driver) down the great U-shaped valley of the Roaring Fork. Try to visualize the creeping glaciers that cut these cirques and rounded these slopes, converging near Lincoln Gulch and slowly but forcefully, over many thousands of years, shaping the valley all the way to Aspen.

Like many other Colorado passes, Independence Pass developed where the rocks were weakened by faults, many of which probably date back to Precambrian time. At 12,095 feet one of the highest highway passes in the United States, Independence Pass lies on the Continental Divide. Waters draining westward reach the Pacific via the Roaring Fork and Colorado Rivers. To the east, Lake Creek (one of a dozen or more Lake Creeks in Colorado) flows into the Arkansas, its waters bound for the Mississippi and the Gulf of Mexico.

The Sawatch Range represents the western half of a large faulted anticline; the eastern half is the Mosquito Range, east of the Arkansas Valley. Sedimentary rocks dip west off the Sawatch Range near Aspen and east off the eastern side of the Mosquito Range near Fairplay. During mid-Tertiary regional uplift, the great anticline split and pulled apart along two large faults now hidden by Arkansas Valley sediments. Between the faults, the central wedge dropped thousands of feet to become the northern end of the Rio Grande Rift, which extends from just north of Leadville southward through New Mexico.

Glaciers shaped the crest of the Sawatch Range, and you can see many glacial and frost-related alpine features near Independence Pass. Look for cirques, rolling uplands smoothed by an ice cap, and frost heaving of rocks and soil on open ground.

Mt. Elbert, at 14,433 feet the highest summit in Colorado, is just visible due east of Independence Pass. Its Precambrian gneiss differs from the rounded granite summit of Mt. Massive to the north of it. Grizzly Peak to the south is the core of an eroded Oligocene volcano.

The Precambrian granite summit of La Plata Peak (14,336 feet) to the southeast is surrounded by a Laramide batholith composed of porphyry, a rock characterized by large feldspar crystals in a matrix of smaller crystals. Precambrian igneous and metamorphic rocks make up several peaks in the Sawatch Range north of Mt. Elbert. The range as a whole claims fifteen of Colorado's fifty-four Fourteeners.

Between mileposts 73 and 74, Monitor Rock juts into Lake Creek Valley from the Mt. Elbert massif. Made of particularly hard Laramide granite, it partly blocked the Pleistocene glaciers that flowed down this valley and still bears the scars of their passing: deep glacial striations visible as parallel horizontal grooves above the highway. Glaciers scour such grooves not with ice itself but with rock fragments embedded in the ice.

From milepost 80, Colorado 82 overlooks the glacier-broadened valley of Lake Creek, floored with glacial outwash. Lying at the very edge of the valley of the Arkansas River, Twin Lakes are dammed by terminal and recessional moraines of the Lake Creek glacier, with recent man-made modifications.

Even during the maximum advance of Ice Age mountain glaciers, the valley of the Arkansas River was not glaciated downstream from Leadville. From time to time, though, it was obstructed by ice tongues from side valleys. Terraces of stream-laid glacial gravel and large coalescing alluvial fans bear witness to the immense amount of material scoured from these mountains by the moving rivers of ice. The Lake Creek glacier pushed the Arkansas River eastward, forcing it to cut a canyon through the hard granite that forms the lower slopes of the Mosquito Range.

V
Volcanic San Juan Mountains

Within the San Juan Mountains, rocks of every phase of Colorado's geologic story are represented: Precambrian to Cenozoic, igneous, sedimentary, and metamorphic. There is plentiful evidence also of faults and folds, and most of all of volcanic violence.

The oldest rocks in the San Juan Mountains are exposed in the Needle Mountains and in deep valleys on the northern and southern flanks of the range. They are similar to metamorphic rocks in most other ranges in Colorado: dark schist and gneiss formed by intense heat and pressure deep within the crust. Their antecedents—the rocks from which they formed—include basalt, light-colored volcanic rocks, and sedimentary strata that lay along the margin of the Precambrian continent about 1.8 billion years ago. Here, as elsewhere in Colorado, they were metamorphosed and intruded by granite during the Early Proterozoic Orogeny, roughly 1.7 billion years ago. Their northeast-trending faults parallel the ancient continental margin.

Newer faults developed around 1.6 billion years ago, rift faults that cut deeply through the crust in a northwest-southeast direction and that seem to show shearing, sideways movement like that on the modern San Andreas fault in California. Two long, narrow crustal blocks dropped downward between pairs of faults, forming deep rift valleys in which more than 8,000 feet of conglomerate, sandstone, and shale—now the Uncompahgre formation—were deposited. Movement tends to repeat on deep rift faults; there were at least four stages of movement along these faults.

Repeated movement on the older faults offset the younger ones, bending them into a broad **S** shape and weakening part of the crust. The Eolus granite and other coarse-grained igneous rocks intruded the weakness about 1.4 billion years ago. These rocks are members of the Berthoud plutonic suite.

There is very little to tell us what happened in the San Juan area during the next billion years: a widespread erosion surface, a little Cambrian quartzite (a beach deposit), and a few small Cambrian intrusions. Oceans may have covered this part of Colorado as they did surrounding regions, but if

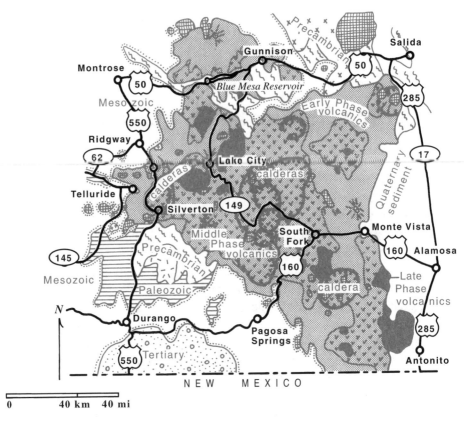

Unlike the linear faulted anticline ranges of the Colorado Rockies, the San Juan Mountains are mostly volcanic—the largest such region in Colorado.

so, the evidence is now gone. During this time interval, living things evolved into more and more complex forms.

Beginning 400 million years ago, Devonian and Mississippian strata—mostly marine limestone—accumulated in shallow seas that extended over most of Colorado.

Early in Pennsylvanian time, around 320 million years ago as North America and Africa collided, the island range of Uncompahgria rose between a pair of northwest-trending faults. Southwest of the new mountains, the Paradox Basin sank roughly 12,000 feet. There, almost isolated from the sea, at least 6,000 feet of salt were deposited, topped with 6,000 feet of brick red sediments washed off Uncompahgria in Permian time.

The textures of Mesozoic sedimentary rocks became finer and finer as Uncompahgria slowly wore down; by Jurassic time the mountains were gone, replaced by near-sea-level plains and swamps. Widespread seas then crept across the entire state in Cretaceous time.

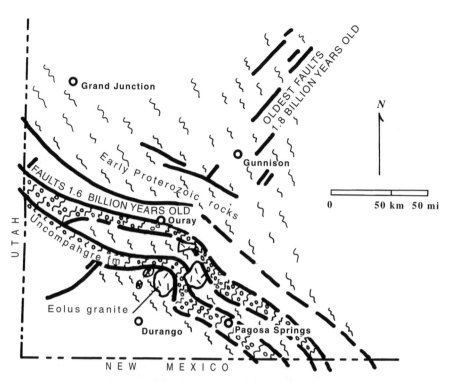

Two intersecting sets of faults influenced the geologic history of the San Juans: northeast-trending faults that formed about 1.8 billion years ago, roughly parallel to the continent's ancient shoreline, and northwest-trending faults created about 1.6 billion years ago.

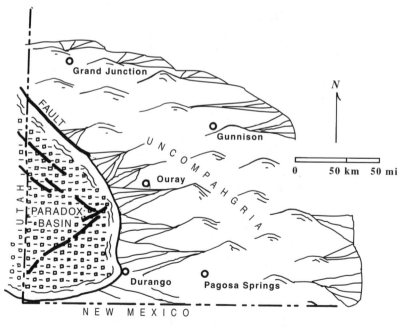

As the Pennsylvanian range of Uncompahgria rose, the Paradox Basin slowly sank, accumulating thick deposits of salt as waters of an inland sea repeatedly evaporated.

During the Laramide Orogeny and the rise of the Rocky Mountains 72 to 40 million years ago, the Uncompahgre Plateau pushed upward as a long narrow fault block about where Uncompahgria had been before. The central San Juan area was faulted and domed even higher around the **S** in the old Precambrian faults. In the San Juan region, erosion almost kept pace with uplift, so by late Eocene time the contours of the land were fairly gentle. A blanket of sediment, the Telluride conglomerate, marks the Eocene surface.

Thirty-six million years ago, near the end of Eocene time, the geologic picture changed abruptly as volcanism burst out in southwestern Colorado. And "burst" is the keyword. Incredible outpourings of lava and volcanic ash lasted, off and on, for 30 million years, through Oligocene and

High valleys of the San Juans, many peppered with mines, have the U-shaped profile that tells of glacial erosion. Steep glacial slopes are scarred by avalanches and rockslides. —Jack Rathbone photo

Miocene time. Lava flows and volcanic ash from many volcanoes merged to build up the San Juan Mountains, the West Elk Mountains, Grand Mesa, the White River Plateau, the Sawatch Range, and South Park. Ash layers spread east to the High Plains. By far the greatest outpourings were in the San Juans, as if the much-faulted weak area in the crust, inherited from Precambrian time, provided pathways for rising magma.

Volcanic Rocks

Two kinds of volcanic rocks are common in the San Juan region: lava flows formed from molten magma and tuff formed from volcanic ash and pellets of fine volcanic cinder. Ashflows avalanche down mountainsides as hot, ground-hugging clouds, becoming welded together by their own internal heat. When thick, they form hard, strong ashflow tuff that stands as cliffs or breaks down into large angular blocks. Thin ashflows and the tops and bottoms of thick ones cool more quickly so they are often not welded but soft. Ashfalls, on the other hand, settle from clouds of volcanic ash rising above exploding volcanoes, and ashfall tuff is more loosely consolidated. In the San Juans, ashfall tuff is rarely seen in outcrops because it slides easily, turns readily into soil, or becomes covered by talus slopes of welded tuff. It can, however, be identified in roadcuts, which must be beveled to low angles to prevent sliding.

Two other kinds of volcanic rocks appear in the San Juan region in lesser amounts: agglomerate, formed from debris flows of volcanic material, and volcanic breccia, containing fragments of lava thrown forcibly from volcanoes or crusting lava that repeatedly broke apart and glued itself together as it flowed.

Unfortunately for those trying to understand them, volcanic rocks rarely form neat parallel layers. Successive flows thin and thicken, flow down valleys, and fill irregularities. And since volcanic rocks rarely contain fossils from which their age can be learned, the San Juan story at first seemed nearly impossible to decipher. Volcanic flows vary, however, in mineralogy and chemistry and contain radioactive elements from which their age may be discovered. Thanks to careful mapping and geochemistry, we can distinguish three phases of San Juan volcanic activity, separated by intervals of faulting, collapse, and erosion.

Early Phase. The Early Phase began about 36 million years ago and lasted about 5 million years. Lava flows, breccia, and thick ash layers built a shieldlike volcanic field 100 miles across and about 4,000 feet high, dotted with volcanic cones whose summits rose much higher. Ash layers extended north to interfinger with volcanic ash and breccia of the same age from the West Elk Mountains. Erosion that followed Early Phase volcanism pared the volcanoes—many of them collapsed—down to mere clustered hills.

Middle Phase. Enormous avalanches of incandescent volcanic ash and numerous lava flows of the Middle Phase began to cover the eroded ruins of the Early Phase about 30 million years ago. For 4 million years, explosive eruptions blasted hot volcanic ash across at least a third of the state: in Colorado, individual sheets of welded tuff are traceable for 50 miles or more from their sources, with volumes of as much as 1,000 cubic miles. As great volumes of magma and gas escaped, the roofs of many magma chambers collapsed, leaving immense circular basins—calderas as much as 50 miles across—outlined by arc-shaped faults. Eruptions continued, filling the calderas with more flows and ash, with volcanic pressures often doming their floors upward as well. Middle Phase volcanic activity ceased about 26.5 million years ago.

None of the San Juan calderas appear as large circular basins today. They can, however, be recognized by careful on-the-ground mapping of arcuate faults around their edges. Many are outlined by streams, which tend to follow the weakened zones along the curving faults.

Volcanic ash from both Early and Middle Phases covered much of Colorado. Where it settled on the high, uneven surfaces of the Rocky Mountains, most of it eroded away. But it contributed greatly to the volcanic pile of the San Juan Mountains.

Late Phase. The Late Phase began 25 million years ago, bringing floods of basalt that spread across many areas of Colorado, along with minor light-colored rhyolite lava flows. Basalt is dark gray or purplish gray, with large amounts of iron- and magnesium-rich minerals—very much the same composition as the mantle, its probable source.

With mid-Tertiary regional uplift the San Juan region and the rest of Colorado rose 5,000 feet or more to its present elevation, and the crust stretched and broke, creating deep faults that reached clear down to the mantle. These faults allowed Late Phase magmas to rise through the crust without mixing much with the crust's light-colored rocks. Late Phase basalt flows are 23 to 6 million years old; they once covered most of the eastern part of the San Juans with a wide lava plain that sloped eastward into the San Luis Valley and south into New Mexico. Much of the southeastern portion of this hard resistant Late Phase basalt plain still forms the land's surface.

San Juan volcanic rocks can be grouped into a few rock types based on their color, which reflects the amount of iron and magnesium they contain.

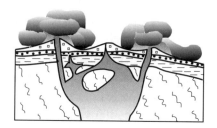

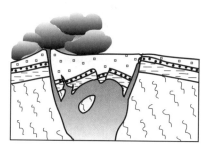

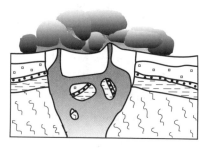

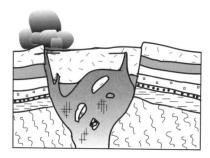

How Calderas Form

Several processes combine in the development of a caldera:

• A large reservoir of magma rises from the mantle, below Earth's crust. The magma eats away at the rock above it, melting and mixing with it until its composition is like the crust around it.

• As the lava erupts, ground above the magma reservoir sinks along faults that ring its edges. These faults, which are the edge of the caldera, remain roughly in the same place throughout the life of the reservoir.

• Many flows of similar magma make up one eruption. Many eruptions, each separated by years of quiet, can come from one caldera.

• Roughly half the lava from each eruption falls or flows into the caldera, to be recycled into more lava in that or later eruptions.

• Rock of the caldera floor behaves like a trapdoor, doming and tilting as the magma below it tries to find an outlet. During and after eruptions, the trapdoor subsides, to make up for the volume of magma erupted.

• Between eruptions, erosion smooths the topography. For simplicity, this step is not shown in the diagram.

• Vent locations change with different eruptions; the molten rock finds the easiest way out.

• As eruption follows eruption, the magma reservoir gradually rises through the pile of volcanic flow rocks, which have deeply buried older rocks. Rocks above the reservoir continue to be recycled into magma.

• Below the caldera there must be a continuing source of magma (not shown in this diagram) to keep the caldera active. When the source is gone, volcanic activity ends, and the flows and reservoir cool and solidify.

The darkest gray, nearly black rocks are basalt. Andesite, dacite, and rhyolite are successively lighter. Weathering changes the appearance of all these types; a freshly broken surface reveals the true color. Molten basalt is quite fluid and flows rapidly across large areas. Andesite, dacite, and rhyolite are successively thicker and gummier and form less extensive flows and taller, steeper volcanoes. Rhyolite lava is so thick that it usually creates small, steep-sided lava domes. Dacite and rhyolite magmas that contain a lot of gas, however, may erupt explosively, creating sheets of tuff and welded tuff—some of the most widespread volcanic deposits in Colorado.

Despite the magnitude of San Juan eruptions, great quantities of magma never reached the surface but cooled below the volcanoes as intrusions—batholiths, stocks, laccoliths, sills, and dikes. Many of theses intrusive rocks cooled fairly near the surface and are now exposed by erosion. Some of them are so fine grained that they look almost like their volcanic counterparts; others are porphyry, rock containing large crystals within a finer matrix. Gravity measurements at the surface indicate that immense batholiths of coarse-grained rock underlie the San Juan area, the hardened contents of large magma chambers below the region's volcanoes.

As the San Juan volcanic rocks cooled, water circulating through them, through surrounding rocks, and through the batholiths that fed them gradually deposited unusual quantities of valuable minerals—gold, silver, lead, copper, and zinc—in faults and joints in the rocks. Most mines of the western San Juan Mountains are near intrusions; some of the most productive lie right at the contacts between intrusions and the rocks that surround them. Most of the deposits are within the Colorado Mineral Belt, the northeast-trending band that contains most of the state's mineral wealth.

Big geologic questions remain concerning the whys and wheres of San Juan volcanism. Some geologists propose that the volcanism was triggered as the North American plate overrode the Pacific plate and part of its mid-ocean ridge by perhaps as much as 1,200 miles. If so, friction would have melted great quantities of crustal rock, which then may have risen along old faults to build volcanoes of the Early and Middle Phases. Eventually, the subducted plate might have run up against the deep roots of the Rockies and been unable to push farther east, creating the huge forces that brought about regional uplift in Colorado and neighboring states. Development of the San Andreas fault in California may then have relieved the pressures and brought on tension, stretching the domed-up area to the breaking point and creating the deep faults associated with the Rio Grande Rift. Those faults, reaching down to the mantle, would have allowed the escape of magma from the mantle—the basalt flows of the Late Phase of volcanism.

A far-fetched story? It may be. It fits the facts, but the facts are too few. We need, among other things, more evidence of what is beneath the continental crust and a better understanding of deeply buried rocks and subducted plates.

Pleistocene to Recent Time

Like other high mountains in Colorado, the San Juans were scoured by glaciers during the Pleistocene Ice Age. Rivers of ice sharpened mountain peaks and carved deep, cliff-walled valleys, while rivers of meltwater shaped canyons and ravines around the margins of glaciated areas. In the western San Juans, such erosion removed most of the volcanic rocks, cutting down through older sedimentary rocks and into the Precambrian foundation below. In the central and eastern parts of the range, steep valleys and tall peaks of volcanic rocks remain and reveal many details of the interiors of volcanoes, calderas, and lava and ash flows. Throughout the San Juans, landslides have shaped new slopes, dammed streams, and created lakes to give us the varied San Juan scenery of today.

U.S. 50
Gunnison—Montrose
66 miles (106 km)

Near Gunnison, rain and river water carved the strikingly castellated, chocolate-colored Palisades in relatively soft breccia of fine volcanic pumice and ash enclosing fragments of harder volcanic rocks. The breccia originated between 34 and 29 million years ago, during Early Phase volcanism in the West Elk Mountains northwest of Gunnison.

Southwest of town, rugged hills of Precambrian granite and gneiss appear south of the river, and from milepost 150 the highway is surrounded by this rock—gray, hard, crystalline, and garlanded with dark and light veins. The country begins to open up here, for we are well west of the main Rocky Mountain ranges.

For several miles, U.S. 50 follows the north shore of Blue Mesa Reservoir, at first in Precambrian rock and then below slopes and cliffs of volcanic breccia. Practically the entire skyline is rimmed with brown volcanic rocks erupted from either the West Elk Mountains or the San Juans. West of milepost 141, the breccia overlies pink and green Morrison shale, which shows up in many roadcuts and is readily recognized by its pastel colors.

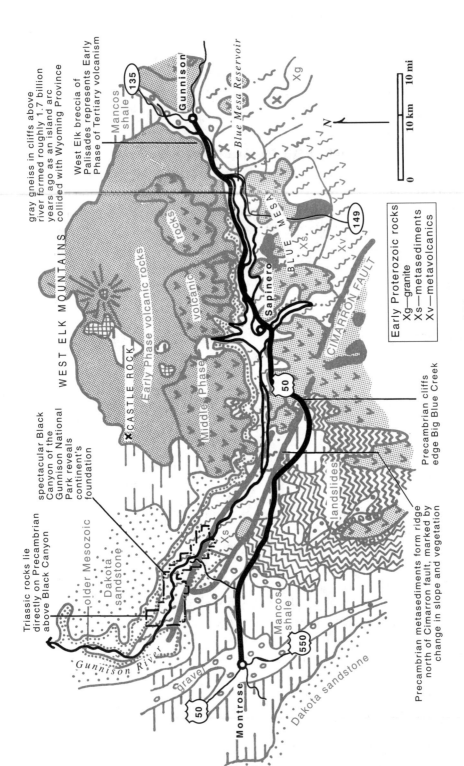

West Elk Mountains

gray gneiss in cliffs above river formed roughly 1.7 billion years ago as an island arc collided with Wyoming Province

West Elk breccia of Palisades represents Early Phase of Tertiary volcanism

Mancos shale

Gunnison

135

Blue Mesa Reservoir

Xg

X

X

N

0   10 km
0   10 mi

149

Sapinero

BLUE MESA

Xs

Xv

Xv

CIMARRON FAULT

Early Proterozoic rocks

Xg—granite
Xs—metasediments
Xv—metavolcanics

Precambrian cliffs edge Big Blue Creek

50

ROCKS

Early Phase volcanic rocks

volcanic

CASTLE ROCK

Middle Phase

landslides

Precambrian metasediments form ridge north of Cimarron fault, marked by change in slope and vegetation

spectacular Black Canyon of the Gunnison National Park reveals continent's foundation

older Mesozoic

Dakota sandstone

Xs

Triassic rocks lie directly on Precambrian above Black Canyon

Gunnison River

gravel

Mancos shale

550

Dakota sandstone

Montrose

50

Geology along U.S. 50 between Gunnison and Montrose.

Above the Gunnison River, the West Elk breccia rests on Mesozoic sedimentary rocks. The craggy Palisades are about 600 feet high.
—J. C. Olson photo, courtesy of U.S. Geological Survey

Erosion of soft shale undermines the breccia, helping to form palisades and cliffs above the highway.

Sedimentary strata between here and Montrose are all Mesozoic. Paleozoic formations were removed by erosion 300 to 200 million years ago when this area was part of the Uncompahgria highland.

Tuff and welded tuff form light-colored hills to the south as well as upper parts of the cliffs to the north. Both are rhyolite, of about the same chemical makeup as granite but with much finer—in this case microscopic—crystals. Welded tuff forms during explosive eruptions when white-hot clouds of volcanic ash swoop down the slopes of volcanoes, often with terrifying rapidity, and weld together as they settle. In 1902 such a cloud erupted from Mt. Pelée in Martinique, destroying the city of St. Pierre and killing an estimated 40,000 inhabitants. More recently, several similar eruptions have been predicted far enough in advance to allow evacuation of endangered areas.

Volcanic fragments in the West Elk breccia protect underlying soft volcanic tuff from wind and rain, helping to form the castlelike spires and cones that edge the West Elk Mountains for many miles. —W. R. Hansen photo, courtesy of U. S. Geological Survey

Thick welded tuff tops some of the West Elk breccia; thinner layers are responsible for funny "hats" and "collars" on its pinnacles. Most of the welded tuff in this area can be traced to caldera sources in various parts of the San Juans.

West of Blue Mesa Dam, Precambrian metamorphic rock, called the Black Canyon gneiss in this region, abruptly becomes the dominant rock along the Gunnison River. Black Canyon really starts just above the dam, where the canyon walls are about 200 feet high. They become more than 2,000 feet high 20 miles farther west in Black Canyon of the Gunnison National Park. The gneiss, dark with biotite, hornblende, and chlorite, is so extensively injected with granite dikes and sills that it is referred to as an injection gneiss.

West of the dam, U.S. 50 climbs away from the river through Mesozoic sedimentary rocks and crosses a bumpy, humpy landslide area scattered with big blocks of tuff. Such hummocky terrain is characteristic of landslides, and you can easily develop a knack for recognizing it. The weak volcanic breccia slides easily, especially during or after heavy rains, when it absorbs water and increases in weight. Wet unwelded tuff below "greases

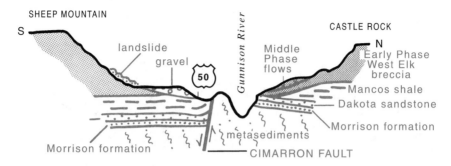

Section across U.S. 50 and the Black Canyon near Big Blue Creek.

the skids." Highway bumpiness proves that this slide still moves. North of the river, large slides are lubricated by slippery Jurassic and Cretaceous shale.

West of milepost 124, U.S. 50 crosses the beveled top of the Precambrian rocks and drops into Blue Creek Canyon. The metamorphic rocks are well exposed, laced with large and small pegmatite dikes, some of which contain oversize, intergrown crystals of pink quartz and feldspar.

West of Big Blue Creek, hummocky landslides cover the Cimarron fault, a major fault that extends at least 30 miles northwest and 30 miles southeast of here. Upfaulted Precambrian rocks north of the Cimarron fault form the high ridge that rises north of the highway for about 10 miles. Cretaceous Mancos shale butts right into these uplifted Precambrian rocks—as you can see from milepost 114—proving that movement on the fault took place after the Mancos shale was deposited in Cretaceous time. The fault parallels northwest-trending faults in the San Juan Mountains that originated about 1.6 billion years ago.

Notice how bumpy the highway is as it crosses the Mancos shale near Montrose. High clay content causes the shale to swell when wet, something neither pavement nor plants readily adapt to. Yellow and gray badlands of Mancos shale edge the Uncompahgre Valley.

Near the Black Canyon of the Gunnison National Park junction, the faulted ridge of Precambrian rock is veneered with Jurassic shale and sandstone and Cretaceous Dakota sandstone. These sedimentary rocks slant up northeastward, bending over the Precambrian fault block.

The northern San Juans are in view to the south near Montrose; tuff layers form somber cliffs in the middle distance. The Uncompahgre Plateau to the west and lava-capped Grand Mesa on the skyline north of Montrose developed as a single broad uplift during the Laramide Orogeny. The same area had been lifted and eroded in Pennsylvanian and Permian time when it was Uncompahgria, an Ancestral Rocky Mountain range.

The steep face of the Gunnison uplift is composed of Precambrian metamorphic rock. Here, it is lifted along the Cimarron fault, a well-defined line sloping across this photo from left to right. Smoother slopes below are Mancos shale. —W. R. Hansen photo, courtesy of U.S. Geological Survey

In just the last 10 million years, the Uncompahgre and Gunnison Rivers have excavated the immense Uncompaghre Valley, carving down through sloping Tertiary sediments and Mesaverde group sandstone, and deep into the Mancos shale.

Black Canyon of the Gunnison National Park
56 mile (90 km) round trip from Montrose

Within the 12-mile stretch set aside as a national park, the Black Canyon displays an unusual combination: profound depth with extreme narrowness. The canyon is between 1,730 and 2,425 feet deep and, at the Narrows, only 1,300 feet from rim to rim. Carved by the Gunnison River, it is walled with very old, very hard, very durable rock capable of standing in near-vertical ramparts. The canyon was excavated sometime after the end of volcanic eruptions in the West Elk Mountains and the San Juans. Regional uplift in mid-Tertiary time markedly increased the river's gradient; then and later during Pleistocene time, when its flow swelled and strengthened

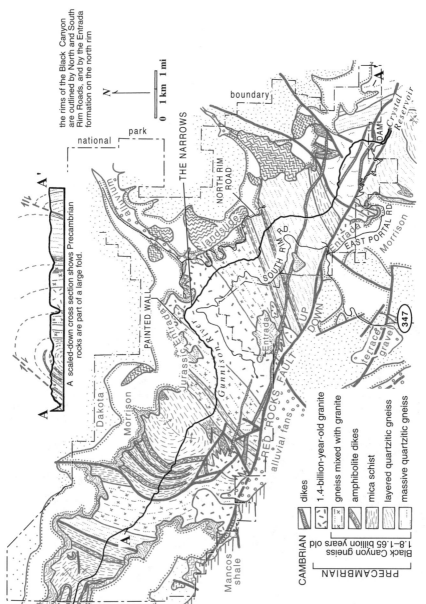

the rims of the Black Canyon
are outlined by North and South
Rim Roads, and by the Entrada
formation on the north rim

N

0   1 km   1 mi

national   park

THE NARROWS

A'

A scaled-down cross section shows Precambrian
rocks are part of a large fold.

A'

NORTH RIM
ROAD

boundary

landslide

PAINTED WALL

SOUTH RIM RD.

Entrada

Dakota

Morrison

Jurassic Entrada

Gunnison River

RED ROCKS FAULT

UP

DOWN

Entrada

East Portal Rd.

DAM
Crystal
Reservoir

A

Mancos
shale

alluvial fans

Terrace
gravel

347

Morrison

A'

CAMBRIAN

dikes

1.4-billion-year-old granite

gneiss mixed with granite

amphibolite dikes

mica schist

layered quartzitic gneiss

massive quartzitic gneiss

PRECAMBRIAN

Black Canyon gneiss
1.8–1.65 billion years old

Geology of Black Canyon of the Gunnison National Park.

with runoff from mountain glaciers, it carried abundant sand, gravel, and boulders, efficient tools for canyon-cutting.

At times the river flowed well north of its present course, across land that now underlies the West Elk Mountains. Each time lava flows from West Elk volcanoes barred its path, the river's course swung south; each time the San Juan volcanoes erupted, their lava flows and ashfalls pushed the river north again. As volcanic activity lessened and died out, the river cut downward through layers of lava and ash, through underlying sedimentary rocks, and into a ridge of tough Precambrian rock that lay across its path. Encountering this hard resistant rock but unable to leave its own valley, the surging water scoured its way downward, pounding, hammering, and deepening a canyon scarcely wider than the river itself.

In the shadowy Narrows of the Black Canyon, the Gunnison River is only 40 feet wide (note figure at lower right). Here, the canyon is about 1,750 feet deep and only 1,300 feet from rim to rim. —W. R. Hansen photo, courtesy of U.S. Geological Survey

Meanwhile, tributaries stripped volcanic and sedimentary rocks back from the canyon's rim. Draining low, unglaciated country rather than the high ranges to the east, they were no match for the Gunnison River, and when they reached the hard Precambrian metamorphic rock they could not cut down as fast. Now, when they flow, they fall into the Black Canyon from unusual hanging canyons high on its walls. Islands, pinnacles, and dark, shadowy clefts within the main canyon were created by weathering and erosion along vertical joints and fissures.

The 2,300-foot Painted Wall is the highest cliff in Colorado—nearly twice as high as the Empire State Building. Little sunlight reaches the canyon depths. Notice along the edges of the canyon how joint systems determine the course of tributary streams.

A short distance back from the canyon rim, Mesozoic sandstone and shale layers lie on the beveled surface of the metamorphic rocks, above an unconformity smoothed off during the long period of erosion at the end of Precambrian time. Early Paleozoic quartzite and limestone once covered the eroded rocks but were washed away when this area rose as part of the Pennsylvanian highland of Uncompahgria. Mesozoic sediments, deposited after Uncompahgria was worn down, are now being beveled off in turn.

Precambrian metamorphic rocks in the Black Canyon, locally called the Black Canyon gneiss, are similar to metamorphic rocks exposed elsewhere in Colorado and that probably underlie most of the state. Roughly 1.8 billion years ago, they were part of an island arc that lay south of the Wyoming Province, the early continent that would become part of North

Metamorphic rocks show intricate contortions caused by flow under great heat and pressure in the depths of Earth's crust. Light bands are mostly feldspar and quartz; dark bands contain quartz and biotite.

—W. R. Hansen photo, courtesy of U.S. Geological Survey

America. As the island arc and the province collided about 1.7 billion years ago, they crumpled into mountains.

With tremendous pressures and intense heat, buried as much as 7 to 10 miles below the surface, the rocks were crunched on a grand scale, partially melted and thoroughly folded and refolded—a process lasting hundreds of millions of years. Most of their original textures were obliterated as individual mineral grains changed shape or melted together, their edges becoming ragged and uneven. Elements in clay recombined and recrystallized into feldspar and muscovite, while dark-colored minerals of basalt and iron-rich sedimentary rocks became dark needles of hornblende and flat plates of biotite. Mineral crystals grew parallel with one another, aligned by the pressure of mountain building, making the rock platy, or foliated, in directions not always akin to the original bedding.

At about the same time, granite magma pushed its way upward in what is now the northwestern part of the canyon. Faintly textured areas in the granite suggest some sedimentary rock melted completely, possibly mixing with magma before recrystallizing. During the next 300 million years, dark dikes penetrated the deeply buried rocks, following northwest-trending faults and joints.

Around 1.4 billion years ago another large granite mass intruded, mostly paralleling the grain of the ancient gneiss. Feldspar crystals growing in the magma and chunks of older rock lined up with the flow direction. Later, lighter-colored granite dikes cut sharply across the metamorphic rocks. Some of the watery magma in these dikes leaked into joints in the gneiss; there it cooled extremely slowly, allowing oversize feldspar and quartz crystals to develop in pegmatite veins.

Far more recently, a mere 510 million years ago, more dark dikes cut the older rocks, again mostly lined up in a northwest-southeast direction.

Despite all the grand-scale folding, melting, and intrusions of ancient rock, there are places in Black Canyon where the original 1.8-billion-year-old bedding can still be recognized. When it and variations in color and chemical composition are mapped and studied, they show that the ancient layers from which the rock formed were once sandstone, shale, basalt lava flows, and dikes.

Leaving the national park, Colorado 347 descends through Mesozoic sandstone and red, purple, and light green shale. The spectacular view outlines the jagged profile of the San Juan Mountains 35 miles to the south and below them the spreading Uncompahgre Valley, edged with humpy gray and yellow badlands of Mancos shale. Westward, resistant sandstone of the Dakota formation crosses the valley beneath the Mancos shale and bends up onto the Uncompahgre Plateau, the forested rise that fills the western skyline.

U.S. 160
Alamosa—Pagosa Springs
89 miles (143 km)

Between Alamosa and Monte Vista, U.S. 160 crosses the western edge of the San Luis Valley. The faults that outline the valley's edge are concealed by thousands of feet of volcanic rock, gravel, and sand. For a discussion of this intermountain depression and the great rift valley of which it is a part, see **U.S. 285: Poncha Springs—New Mexico.**

West of Monte Vista, U.S. 160 crosses the alluvial fan of the Rio Grande and follows the river's valley into the eastern San Juan Mountains. Tongues of volcanic rock reach out into the valley near Del Norte. Volcanic necks, conduits of former volcanoes, jut up sharply around the town, and two small Tertiary intrusions, surrounded by radiating dikes, are a short distance northwest.

In 1955 a geologist pointed out the possibility of oil-bearing sedimentary rocks beneath the southeastern half of the San Juan volcanic field. But sedimentary rocks, if there, lie under thousands of feet of volcanic rock, making seismic surveys, a favored tool for oil exploration, impossible. A later study predicted a large batholith beneath the volcanic pile, and the possibility of sedimentary rocks was forgotten. Nine years later while searching for gold and silver north of Del Norte, geologists noticed oil in their drill cores. This led to the discovery of oil seeps north and south of Wolf Creek Pass and a little oil in discarded drill cores from Summitville. By 1984 an exploratory well that penetrated 4,500 feet of volcanic rocks and nearly 5,000 feet of underlying strata demonstrated the presence of oil-bearing sedimentary rocks beneath this part of the San Juan volcanic field. A producing well is now operating a few miles west of Del Norte.

Late Phase basalt lava flows appear on Del Norte Peak and Hogback Mesa southwest of Del Norte. Early Phase andesite flows appear on hills north and south of the highway between Del Norte and South Fork, as do many layers of Middle Phase tuff.

U.S. 160 follows the Rio Grande as far as the town of South Fork before ascending the South Fork of the Rio Grande to Wolf Creek Pass. The headwaters of the Rio Grande are on the Continental Divide about 50 miles due west of the town of South Fork, well north of this highway.

Massive pinkish gray ashflow tuff takes over the scenery along the South Fork, its irregular layers now forming spires and precipices close to the river and precipitous cliffs and ledges high above. This a good place to think about the monstrous explosive eruptions, the gigantic clouds of incandescent ash, the darkened skies, and the choking air that created and accompanied the expulsion of these impressive tuffs. The many layers of Middle

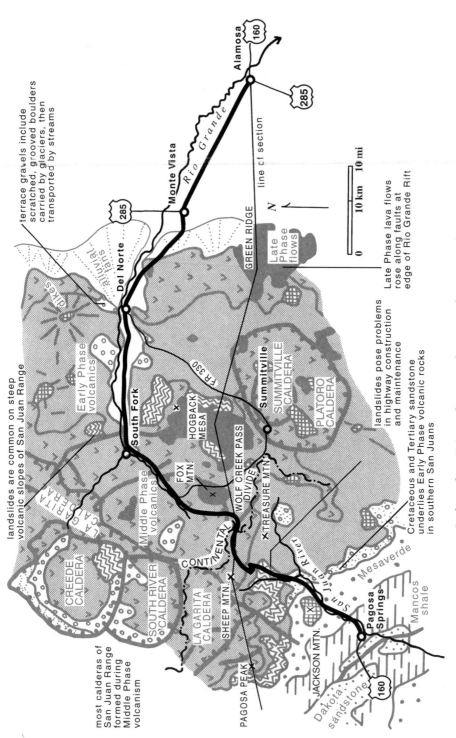

Geology along U.S. 160 between Alamosa and Pagosa Springs.

terrace gravels include scratched, grooved boulders carried by glaciers, then transported by streams

landslides are common on steep volcanic slopes of San Juan Range

most calderas of San Juan Range formed during Middle Phase volcanism

Late Phase lava flows rose along faults at edge of Rio Grande Rift

landslides pose problems in highway construction and maintenance

Cretaceous and Tertiary sandstone underlies Early Phase volcanic rocks in southern San Juans

Alamosa
160
285
Monte Vista
Rio Grande
line of section
GREEN RIDGE
N
0   10 km   10 mi
0   10 mi
Late Phase flows

285
Del Norte
alluvial fans
dikes
Early Phase volcanics

South Fork
Middle Phase volcanics
FR 330
HOGBACK MESA
FOX MTN
×
SUMMITVILLE
Summitville
PLATORO CALDERA
SUMMITVILLE CALDERA

CREEDE CALDERA
LA GARITA CALDERA
SOUTH RIVER CALDERA
SHEEP MTN
×
CONTINENTAL
WOLF CREEK PASS
DIVIDE
TREASURE MTN
×
×

PAGOSA PEAK
×
JACKSON MTN.
×
San Juan River
Mesaverde
Mancos shale
Pagosa Springs
160
Dakota sandstone

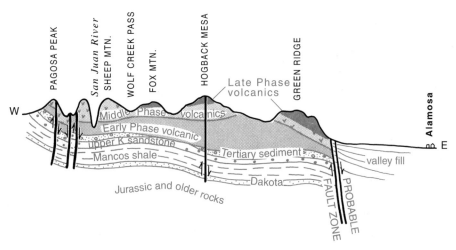

Section parallel to U.S. 160 between Alamosa and Pagosa Peak.

Phase welded tuff, each from a different eruption, burst from caldera centers scattered throughout the San Juans. Individual flows and tuff sheets may be several hundred feet thick; piled together they are thousands of feet thick.

The old gold mining town of Summitville, founded in 1870, lies about 30 miles south of this highway. Soaring prices for gold in the 1970s led to renewed exploration and the discovery of a large low-grade gold deposit there. Mining began in 1985. Gold was extracted from low-grade ore by heap leaching—proven successful at many other mines. Crushed ore, heaped on a waterproof pad, was soaked repeatedly with a cyanide solution to dissolve gold and other metals. The solution was then processed to recover the gold. Over several years roughly 260,000 ounces of gold were recovered.

Unfortunately, just before this mine began processing its ore, the Colorado Legislature temporarily eliminated the state's mine enforcement budget, so the mine was not inspected or required to comply with standard regulations. By 1989 dissolved metals and cyanide had been released several times into nearby creeks, killing fish for 20 miles downstream and contaminating irrigation water in the San Luis Valley. Since 1993 the Environmental Protection Agency has spent more than $100 million cleaning up the now-abandoned site—an expensive environmental lesson.

Above the massive ashflows and visible as U.S. 160 climbs toward Wolf Creek Pass, dark Late Phase lava flows cap several summits, including Fox Mountain southeast of the highway. Bands of yellow, white, and light gray Middle Phase tuff show among the trees. A few miles east of the pass are excellent massive exposures of a Middle Phase dacite flow that was a precursor to the eruption that formed La Garita Caldera.

Evidence of glaciation is apparent in the steepened, slide-prone walls of side valleys and the cirques around Wolf Creek ski area. The tunnel at milepost 168 protects the highway from avalanches, notoriously severe here where glaciers over-steepened the slopes. Near Wolf Creek Pass, watch for signs of landslides: hummocky surfaces lying below steep, scarred slopes or cliffs. Slides have plagued highway construction and maintenance crews here for years. Deep highway cuts in these unstable volcanic rocks almost inevitably start new slides or reactivate old ones.

The southern end of La Garita Caldera is a few miles west of Wolf Creek Pass. The caldera, the largest known in the San Juans, is roughly oval, 30 miles across from east to west and 50 miles long from north to south. It apparently formed during one immense Middle Phase eruption that moved gradually from south to north, spewing more than 1,000 cubic miles of hot

Horizontal layers on Sheep Mountain are Middle Phase welded tuff sheets erupted from Platoro Caldera. —Felicie Williams photo

volcanic ash over surrounding country as much as 60 miles out from it, building an enormous welded layer known as the Fish Canyon tuff.

U.S. 160 zigzags steeply down the slopes of Sheep Mountain west of the pass. Landslides along this and adjacent mountains have exposed volcanic rocks of the Early Phase. At the viewpoint at milepost 161, coarse volcanic breccia—the result of a mudflow—has eroded into weird dark pinnacles. The breccia, as well as volcanic rocks in the canyon below, came from a volcano whose intrusive core makes up Jackson Mountain a few miles southwest of the viewpoint.

The West Fork of the San Juan River, below the viewpoint, winds across a flat valley floor underlain by outwash from Pleistocene glaciers. The highway soon drops to this level, crossing white Tertiary sandstone (milepost 153) and gray Cretaceous shale, both of which appear in roadcuts.

A large dike that extends from Jackson Mountain appears in a roadcut near milepost 150, offering a close look at the intrusive rock. Clusters of small feldspar and quartz crystals are scattered through the otherwise very fine-grained gray rock, which is called porphyry because of this coarse-and-fine texture. The green mineral that coats some quartz crystals is chlorite. Minute pyrite crystals shine like tiny dots of gold.

Just west of milepost 148, a sandy ledge appears in gray Cretaceous shale exposed across the river. This sandy zone thickens westward to become part of the Mesaverde group, several hundred feet thick at Mesa Verde National Park.

Pagosa Springs, near the western edge of a shale-floored valley, is surrounded by hills and cuestas of tilted Dakota and Mesaverde sandstones of Cretaceous age. The town gets its name from a cluster of hot springs just south of the San Juan River: *pagosa* means "boiling water" in the Ute language. Some of the hot water is piped to swimming pools, spas, and fountains, and some is used to heat a nearby motel, but plenty remains in a large morning-glory-shaped pool and several small seeps. In this area, characterized by volcanism of Tertiary and even Quaternary age, the temperature of the Earth's crust increases rapidly downward. Groundwater working its way through the rock becomes heated and rises quickly without complete cooling.

The water in these springs contains many minerals in solution, especially silica. As it cools, the chemicals precipitate around the springs in the form of siliceous sinter. The springs have gradually built a shelf of sinter wide enough to deflect the course of the San Juan River so that it swings westward around them. The rapidity with which the sinter is deposited is evident on man-made fountains, where several feet of sinter have formed in just a few decades.

An eroded mass of intrusive igneous rock, Mt. Sneffels projects through surrounding volcanic layers. Slanting talus slopes are amply supplied with rocks falling from the summit cliffs and in turn nourish a rock glacier in the cirque below the peak. —W. Cross photo, courtesy of U.S. Geological Survey

U.S. 550
Montrose—Silverton
60 miles (97 km)

U.S. 550 gradually ascends the Uncompahgre River between Montrose and Ouray, crossing terraces of stream-deposited glacial gravel. Visible at valley edges for the first few miles are barren hills of Mancos shale, gray grading up into yellow. Laid down in horizontal layers in a shallow Cretaceous sea, the shale, along with older rocks below it, now tilts up southwestward toward the Uncompahgre Plateau and the San Juan Mountains. Flat tops on buttes are remnants of old valley floors, the shale below them weathered yellow. High clay, salt, and selenium content in the Mancos shale, and its tendency to swell when wet, discourage all but a few extra-hardy plants.

Almost all of the San Juan skyline is composed of Early Phase volcanic rocks. However, Mt. Sneffels, the most prominent peak to the south, is a small Tertiary intrusion, and Potosi and Wetterhorn Peaks east of it are topped with Middle Phase welded tuff.

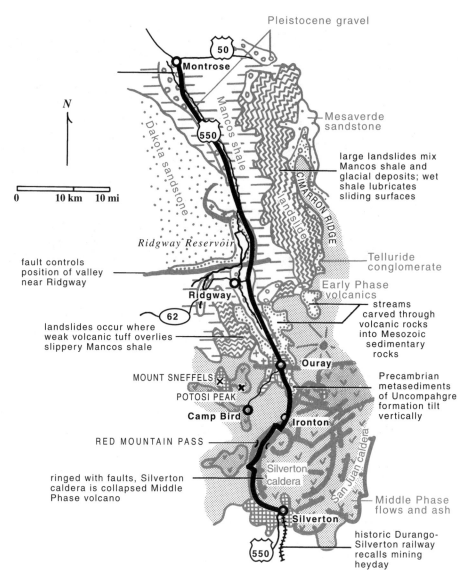

Pleistocene gravel

Mesaverde
sandstone

large landslides mix
Mancos shale and
glacial deposits; wet
shale lubricates
sliding surfaces

Telluride
conglomerate

Early Phase
volcanics

streams
carved through
volcanic rocks
into Mesozoic
sedimentary
rocks

Precambrian
metasediments
of Uncompahgre
formation tilt
vertically

Middle Phase
flows and ash

historic Durango-
Silverton railway
recalls mining
heyday

fault controls
position of valley
near Ridgway

landslides occur where
weak volcanic tuff overlies
slippery Mancos shale

ringed with faults, Silverton
caldera is collapsed Middle
Phase volcano

MOUNT SNEFFELS

POTOSI PEAK

Camp Bird

RED MOUNTAIN PASS

Montrose

Ridgway

Ouray

Ironton

Silverton

Geology along U.S. 550 between Montrose and Silverton.

At milepost 116, the Dakota sandstone comes to the surface, forming ridges on both sides of the river. Below it, especially farther south near milepost 113, is easy-to-recognize green and purple Jurassic Morrison shale. Just north of Ridgway, the highway crosses a fault and the Mancos-Dakota-Morrison sequence is repeated southward as the layers bend upward in a large monocline. Ridgway's wide valley is floored with glacial outwash gravel and Mancos shale.

On the skyline east of Ridgway, the jagged crest of Cimarron Ridge consists of Early Phase lava flows and ash layers. Rounded slopes below are deeply weathered Mancos shale surfaced with glacial moraines. —Felicie Williams photo

Older Jurassic rocks appear farther south, among them the distinctive light-colored Slick Rock member of the Entrada sandstone, an ancient dune deposit that forms rounded smooth-surfaced outcrops easily identified all over southwest Colorado and adjacent parts of Utah.

Triassic, Permian, and then uppermost Pennsylvanian rocks appear below the Jurassic Slick Rock member. The dark red sandstone and shale layers, forming slopes and ledges, appear as layer-cake cliffs near Ouray. They are difficult to date exactly because they contain very few fossils. The whole sequence spans about 120 million years, during which time the Pennsylvanian highland of Uncompahgria yielded gradually to weathering and erosion, its streams depositing these many layers of mud and sand on deltas and floodplains around its flanks.

Ouray was settled in 1876 when gold and silver were discovered near Mt. Sneffels. At first glance it seems a logical and strikingly beautiful place for a town, but avalanches and landslides do not make it the safest place to live.

This district has produced over $125 million in gold, silver, lead, copper, and zinc. The richest deposits were near Camp Bird Mine about 6 miles southwest of Ouray and at Red Mountain 12 miles south of the town.

Between Ouray and Silverton, U.S. 550 crosses the Colorado Mineral Belt, the northeast-trending zone that contains most of the state's mining riches. Mining continues in the district still, though not on as great a scale as in the past. On cliffs northeast of Ouray, a rusty colored, jagged

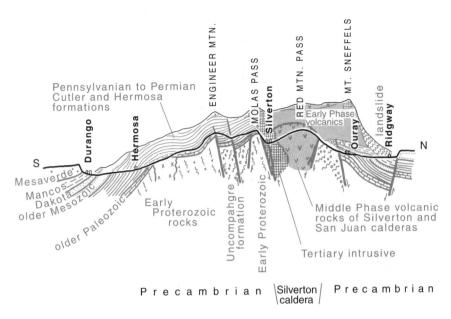

Section across the San Juan Mountains parallel to U.S. 550 between Ridgway and Durango.

Laramide stock known as the Blowout seems to spread out above the Dakota sandstone. It was probably the feeder for the laccolith that gently domes overlying Mancos shale.

Ouray's municipal swimming pool is supplied with water piped from hot springs south of town in Box Canyon. The water is heated as it flows along faults in rocks still hot from San Juan volcanism. The word *uncompahgre* means "hot water spring" in the Ute language.

Below the Triassic, Permian, and Pennsylvanian redbeds in the Ouray cliffs, and also rising southward onto the Uncompahgre uplift, are early Paleozoic marine sedimentary rocks. The most distinctive of these is the massive gray Leadville limestone. Below it are older, browner Devonian limestone layers of the Ouray formation.

Ouray's Box Canyon Falls, just south of town and well worth a visit, formed where Canyon Creek, in a hanging valley, dissolved a deep, narrow cleft in fault-weakened limestone. Rock walls overhang the falls by nearly 100 feet. Formed of the mineral calcite, limestone is soluble in water that has picked up acids as it percolates through forest soil and volcanic rock. Upstream hot springs may have helped it to dissolve the limestone.

Just above the falls and south of another fault, gently sloping Devonian sandstone lies across Precambrian rocks on the erosion surface formed during a long interval at the end of Precambrian through Cambrian time. Here, the Precambrian rocks, called the Uncompahgre formation, don't

Near Box Canyon Falls, Devonian sandstone layers lie almost horizontally across the beveled surface of nearly vertical Precambrian metasedimentary rocks. Far younger Tertiary Early Phase volcanic rocks form the crags in the background. —Jack Rathbone photo

look like the gneiss and granite elsewhere in Colorado. They are younger than the 1.7-billion-year-old gneiss and Routt plutonic suite granite and older than the 1.4-billion-year-old Berthoud plutonic suite granite found in the Colorado Rockies and Black Canyon of the Gunnison. Their layered conglomerate, sandstone, and shale, deposited in a downfaulted valley sometime after the Early Proterozoic Orogeny, were tilted vertically late in Precambrian time, but only slightly metamorphosed into quartzite, slate, and schist.

The oldest Tertiary volcanic deposits near Ouray, a thick widespread layer of Early Phase volcanic ash and breccia, wall the large amphitheater above the town.

South of Ouray, U.S. 550 follows a route pioneered by the historic Million Dollar Highway, famous because when built it took the then-enormous sum of $1,000,000 to construct 50 tortuous miles of crude mountain road. The highway was blasted through the Uncompahgre formation, steeply dipping and in many places rounded by Pleistocene glaciers. In winter, avalanches are common here; their paths show up as treeless chutes in summer. Near mile 91.7, glacial striations, scratches gouged by glacier-carried rocks, appear just above the highway. Beyond milepost 90, gray volcanic rocks overlie an irregular Precambrian surface; all Paleozoic rocks were eroded off this part of the San Juans before the volcanic rocks were added.

Near Ironton the highway emerges onto a flat lake floor filled with sediments deposited when the valley was dammed by a landslide. Gaudy colors of surface rocks result from oxidation of iron-bearing minerals. Though not concentrated enough to be mined, such "color" helped lead prospectors and geologists to more valuable mineral deposits.

The many mines along the highway near Red Mountain Pass worked small, pipelike ore bodies rich in silver, copper, and lead. Underground, this area is honeycombed with mine tunnels. Water seeping from abandoned mines contains toxic minerals, though the groundwater may have been laced with them before the area was mined. Cleanup has been difficult in such a large, partly boggy area. Plastic-lined settling ponds now trap water so that toxic substances can be removed.

At Ironton the highway enters the Silverton Caldera, one of many formed in the San Juans during Middle Phase volcanism. Calderas develop during or after explosive volcanism, when a roughly circular area collapses or subsides into a partly emptied magma chamber below a volcano. Later filled in and covered by more volcanic outpourings, then reshaped by stream and glacial erosion, the Silverton Caldera no longer looks like famous calderas at Crater Lake in Oregon or Haleakala Volcano in Hawaii. However, curving faults outline it, and in places streams follow the weak faulted zones. From Ironton to Silverton, Mineral Creek and U.S. 550 follow gently curving faults around the west side of the caldera; the curving upper valley of the Animas River similarly defines its eastern and southern margins.

Patterned with marginal and radial faults and dikes and impregnated with minerals, the Silverton Caldera has produced more than $150 million in silver, gold, lead, copper, and zinc. As Middle Phase volcanic and intrusive rocks cooled, water circulated through them and through even older rocks below, becoming heated and acidic. The water dissolved traces of metals from the rocks and carried them upward, depositing them along faults and cracks in cooler rock above.

Mining began around Silverton in 1870, when gold was discovered on what was then Ute Indian territory. Legally, mining began in 1874 after a treaty was signed and ratified, but nearly 4,000 claims had already been recorded! Despite the remoteness of the area and hardships involved in high-altitude mining, Silverton flourished even before the narrow-gauge railway came through from Durango in 1882. In summer the narrow-gauge trains still wind several times a day between Durango and Silverton through rugged, lonely Animas Canyon, where mine-dotted crags testify to the lure of silver and gold and the tenacity of fortune hunters.

In Silverton, the San Juan Historical Society Museum displays local minerals. Donated to the historical society by Sunnyside Gold Corporation, the Mayflower Mill, 2 miles northeast of town, is open in summer. Underground mine tours are offered, and jeep tours carry visitors to old mines in the district.

Between Red Mountain Pass and Silverton, U.S. 550 descends the U-shaped glaciated valley of Mineral Creek, its glacial profile modified by red and yellow talus slides and avalanche chutes. The valley curves eastward, following curving faults that edge the Silverton Caldera. Above timberline just left of center in the photograph, a large rock glacier creeps down a steep ravine. —L. C. Huff photo, courtesy of U.S. Geological Survey

U.S. 550
Silverton—Durango
51 miles (82 km)

South of Silverton, U.S. 550 passes through about a mile of gray Tertiary intrusive rock, then climbs toward Molas Pass through Paleozoic strata:

- Cambrian quartzite, a beach sand deposited above Precambrian rocks;
- brownish Devonian shale and limestone, products of a shallow sea;
- massive, red-stained Leadville limestone, a widespread Mississippian marine deposit with thin beds of black chert, a hard, dense mineral made up of microscopic grains of quartz.

The top of the Leadville limestone is quite uneven, a karst surface caused by solution of the limestone by rainwater and groundwater as the land rose above sea level. Karst areas exist today in the southeastern United States and other places where limestone is exposed in a wet, warm climate. Such regions typically contain solution caves, sinkholes, and irregular, broken land surfaces coated with bright reddish soil.

Here, the irregular layer of fragmented limestone and reddish soil is known as the Molas formation. It includes nodules of black chert containing Mississippian fossils, derived from the Leadville limestone as it dissolved away. Pebbles of this chert are common in overlying Pennsylvanian rock as well.

Molas Lake and several smaller ponds fill hollows in the old karst surface on top of the Leadville limestone. The Needle Mountains, visible across the lake, are Precambrian granite and gneiss. —L. C. Huff photo, courtesy of U.S. Geological Survey

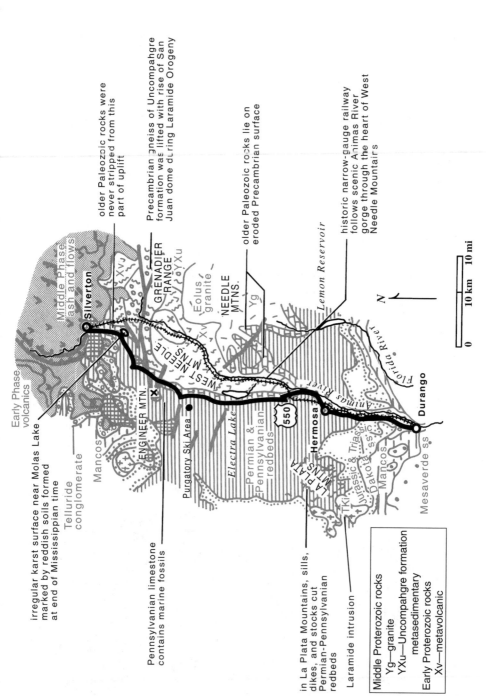

Geology along U.S. 550 between Silverton and Durango.

Early Phase volcanics

irregular karst surface near Molas Lake marked by reddish soils formed at end of Mississippian time

Telluride conglomerate

Mancos

ENGINEER MTN.

Pennsylvanian limestone contains marine fossils

Purgatory Ski Area

Electra Lake

Permian & Pennsylvanian redbeds

in La Plata Mountains, sills, dikes, and stocks cut Permian-Pennsylvanian redbeds

Laramide intrusion

LA PLATA MTNS.

Tri/Jurassic & Triassic

Dakota ss.

Mancos

Mesaverde ss.

Middle Phase ash and flows

Silverton

XVj

GRENADIER RANGE   YXu

WEST NEEDLE MTNS.

Eolus granite

XV

Yg

NEEDLE MTNS.

older Paleozoic rocks were never stripped from this part of uplift

Precambrian gneiss of Uncompahgre formation was lifted with rise of San Juan dome during Laramide Orogeny

older Paleozoic rocks lie on eroded Precambrian surface

Lemon Reservoir

Florida River

historic narrow-gauge railway follows scenic Animas River gorge through the heart of West Needle Mountains

550

Hermosa

Animas River

Durango

N

0   10 km   10 mi

Middle Proterozoic rocks
Yg—granite
YXu—Uncompahgre formation metasedimentary
Early Proterozoic rocks
Xv—metavolcanic

Slopes above Molas Lake are marked with ledge-forming horizontal limestone containing abundant fossil shellfish. These Pennsylvanian rocks, collectively called the Hermosa group, are cyclic, with repeated sequences of sandstone, shale, and limestone. Strangely, cycles typify Pennsylvanian sediments all over the world. In the eastern United States and Europe the cycles include layers of coal formed in ancient swamps. We don't yet know the whys and wherefores of the Pennsylvanian cycles, though they may be due to worldwide fluctuations in sea level caused by repeated ebb and flow of arctic or antarctic glaciation.

The flattened tops of some of the peaks near Molas Pass are Early Phase lava and ash flows. Surrounding country, high and rolling, was smoothed by a Pleistocene ice cap, leaving only a few jagged peaks that projected through the ice. There is abundant evidence of glaciation in this region. Look for cirques, U-shaped valleys, and smoothly rounded rock surfaces marked with glacial striations.

The Grenadier Range, about 10 beeline miles east-southeast of Molas Lake, is composed of steeply dipping layers of Precambrian quartzite and slate of the Uncompahgre formation, the same metasedimentary rocks that appear near Ouray. The West Needle Mountains, visible to the south, consist of older Precambrian metamorphic rocks, primarily of volcanic origin, locally named the Twilight gneiss. Both these units were intruded by the Eolus batholith, the 1.4-billion-year-old granite that makes up most of the Needle Mountains, visible in the far distance behind and between the other two ranges. The three ranges lie at the heart of the dome-shaped uplift that underlies the San Juan Mountains at the position of the S-shaped weak zone where northeast-trending and northwest-trending Precambrian faults intersect. The Needle Mountains include Mt. Eolus, Sunlight Peak, and Wisdom Peak, all Fourteeners.

Between mileposts 61 and 60, U.S. 550 crosses a narrow slice of the Uncompahgre formation that extends from the west end of the Grenadier Range. Watch for limestone recrystallized into marble, shale compressed to slate, and sandstone cemented into quartzite.

Crossing another fault, the highway cuts into a small Tertiary intrusion near milepost 60, with gray granite containing scattered needles of black hornblende. The road then crosses more Pennsylvanian cyclic sediments, which include black shale (easily mistaken for coal) near Coal Creek. A fold in these cyclic layers is visible at milepost 59.

Older Paleozoic sediments appear in a roadcut at milepost 54: the Leadville limestone with its irregular karst surface, brownish Devonian limestone below it, and then Cambrian quartzite. There are no Silurian or Ordovician rocks here. The sedimentary rocks lie on dark Precambrian gneiss; they dip southwest off the south side of the Needle Mountains. In places,

Cyclic Pennsylvanian marine sediments of the Hermosa group form the lower slopes of Engineer Mountain. Pennsylvanian and Permian redbeds on higher slopes make up the Cutler formation, composed of sand and mud washed from the Pennsylvanian highland of Uncompahgria. The cliff on the right side of the mountain is an igneous sill, probably of Tertiary age. —Jack Rathbone photo

the highway runs right on the Mississippian karst surface, recognizable by its uneven topography and bright red soils.

Between mileposts 60 and 50, the highway stays in Paleozoic rocks, skirting the West Needle Mountains and their Precambrian gneiss. On the far eastern side of the range, the Animas River cuts down through this gneiss. The narrow-gauge railway between Durango and Silverton threads through the Animas River gorge, in places clinging precariously to precipitous canyon walls, giving its passengers a good look at these ancient rocks.

U.S. 550 crosses more Mississippian karst at milepost 41, on particularly fossiliferous, cherty limestone. Then it drops through older Paleozoic rocks mostly concealed by vegetation.

As the highway grade decreases, the road returns to younger rocks. Starting at milepost 36, the highway crosses an entire Paleozoic-Triassic-Jurassic sequence, from oldest to youngest:

- whitish Cambrian Ignacio sandstone;
- brown Devonian Ouray limestone with twinkling calcite crystals;
- gray Mississippian Leadville limestone;

- Pennsylvanian sandstone-shale-limestone cycles of the Hermosa group (2,000 feet thick near Hermosa, including greenish gray sandstone and conglomerate, thin fossil-bearing gray and black limestone, and thin beds of gypsum);
- Pennsylvanian and Permian redbeds, the Cutler formation (near milepost 28);
- Triassic redbeds, with the boundary between Permian and Triassic hard to determine since the rocks contain few fossils;
- Jurassic Slick Rock member of the Entrada formation;
- mottled green and purple shales of the Morrison formation.

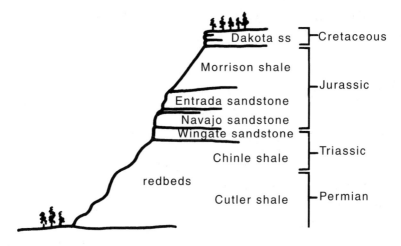

Permian and Mesozoic rocks appear above U.S. 550 at milepost 26. The Dakota sandstone forms the canyon rim.

Durango lies at the edge of the Plateau Country of southwestern Colorado. Sedimentary layers flatten out here, forming cuestas, mesas, and plateaus, all dipping gently south into the San Juan Basin.

Durango is Silverton's chief contact with the outside world. The narrow-gauge railway that connects these towns once transported mining supplies and mineral shipments southeast to Chama, New Mexico, and then northeast to Antonito and the standard-gauge. Today's boom is in the tourist trade, so it now carries sightseers and vacationers instead.

Colorado 135
Gunnison—Crested Butte
27 miles (43 km)

For a few miles Colorado 135 follows the Gunnison River upstream, with a view to the north of Red Mountain and Flat Top, both soft Cretaceous Mancos shale capped by 10-million-year-old Late Phase basalt flows. The open alluvium-filled valley north of Gunnison is underlain by the same shale, deposited in a shallow sea that flooded North America in Cretaceous time. Nearly horizontal Dakota sandstone and multicolored Morrison shale appear along the valley's edges. A massive white sandstone at the base of the Morrison formation, all that is left of some Jurassic sand dunes, lies directly on eroded Precambrian rocks, which appear at the surface at milepost 9.

Precambrian granite walls the canyon south of Almont; metamorphic rocks appear just north of Almont, their indistinct vertical light and dark bands probably originally horizontal layers of sandstone and shale. The road remains close to the contact between the Precambrian and Jurassic rocks; just beyond Roaring Judy Fish Hatchery it crosses a fault and reenters the Mancos shale. This thick gray shale also underlies the valley of the East River north of Almont, draping in a syncline that is part of the downwarp known as the Piceance Basin between the Rockies and the Uncompahgre uplift.

Symmetrical Round Mountain, almost straight north of milepost 12, is a laccolith, an intrusion that formed as magma oozed between sedimentary strata, doming the layers above, which have since eroded away. There are many more laccoliths in this area: Crested Butte northwest of Round Mountain, quite deeply eroded; Mt. Whetstone farther west, its summit high enough to be indented by two glacial cirques; and behind Mt. Whetstone about thirty more summits in the northern part of the West Elk Mountains.

Wide valleys near Crested Butte are floored with stream-deposited glacial outwash from retreating glaciers. Horizontal terraces of similar but older stream-deposited gravel edge the valleys. At milepost 21, Colorado 135 crosses the lower limit of glaciation, marked by an inconspicuous terminal moraine scattered with boulders.

Glaciers carved deeply into the mountains surrounding Crested Butte. In addition to their usual fingerprints—cirques, U-shaped valleys, and hummocky moraines—they left a legacy of unstable ground and recurrent landslides. In the West Elk Mountains, laccoliths rest on gently tilted shale and sandstone and relatively unconsolidated Tertiary sediments, all easily

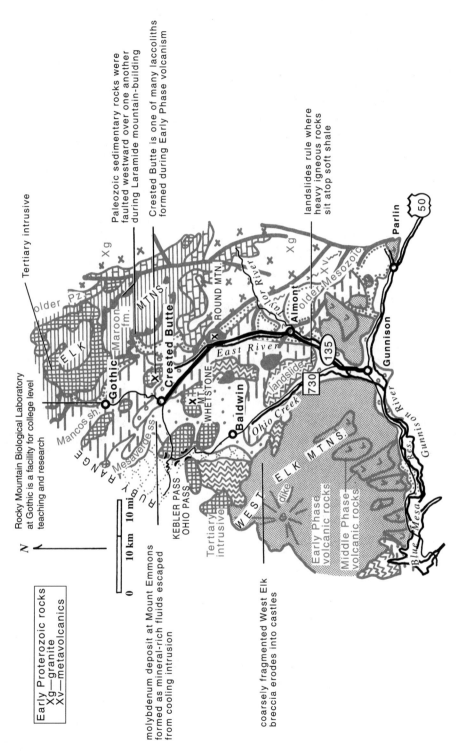

Early Proterozoic rocks
Xg—granite
Xv—metavolcanics

N

0   10 km   10 mi

Rocky Mountain Biological Laboratory
at Gothic is a facility for college level
teaching and research

Tertiary intrusive

Paleozoic sedimentary rocks were
faulted westward over one another
during Laramide mountain-building

Crested Butte is one of many laccoliths
formed during Early Phase volcanism

landslides rule where
heavy igneous rocks
sit atop soft shale

molybdenum deposit at Mount Emmons
formed as mineral-rich fluids escaped
from cooling intrusion

coarsely fragmented West Elk
breccia erodes into castles

older Pz

Xg

ELK

Maroon fm.

MTNS.

ROUND MTN.

Gothic

Crested Butte

Mancos sh.

Mesaverde ss

RANGE

RUBY

KEBLER PASS
OHIO PASS

Tertiary
intrusive

WEST ELK MTNS.

dike

Early Phase
volcanic rocks

Middle Phase
volcanic rocks

MT.
WHETSTONE

Baldwin

Ohio Creek

East River

landslide

730

135

Almont

Taylor River

Xg

older Mesozoic

Xv

Gunnison

Gunnison River

Parlin

50

Blue Mesa

Geology along Colorado 135 between Gunnison and Crested Butte.

incised by the glaciers. When the ice retreated, steepened slopes were left unbuttressed and prone to collapse.

Moraines and landslides may look similar, but the bare steep source areas above landslides distinguish them. A large slide on Mt. Whetstone, for instance, left a steep scarplike hollow where it broke away, and typically bumpy terrain where it came to rest.

The town of Crested Butte was founded in 1879 to supply gold and silver mines farther up in the mountains. Metallic ores associated with Tertiary volcanism were discovered in the Ruby Range northwest of Crested Butte in about 1860 and in the Elk Mountains somewhat later. Mining

Crested Butte, a Miocene laccolith, rests on Cretaceous Mancos shale, with a gentle shale slope below the steep-sided igneous rock. Numerous long open cracks weaken the igneous mass, a warning that it is not very stable. —Felicie Williams photo

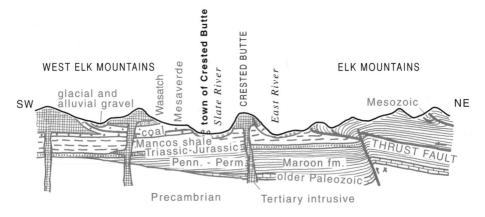

Southwest-northeast section through Crested Butte.

started in the 1870s. Placer gold was discovered up nearby Washington Gulch, and gold- and silver-bearing veins were located among the high peaks to the north and west.

By 1881 the emphasis had shifted to coal. Sandstone layers above the Mancos shale are part of the Mesaverde group—shore, swamp, and lagoon deposits of the receding Cretaceous sea. Thick coal beds lie between the sandstone layers—hence the black mine dumps partway up slopes south and west of town. Mesaverde coal is mostly soft or bituminous coal, but where it has been baked by mid-Tertiary intrusions it has hardened to anthracite, or hard coal. The mines closed in 1952 for economic reasons, not for lack of coal.

Dinosaurs left deep footprints on the matted, peatlike vegetation that later became coal. Many of their tracks later filled with sand, so as the miners removed the coal they found casts of the huge footprints, some of them almost a yard across, hanging from the mine ceilings! Some of these tracks show us that their makers walked in herds with their offspring among them.

In the 1970s geologists discovered a sizable molybdenum deposit beneath Mt. Emmons, a few miles west of town, where iron-rich seeps had indicated the presence of mineralization.

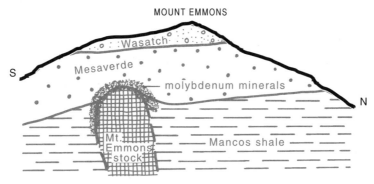

The Mt. Emmons molybdenum deposit formed where a Late Phase granite porphyry stock cooled and crystallized deep below the surface. Hot mineral-bearing fluids escaping from the stock precipitated molybdenite in joints near the contact with overlying sedimentary rocks.

Buildings of the Crested Butte ski area lie on soft Mancos shale anchored in place by thick dikes of Early Phase igneous rock. Some of these dikes may have been feeders for the Crested Butte laccolith.

Paleozoic strata that pre-date and postdate the Ancestral Rockies overlie Precambrian rocks in the Elk Mountains east and north of Crested Butte. In Pennsylvanian time, as Uncompahgria began to rise to the west and Frontrangia to the east, older Paleozoic rocks were preserved in the sheltered basin between the two ranges. Gradually they were covered by nearshore marine deposits such as tidal mud, shoal and shore sand, reef limestone, salt, and gypsum. These deposits were in turn overlain by fans of coarse sand and mud eroded from Uncompahgria and Frontrangia, now hardened into the deep red rocks of the Maroon formation.

In the Elk Mountains, high ridges of this formation mark the western boundary of the present Rocky Mountains. During the Laramide Orogeny the Maroon formation was folded and thrust westward over itself and over younger strata. Mesozoic rocks that lay above the Maroon formation have mostly eroded away.

Colorado 145
Telluride—Cortez
72 miles (116 km)

In most parts of Colorado, elemental or native gold first lured prospectors and miners. This was true in the western San Juans, too, but later prospecting revealed that this area also contained a gold-bearing tellurium compound called telluride. The ores for which the town of Telluride was named occur in veins associated with many small Tertiary intrusions, along with native gold and compounds of lead, silver, zinc, and copper.

Some Telluride veins have been mined vertically for nearly 3,000 feet; others have been followed horizontally for close to 7 miles. Mining towns high between the peaks, at elevations up to 12,000 feet, were occupied year-round. Many mine tunnels eventually interconnected, so quite a few could be reached from the Idarado Mine at the upper end of the valley—a boon in winter when it was warmer inside the mountain. About $250 million worth of metals were mined here.

The only surface evidence of the extensive tunnels are mine dumps and mill tailings. Tailings that once trickled with metal-contaminated water have been sealed off and reclaimed. Acid mine drainage among the peaks far above town is channeled and treated. Most of the original mines are now closed off, their portals concealed by second-growth forest. For more information about Telluride mining, visit the San Miguel County Museum.

West of Telluride, Colorado 145 passes Society Turn, where Telluride's carriage set paraded on Sunday afternoons, and proceeds up the glaciated valley of the South Fork of the San Miguel River. An 800-foot-thick sill forms the cliff on the other side of this valley; it may have once connected with the Tertiary stock at Ophir and the Mt. Wilson intrusions to the southwest. The sill lifts overlying Dakota sandstone well above its normal position.

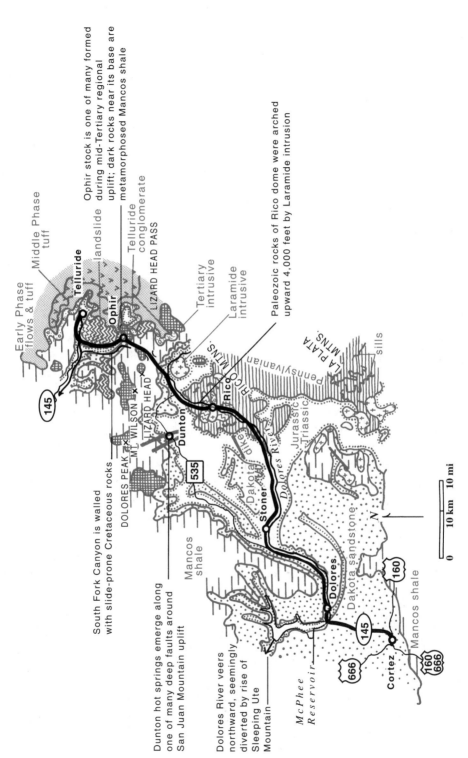

Early Phase
flows & tuff

Middle Phase
tuff

Telluride

landslide

Telluride
conglomerate

LIZARD HEAD PASS

Ophir stock is one of many formed
during mid-Tertiary regional
uplift; dark rocks near its base are
metamorphosed Mancos shale

145

Ophir

Tertiary
intrusive

Laramide
intrusive

DOLORES PEAK

MT. WILSON

LIZARD HEAD

Dunton

535

Paleozoic rocks of Rico dome were arched
upward 4,000 feet by Laramide intrusion

RICO MTNS.

Rico

Pennsylvanian

LA PLATA MTNS.

sills

South Fork Canyon is walled
with slide-prone Cretaceous rocks

Dunton hot springs emerge along
one of many deep faults around
San Juan Mountain uplift

Dakota dike

Stoner

Dolores River

Jurassic
Triassic

Mancos
shale

Dolores River veers
northward, seemingly
diverted by rise of
Sleeping Ute Mountain

Dolores

Dakota sandstone

160

N

0   10 km   10 mi

McPhee
Reservoir

145

Mancos shale

666

Cortez

160
666

Geology along Colorado 145 between Telluride and Cortez.

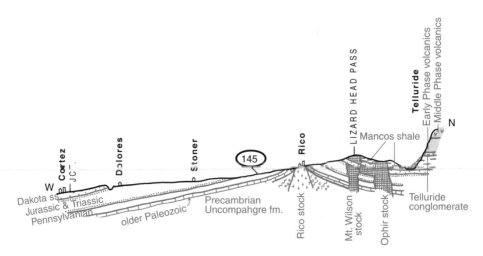

Section along Colorado 145 from Telluride to Cortez.

Glacially polished, grooved, and striated rock appears along both sides of the 3-mile side road to Ophir. Cabins in and around town rest on hummocky glacial moraines. The glaciers that occupied this smaller valley during the Ice Age couldn't carve downward as rapidly as the large South Fork glacier, partly because of their smaller size and partly because of the hardness of the Ophir stock, so their valley remains a hanging valley high above the South Fork.

Landslides below the highway near milepost 64 destroyed tracks and trestles of a narrow-gauge railway that ran from Telluride to Durango. Landslides are also responsible for Trout Lake, as timber-covered slopes east and northeast of the lake are blocks of tuff that slid down from the main ridge line, skidding across slippery Cretaceous shale.

At Lizard Head Pass a roadside exhibit points out surrounding peaks as well as some of the area's history. The rugged ridge of Yellow Mountain to the northeast is sloped with Early Phase tuff colored by oxidation of iron sulfide produced by igneous vapors seeping along veins and fractures. The rust-colored stains are a reminder that this area is part of the Colorado Mineral Belt. The tuff lies on Cretaceous shale, which surfaces near the pass and forms the smooth green slopes of nearby Sheep Mountain.

Yellow Mountain's jagged skyline is eroded from younger lava flows belonging to the Middle Phase of volcanism. No Late Phase volcanic rocks occur in this part of the San Juans. Smooth shale slopes rise toward the large, irregular intrusion that forms the high peaks of the San Miguel Mountains, northwest of milepost 58, and peaks west of the narrow spine of Lizard Head.

Sedimentary and volcanic rocks, shattered and tilted vertically by intrusion of the Ophir stock, have eroded into the spectacular pinnacles of Ophir Needles. The granite porphyry of the stock contains large pink and white feldspar crystals amid smaller crystals of quartz, hornblende, and mica. The Ophir district produced silver, lead, zinc, gold, and tungsten.
—Felicie Williams photo

Glaciers smoothed much of the high country around Lizard Head Pass, but the sharp, pinnacled silhouettes of many peaks tell us they jutted up through the ice as nunataks. Evidence of glaciation continues almost to Rico. Its lower limit, at about 8,000 feet in most of Colorado, is above 9,000 feet here because of the southern exposure. Terraces of stream-deposited glacial outwash gravel extend downstream.

Between Lizard Head Pass and Rico, Colorado 145 runs through older and older sedimentary rocks that bow upward across the Rico dome. From northeast to southwest, they are:

- Dakota sandstone (milepost 56), the lowest Cretaceous rock in most of Colorado, resistant, blocky sandstone that often tops cuestas and hogbacks;

- about 1,000 feet of Jurassic floodplain, lake, tidal flat, and dune deposits, from Morrison shales with their Easter egg colors to the Slick Rock member, the unmistakable smoothly rounded ledge at the base of the Entrada sandstone (milepost 54);

- Triassic red sandstone and shale, probably floodplain, delta, and estuary deposits;

Lizard Head is an eroded remnant of pre-volcanic conglomerate below reddish gray Early Phase welded tuff. —Felicie Williams photo

- Permian Cutler formation, sandstone, shale, and conglomerate with pebbles and cobbles of Precambrian igneous rock, proof that Uncompahgria had lifted high enough late in Pennsylvanian time for erosion to cut into its core;
- Pennsylvanian marine shale and fossiliferous limestone of the Hermosa group.

In the rugged Rico Mountains surrounding the town of Rico, many Laramide sills, dikes, and irregular intrusions penetrated an area about 5 miles across and created the Rico dome, pushing up overlying strata at least 4,000 feet. At 70 million years old, much older than the igneous rocks that dominate the San Juans, the intrusions lie along the Precambrian faults of the Colorado Lineament, the path followed during the Laramide Orogeny and later in mid-Tertiary time by mineral-rich fluids that created the Colorado Mineral Belt.

Near the town of Rico, landslides hide most of the intrusive rocks; Rico itself is built right on a slide. For many years this district produced incredibly rich silver ore from Pennsylvanian rocks near their contacts with nearby intrusions. More recently, lead, zinc, copper, gold, and pyrite have been mined, and a molybdenum deposit has been found.

South of Rico, strata dip south off the Rico dome. As the highway travels *down* the Dolores River it goes *up* the rock sequence, reversing the order north of Rico. Laramide intrusive rock, porphyry with big feldspar and hornblende crystals, is right next to the road near milepost 44. South of milepost 22 a quarry produces sand and gravel from glacial outwash. Low tan cliffs of Dakota sandstone surround Dolores.

The Slick Rock member of the Entrada sandstone displays the distinctive crossbedding of a dune deposit. —E. B. Eckel photo, courtesy of U.S. Geological Survey

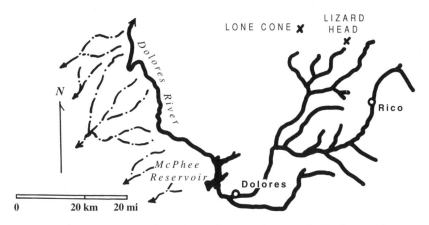

The Dolores River turns abruptly north near Dolores, probably diverted from an original southwestward course by intrusion of the Sleeping Ute Mountain laccolith. Erosion since the laccolith formed caused small tributary streams to turn southwest again, but the Dolores, trapped in a deeper channel, still flows north into the Colorado River in Utah.

South of McPhee Reservoir, Colorado 145 crosses a gently sloping surface of Dakota sandstone to join U.S. 160 near Cortez. The town of Dove Creek, about 35 miles to the northwest, bills itself as "the pinto bean capital of the world" for a geologic reason. The Dakota surface between Colorado 145 and Dove Creek is veneered with a layer of glacially ground, windblown Pleistocene silt particularly well suited to growing beans—the fine soil is a relic of a very different climate.

Colorado 149
South Fork—Blue Mesa Reservoir
123 miles (198 km)

Colorado 149 crosses the center of the San Juan Mountains, traversing a cluster of calderas and the flows that erupted from them. Deeply eroded, the calderas are hard to discern. However, by mapping their ring-shaped fault systems and types of volcanic rock, and studying rock samples in the laboratory, geologists have identified individual tuff layers and correlated them with their source calderas.

The community of South Fork is surrounded by hills of welded tuff of Middle Phase volcanism, the time of huge caldera-forming eruptions. The uppermost layer, the Fish Canyon tuff, is exposed northeast of the highway beyond milepost 4. Erupted from La Garita Caldera, it is distinguished from other welded tuff by the presence of tiny crystals of green hornblende. Below it are thick layers of tuffs from an older caldera that was completely obliterated by La Garita eruption.

La Garita Caldera is the largest (30 by 50 miles) and oldest caldera exposed in the central San Juans, though at least one older one lies buried beneath it. Geologists think La Garita eruption started several miles west of Wolf Creek Pass, which is south of South Fork on U.S. 160. During a few days or weeks, the eruption moved north through three eruptive centers, each of which collapsed along roughly circular faults, together forming a huge, compound, oval caldera. A gigantic outpouring of volcanic ash—more than 1,000 cubic miles of it—ponded over a mile deep in the caldera and blanketed the surrounding country as far as 60 miles from the caldera rim. Known as the Fish Canyon tuff, the ash layer is as much as 1,000 feet thick. (For comparison, in 1980 Mt. Saint Helens blew off a mere 0.06 cubic mile of volcanic ash.) Several younger calderas clump along the western margin of La Garita Caldera, their magmas rising along the faults that edged the older structure, a common situation as calderas go.

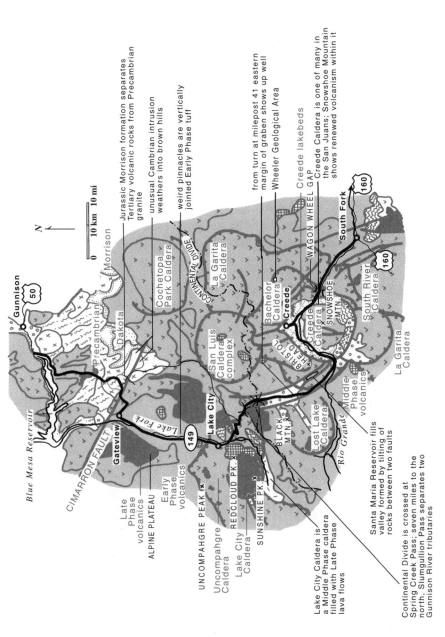

Geology along Colorado 149 between South Fork and Blue Mesa Reservoir.

Colorado 149 follows the Rio Grande upstream for some distance through a graben, a valley dropped downward between pairs of more or less parallel faults. The V-shaped canyon between South Fork and Creede was shaped by huge rockfalls where the canyon walls were undermined by the river and the road. Rockfalls here pose a major problem for highway maintenance crews.

Just north of Wagon Wheel Gap, a road to the north leads to Wheeler Geologic Area, a wonderland of pink and white spires, cones, and gnomelike figures eroded in three successive sheets of relatively soft Middle Phase tuff. A four-wheel-drive vehicle is a must for this rocky 24-mile trip. Though the variety of form and color of the eroded figures is unusual, small clusters of similar spires have also developed in other parts of the San Juans.

The best-preserved caldera in the San Juans, the Creede Caldera formed during Middle Phase volcanism. Nearly 10 miles across, it is centered well south of the town of Creede. After it formed, magma rising from the underlying reservoir produced a 2,000-foot-high dome in its center, leaving a crescent-shaped valley between the caldera wall and the dome.

At Wagon Wheel Gap, the Rio Grande cuts through the remnants of a thick, gently sloping lava dome draped against the wall of the Creede Caldera. Irregular columnar jointing makes the dome look like a log stockade. This photograph was taken in 1874, before the present highway was built. —W. H. Jackson photo, courtesy of U.S. Geological Survey

After the Creede Caldera collapsed, a broad new volcanic dome developed in its center, almost filling the caldera. This dome now appears as Snowshoe Mountain south of Creede. —P. W. Lipman photo, courtesy of U.S. Geological Survey

From Wagon Wheel Gap, Colorado 149 follows this flat-floored valley, which also determined the course of the Rio Grande. As volcanic ash from later eruptions fell into a lake that for a time filled the crescent valley, it preserved many fossil leaves and insects, as in some of the white tuff near the highway west of Creede. The valley also received large amounts of glacial outwash gravel in Pleistocene time.

Creede Caldera lies inside the older, larger Bachelor Caldera. Near the town of Creede, both calderas are cut by several parallel faults that run just west of north, with slivers of land dropped downward between them, forming grabens. Mineralization in shattered volcanic rocks near these faults made the Creede district one of the most productive silver-mining areas in the United States. Discovered in 1889, the ores occurred in quartz and amethyst veins north of the town. The district produced gold, lead, and zinc as well as silver. The Underground Mining Museum in Creede has brochures for the Bachelor Historic Loop, a 17-mile drive up the canyon to see the mines, some of them literally "cliff-hangers."

The last mines closed in 1985. Since then, mills have been dismantled and waste piles sealed for safety, with a citizens' group managing reclamation.

Three or more easily identified terrace levels show up along the Rio Grande west of Creede. Each level represents a former river floodplain and a time of stability in the history of the river, when it was abundantly supplied with both gravel and water. The uppermost terrace is the oldest.

Higher terracelike benches visible on slopes in the distance across the valley are ancient lake shorelines carved into surrounding volcanic rocks.

West of milepost 29, the road leaves the caldera deposits and reenters Middle Phase volcanic rocks. Andesite lava flows of both Early and Middle

Phases are well exposed on the cliffs of Bristol Head, visible to the north between mileposts 32 and 33. There, the highway crosses the eastern of two faults that define the Clear Creek graben, a downdropped block about 20 miles long and 1 to 5 miles wide. It is faulted along both sides, its central block 2,000 feet lower than rocks on either side. Bristol Head and the cliffs running north from it form the eastern edge of the graben, more gently sloped Baldy Mountain its western edge. The central downdropped block is tipped down to the west, its upper edge forming a central ridge. Clear Creek and Santa Maria Reservoir lie east of this ridge; the Rio Grande and Colorado 149 run west of it.

Where Colorado 149 swings into the western half of the Clear Creek graben, the Rio Grande flows in lazy meanders on the almost horizontal valley floor. Near milepost 41 the highway leaves the river, whose headwaters lie some 35 miles west of here, and continues up the Clear Creek graben, crossing to its eastern side between mileposts 44 and 45.

The road crosses three groups of fault-dropped grabens between South Fork and milepost 45. Geologists think these structures formed over the

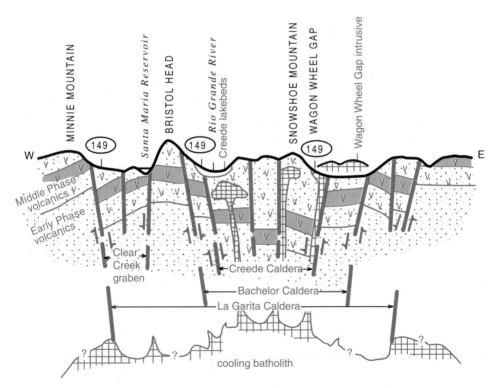

Section from east of Wagon Wheel Gap to Minnie Mountain across three nested calderas and Clear Creek graben. Darker shading marks tuff erupted from La Garita Caldera. Tuffs and flows erupted from other calderas are less extensive.

collapsing roof of an immense magma chamber that lies below the cluster of calderas.

The overlook near milepost 47 provides a remarkable view of San Juan high country, looking southwestward up the valley of Hermit Creek toward the Rio Grande Pyramid and other summits along the Continental Divide. A series of small lakes occupies the long, straight valley, its U-shape gentled by rockslides since the glacier that sculpted and supported the valley walls melted away.

As the highway continues northward toward Spring Creek Pass, basalt flows of Late Phase volcanism are visible topping summits to the north and northeast. Considerably harder and darker than underlying tuffs of the Middle Phase, they form smooth resistant caprock over distances of several miles. Spring Creek Pass at 10,901 feet is a low spot on the Continental Divide, the generally north-south line (east-west just here) which divides Atlantic and Pacific drainage. Colorado 149 then climbs fairly steeply to Slumgullion Pass, 11,361 feet in elevation. Visible from Slumgullion Pass, the Continental Divide runs eastward from Spring Creek Pass along a ridge of high summits.

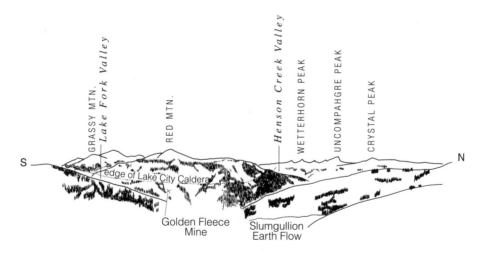

The view westward from Windy Point overlook includes the Lake City Caldera, peaks within and around the caldera, and part of Slumgullion Earth Flow.

Windy Point overlook near milepost 65 affords a spectacular view of the surrounding mountains. Grassy Mountain and Red Mountain rise to the west, and the valley of Lake Fork of the Gunnison River curves below them. Henson Creek swings around their northern flank. These two valleys outline the Lake City Caldera, formed during the explosive Middle Phase of volcanism. Peaks within the caldera developed during the Late Phase. Lake City Caldera lies entirely within the larger Uncompahgre Caldera, whose

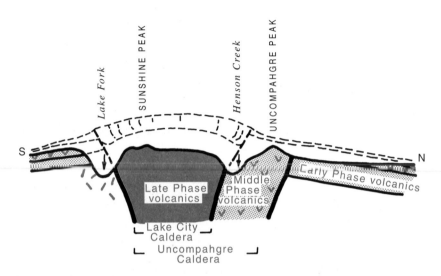

*Lake City Caldera was originally 3,300 feet higher. Since the ring of faults surrounding the caldera slopes inward, creeks and glaciers around its margin ended up outside the caldera when they cut straight down (**arrows**).*

older, less regular boundaries are not as well defined by today's topography. Uncompahgre Peak, north of Lake City Caldera but within Uncompahgre Caldera, is topped with Middle Phase lava flows. Sharp-pointed Wetterhorn Peak is outside both calderas.

Windy Point overlook also provides a view of part of the Slumgullion Earth Flow and the source from which it came. Annual surveys and trees at drunken angles indicate the flow still moves sporadically. The highway crosses the source area and then skirts the south side of the flow; watch for a large turnout near milepost 66 for the best place to see the slide itself. It dams San Cristobal Lake near Lake City—a good place for another look.

Lake City mines produced both gold and silver, mostly from Middle Phase veins outside the Lake City Caldera. The Golden Fleece Mine, visible on the lower slopes of Red Mountain, was the largest in the area. Red Mountain is a large volume of rock altered to red clay by sulfurous fumes and percolating rainwater as the volcano cooled. It is considered a promising source for aluminum. Middle Phase volcanic rocks that filled the Uncompahgre Caldera—including tuff from the giant La Garita eruption—surround Lake City. The one-time mining town now welcomes tourists with Hinsdale County Museum and with jeep tours to mining sites and old ghost towns.

North of Lake City, Colorado 149 crosses another earthflow and reaches bluffs of tuff and breccia of the Early Phase of volcanism. The rock is gray or purplish gray and fairly coarse, with numerous small fragments embedded in it. Where the volcanic ash fell into lakes and streams, it was washed and sorted by water and shows strong horizontal banding.

The area around Lake City and Gateview is one of the few places where Early, Middle, and Late Phase volcanic rocks appear together: Late Phase ashflows atop Alpine Plateau northwest of milepost 80; Middle Phase layers on slopes and ledges below them; and Early Phase tuffs edging the Lake Fork of the Gunnison River.

In many places, tuffs and flows shrank as they cooled, fracturing into vertical columns. Weathering along the vertical fractures eroded the columns into pointed cones often described as "tent rocks." You can see some of these along the Lake Fork north of Lake City. They look more like skinny tepees or the one-pole tents of nineteenth-century geologists.

Between mileposts 103 and 104, Colorado 149 crosses the base of the volcanic rocks, where breccia and tuff rest on colorful Jurassic shale that in turn lies directly on dark Precambrian granite and gneiss. Prospect pits for uranium dot the shale outcrops. In late Paleozoic time, when this area was part of Uncompahgria, earlier strata eroded away.

Near Powderhorn several small mineral deposits contain zinc, copper, lead, silver, and gold disseminated through Precambrian sediments as they were deposited about 1.8 billion years ago. Southeast of Powderhorn, some unusual Cambrian igneous rocks form treeless reddish hills. Vermiculite, a platy clay mineral derived from mica, has been mined from them. When heated, this mineral expands and can be used for insulation or, because of its ready absorption of water, a soil additive.

The West Elk Mountains on the northern skyline are another volcanic center, with lava flows and ashflows similar to those of the Early Phase in the San Juans. Some of the tallest peaks are volcanic conduits surrounded by radiating dikes. Southward flows from these conduits interlayer with welded tuffs from the San Juan Mountains. During the eruptions, the Gunnison River, flowing westward between the West Elks and the San Juans, was pushed alternately south and north by ashflows and lava flows from first one and then the other volcanic field.

Far western Colorado is sharply carved, with deep canyons and ravines separating flat-topped mesas, buttes, and plateaus. Erosion of horizontal rock layers—some hard, some soft—governs the shape of the land. Independence Rock, Colorado National Monument.
—Ray Strauss photo

VI
Plateau Country

West of the Rocky Mountains is a region of flat-lying sedimentary rock known as the Colorado Plateau. Named for the great river that runs through it, it extends far into Utah, Arizona, and New Mexico. This region escaped the wave of unrest that swept across western North America when the continent collided with a subcontinent along its western coast. The collision thrust up the Sierra Nevada, the ranges of Nevada and Utah, and then the Rockies. The Colorado Plateau bowed and buckled, it is true, and was rotated a bit and uplifted many thousands of feet, but the sedimentary rock layers remained nearly horizontal.

The entire Colorado Plateau lies within the drainage basin of the Colorado River, whose tributaries carve its innumerable canyons. Simple folds and faults in many places control the drainage, blocking established streams and forcing them to detour or to cut deep into hard, intractable Precambrian rock. These same faults and folds give different elevations to different portions of the Plateau Country. The Uncompahgre Plateau, for instance, reaches 9,000 feet above sea level, Mesa Verde is 7,000 feet, and the Roan Plateau averages 8,000 feet. None of the component plateaus show the intense folding and thrust faulting that characterize the Colorado Rockies.

This is a colorful land in which many of the sedimentary layers, tinted with shades of salmon, pink, or dark rust red, contrast sharply with the blue of high-altitude sky and the soft greens of piñon, juniper, and sage. It is a desert land now, as it has been before, with average annual rainfall below 10 inches. The warm-hued rocks are often bare of soil, and vegetation is usually scanty, making it a scenic and geologic wonderland whose history is more easily read than that in many other parts of the world.

In the Plateau Country, Precambrian rocks are exposed only in the hearts of the deepest canyons and along the margins of a few fault-block uplands. These sparse outcrops, however, are enough to tell us that the western part of Colorado experienced the same Precambrian events as the rest of the state. Let's look a moment at western Colorado's story.

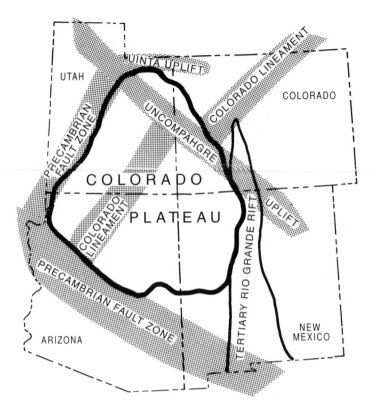

The Colorado Plateau covers parts of Colorado, Utah, Arizona, and New Mexico. Its boundaries are fault zones discernible on maps and satellite images, most of them first activated in Precambrian time.

Precambrian Time

Sedimentary and volcanic rocks deposited along an ancient continental margin some 1.8 billion years ago were folded into the roots of mountains and intruded by granite 1.7 and again 1.4 billion years ago. Each period of mountain building was caused by a collision along the continent's southeastern coastline, first in what is now New Mexico, then in Oklahoma and Texas. Between orogenies, a broad valley sank between deep rift faults in the southwestern corner of the state, seemingly in response to crustal tension. Well after the second orogeny, a similar valley developed in Colorado's northwestern corner.

Paleozoic Time

Cambrian sandstone, now mostly metamorphosed into quartzite, was preserved in the same two corners of western Colorado. Ordovician and

Silurian rocks, if they were ever deposited there, were removed by erosion before Devonian and Mississippian limestones accumulated in shallow seas above the Cambrian layers.

Uncompahgria, an isolated range that was part of the Ancestral Rocky Mountains, rose in Pennsylvanian time in southwest Colorado, faulted upward and westward as a result of North America's collision with Africa. West of Uncompahgria, shallow seas repeatedly flooded a nearly landlocked basin and, through repeated evaporation, left behind thick deposits of salt. Erosion gradually removed earlier Paleozoic strata from Uncompahgria, leaving bare the old horizontal surface of the Precambrian rocks.

Mesozoic Time

Triassic rivers spread floodplain and delta sediments across the horizontal surface. From time to time and place to place, dunes drifted across the land.

North America's collision with a Pacific subcontinent in Jurassic time created the Sierra Nevada in California and Nevada. This immense range blocked sea-moistened winds from reaching the interior, creating a broad dune-swept region as large as today's Arabian and Saharan Deserts. The region that eventually became the Colorado Plateau lay at about the same latitude as today's major desert areas, 20 to 30 degrees from the equator. Gradually, the scattered dune fields of early Triassic time were replaced by sweeping sand seas that became the Wingate, Navajo, and Entrada sandstones.

As time went on, the climate changed: the deserts were replaced by lush, swampy lowlands watered by meandering rivers. Mud, sand, and volcanic ash deposited in these lowlands late in Jurassic time is now the Morrison formation, famous for its abundant and varied fossil dinosaur fauna.

As the Cretaceous period began, streams and rivers deposited discontinuous patches of the Burro Canyon formation, more mud and sand. Then, a short-term advance and retreat of an eastern sea left behind the thin but widespread shore deposits that are now the Dakota sandstone.

As the central part of the continent slowly sank, it was finally inundated by a shallow sea that blanketed the Dakota sandstone with a thick, uniform layer of mud known today as the Mancos shale in the Plateau Country, the Pierre shale in eastern Colorado and adjacent states. This thick gray shale preserved the shells and skeletons of innumerable marine animals, among them coiled ammonites, giant oysters, clams, and swimming reptiles.

Late in Cretaceous time the wave of mountain building that started in California in Jurassic time arrived in Colorado, and the land began to rise, forcing the sea to retreat eastward. The receding sea left sandy beach deposits, river deltas, and the muds of bays and swamps—now layers of sandstone, shale, and coal collectively called the Mesaverde group—in western

Mt. Garfield near Palisade, with slopes of Mancos shale and a cap of Mesaverde group sandstones, displays the cliff-slope-cliff topography characteristic of the Plateau Country. —Jack Rathbone photo

Colorado. Because the shoreline shifted eastward as the sea retreated, the western part of this unit is somewhat older than the eastern part. Such time-crossing is quite common in shoreline sediments.

We know that the Cretaceous period came to an abrupt end with the violent impact of an asteroid or comet around 65 million years ago. Although the continent suffered little physical harm—the rise of the Rockies was not related to it—biologically Earth changed forever. The dinosaurs, along with a large proportion of other life forms, were gone. Freed of fearsome predators, mammals gradually took over the world.

Cenozoic Time

Strangely, the entire Colorado Plateau escaped the Laramide Orogeny almost unscathed, like a well-built raft floating in a stormy sea. Mesozoic sediments between the Wasatch Range in Utah and the Rockies in Colorado remained nearly flat-lying. Gently warped or faulted structures of this land—including the many lesser plateaus—were hardly more than ripples on the foreland of the Rockies.

During and since Tertiary time, great rivers flowing from the Rocky Mountain and Uncompahgre highlands have played major roles in shaping the land. When drainage was blocked by surrounding uplifts, an immense basin formed in the area where Colorado, Utah, and Wyoming join. Many thousands of feet of silty and sandy mud were swept from the rising mountains and deposited on river floodplains and deltas in the basin, forming the Wasatch formation. Eventually, a lake known to geologists as Lake Uinta filled part of the basin south of the Uinta uplift. During its 6.5-million-year existence, the lake filled with fine muds that became the Eocene Green River shale—some of it oil shale, one of the greatest undeveloped fossil fuel sources in the world.

Then for many millions of years, possibly fueled in some way by the subduction of the Pacific plate, volcanoes reigned supreme. Lava flows and welded ash sheets spread from large volcanic centers around the margins of the Colorado Plateau, while volcanic ash drifted across much of the western part of the state. As volcanic activity decreased, uplift began again, the regional lifting and stretching of mid-Tertiary time that brought Colorado and parts of adjacent states to their present elevations. Breaking its ties with the Rockies, the Colorado Plateau pulled away, its southern edge swinging westward, opening a long, narrow rift valley that extended from central Colorado south through New Mexico. Basalt spread from deep faults along the rift and along faulted edges of the Colorado Plateau.

Boosted by the renewed uplift as well as by increased rainfall and snowfall during Pleistocene time, the Colorado River and its tributaries assailed the Plateau Country, carving intricate canyons, mesas, cuestas, and badlands, blending the scenery of western Colorado with that of adjacent parts of Utah, Arizona, and New Mexico.

The Plateau Country possesses not only superb scenery but a wealth of important energy resources—oil and gas in Paleozoic, Mesozoic, and Cenozoic rocks, uranium in Triassic and Jurassic sandstones, coal in Cretaceous strata, oil shale in the Tertiary Green River formation, and sunshine, gift of its desert climate.

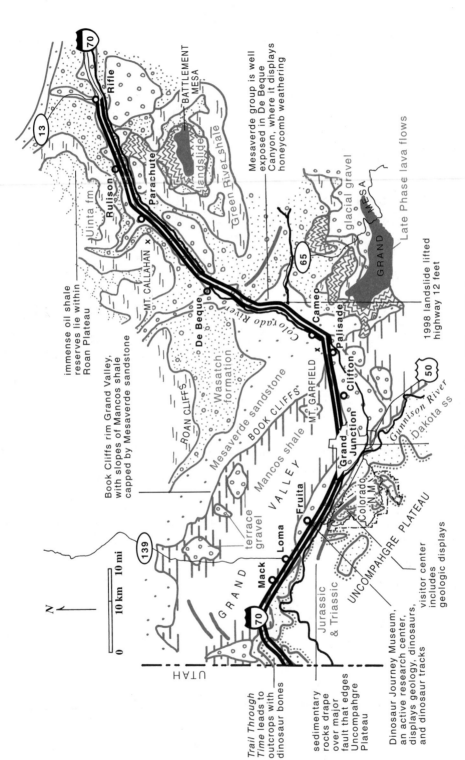

Geology along I-70 between Rifle and Utah.

Interstate 70
Rifle—Utah
88 miles (142 km)

Interstate 70 curves southwestward from Rifle, following the Colorado River through Tertiary and Cretaceous sedimentary rocks. At one time called the Grand, the river gave its name to places like Grand Valley, Grand Junction, and Grand Mesa.

The Roan Cliffs north of the interstate, as well as the cliffs of Battlement Mesa and Grand Mesa to the south, expose flat-lying Paleocene and Eocene sandstone and limy shale of the Wasatch and Green River formations. The pinkish tan Wasatch formation, about 5,000 feet of fine silt and sand that washed off the newly rising mountains in Paleocene time, was deposited as alluvial fans and river floodplains. The Green River shale, 3,500 feet thick, forms the grayish tan upper part of the Roan Cliffs. In Eocene time its muds accumulated in Lake Uinta, which occupied a large, irregular basin west of the Rockies and the White River uplift, between the Uinta uplift to the north and the Uncompahgre uplift to the southwest.

Above these two formations, but out of sight from I-70, lies the Uinta formation, about 400 feet of pale delta sandstone and shale that filled in the remainder of the Uinta Basin.

Within the Green River shale lie thick beds of oil shale. The richest oil shale beds known, they hold more than 1.8 trillion barrels of oil in the form of a waxy compound called kerogen. From I-70 you can see the dark brown layer known as Mahogany Ledge, which averages more than 27 gallons of oil per ton of rock. Oil-bearing beds of the Green River shale thicken

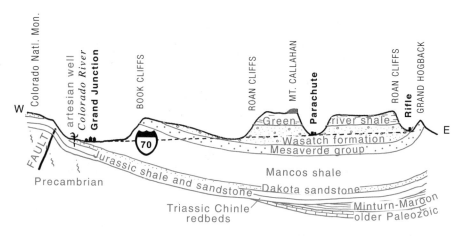

Section parallel to I-70 from Rifle to Colorado National Monument.

northward, so most oil shale prospects are north of here in remote parts of the Roan Plateau.

Unfortunately it's not easy to get the oil out of the shale. It can be freed only by crushing and heating the rock to 900 degrees Fahrenheit, essentially distilling it—an expensive process that would balance the books only during times when oil prices are really high. The search is on, but fitfully, for a recovery method that is both economical and environmentally sound. The area saw a short-lived boom in the early 1980s, when oil prices were high and production was thought to be just around the corner.

Oil and gas wells along the interstate reach down into lower Tertiary and Cretaceous rocks. In 1969, not far from Rulison, an experimental well was drilled over a mile deep into a gas-bearing sandstone of the Mesaverde group. A nuclear bomb was exploded in the bottom of the well to see if it would fracture the rock and cause an increase in release of natural gas. The rock was fractured and gas pressure went up, but gas from the well had become slightly radioactive and couldn't be used. The well is now sealed off, and no drilling is allowed near it.

North of Parachute, a mineral called nahcolite was recovered for a time from the Green River shale. Soluble in hot water, nahcolite was mined by circulating extremely hot water down wells to dissolve it, and then pumping the solution through a pipeline for processing near Parachute. It was a source for soda ash and sodium bicarbonate, to be used in foods, detergent, and the manufacture of glass. The process was not profitable.

Most of the downcutting along the Colorado River accompanied and followed mid-Tertiary regional uplift, which steepened stream gradients and increased the erosional strength of the river and its tributaries. Erosion was also augmented in Pleistocene time when runoff and stream load were many times greater than they are today. Terraces in the valley of the Colorado River indicate stages at which the river stabilized for a time and widened its floodplain before beginning another cycle of downcutting.

West of Parachute, I-70 descends gradually through the lowest part of the Wasatch formation. In De Beque Canyon it drops through stairstep cliffs of the Mesaverde group, coarse floodplain deposits that overlie sand and clay deposited on beaches and bars along the edge of the retreating Cretaceous sea. Shale and coal layers in the Mesaverde group show that muddy lagoons and swamps often festooned the shore, persisting even after the sea retreated. Small springs and seeps are surrounded by bitter white salt deposits brought to the surface by spring water that leached the salts from marine shales below.

Landslides are frequent along the river here, and toppled blocks of sandstone form chaotic debris below freshly scarred cliffs. Erosion of soft shale layers in and below the Mesaverde group undermines the sandstone layers.

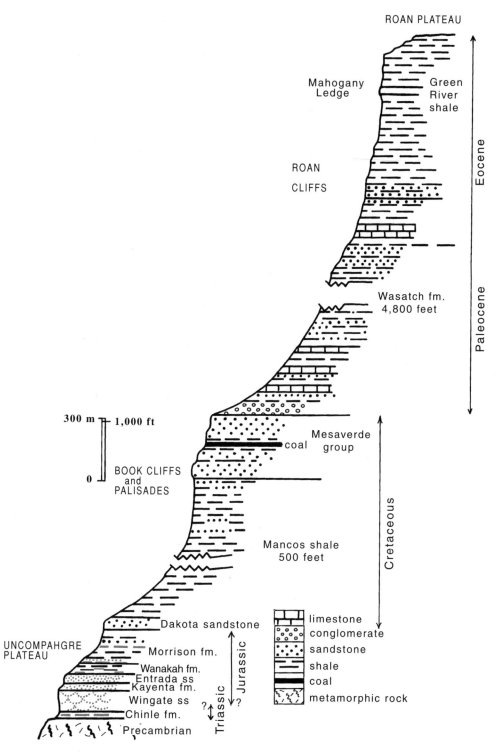

ROAN PLATEAU

Mahogany
Ledge

Green
River
shale

ROAN
CLIFFS

Eocene

Wasatch fm.
4,800 feet

Paleocene

300 m — 1,000 ft

Mesaverde
group

coal

BOOK CLIFFS
and
PALISADES

0

Mancos shale
500 feet

Cretaceous

Dakota sandstone

limestone

conglomerate

UNCOMPAHGRE
PLATEAU

Morrison fm.

sandstone

Wanakah fm.

shale

Entrada ss
Kayenta fm.

Jurassic

coal

Wingate ss

metamorphic rock

?

?

Chinle fm.

Triassic

?

?

Precambrian

Stratigraphic units exposed between Rifle and Grand Junction.

Mt. Callahan, rising 3,600 feet above the Colorado River, is capped by basalt flows probably once continuous with those on Battlement and Grand Mesas. During the 10 million years since the flows formed, the river has cut 3,600 feet downward. —Jack Rathbone photo

Floods formerly swept away the landslide debris, but now that upstream dams control spring floods, the rocky rubble accumulates in the canyon.

Grand Valley Diversion Dam near milepost 50 removes Colorado River water for irrigation. At milepost 44 a large mass of rocky rubble marks a 1950 slide that destroyed a tunnel carrying irrigation water from the dam to orchards near Grand Junction. In an all-out effort the tunnel was rebuilt in nineteen days—in time to save crops and fruit trees.

Several coal seams in the Mesaverde group are thick enough to mine. The mine at Cameo produced coal for the power plant across the river until 1999, when all the profitable coal was mined out. Mesaverde coal is soft, or bituminous, coal with a low sulfur content.

At the mouth of De Beque Canyon, I-70 is well below the base of the Mesaverde group, which caps towering palisades of Mancos shale. The gray shale, yellow where weathered, was deposited on the floor of the shallow

Near milepost 52, a 74-year-old landslide, reactivated by a wet spring, closed I-70 for several days in 1998. Clean, freshly broken rock surfaces and fewer bushes mark the reactivated slide, which lifted the highway 12 feet. —Felicie Williams photo.

Honeycomb weathering, small deep hollows carved by wind, moisture, and frost, frequently appear on Mesaverde group sandstones. —Jack Rathbone photo

Cretaceous sea. It contains clay that swells when wet and shrinks when dry, and the loose, constantly eroding soil that results doesn't support much vegetation. Salt and selenium in the shale add to its uninviting attributes. Where not protected by the Mesaverde caprock, the Mancos shale erodes into humpbacked gray and yellow badlands.

Across the valley to the west, the Uncompahgre Plateau rises as a broad, almost flat-topped faulted anticline. The carved pink cliffs of Triassic-Jurassic Wingate sandstone are in Colorado National Monument. In this area, dark Precambrian rocks appear below the sedimentary layers. A loop road through the national monument, described in the next road guide, rejoins I-70 near the Dinosaur Journey Museum in Fruita.

Grand Junction lies in Grand Valley at the confluence of the Colorado and Gunnison Rivers. The entire valley is underlain by Mancos shale, mostly veneered with terrace deposits. Because of the shale's fine texture and high clay content, the valley has no usable shallow groundwater. Farmers divert irrigation water from the rivers, while most household water comes from reservoirs high on Grand Mesa. Artesian wells supply drinking water for a few areas, mostly west of the rivers. Drilled through Mancos shale into water-bearing Jurassic sandstone, these wells flow without pumping because rain and snowmelt entering the Jurassic aquifer at higher elevations is held in the sandstone by impermeable shale layers above and below it. As the water flows down the sloping sandstone layers, percolating between sand grains, the sandstone, shale, and water well combination works just like a city water system: hydrostatic pressure from a high-up tank or reservoir forces water through pipes and up to faucets on the second or seventh floor.

Along the edge of Grand Valley east and north of Grand Junction, small flat-topped hills of Mancos shale indicate the level of the valley floor during Pleistocene time. The Book Cliffs, with slopes of Mancos shale capped with sandstone of the Mesaverde group, curve westward into Utah. Above them, far in the distance, gray Tertiary strata wall the Roan Plateau. To the west, horizontal strata that top the Uncompahgre Plateau either break off abruptly or curve down dramatically to plunge beneath the valley floor.

The interstate runs close to the Colorado River between mileposts 26 and 16. The river has a low gradient here, and divides and rejoins repeatedly, creating gravel bars and changing its course as it goes.

Near milepost 16, the river curves sharply away from the interstate, plunging into strata that slant down at the north end of the Uncompahgre Plateau. Part of this scenic, little-known canyon is traversed by the Denver & Rio Grande Railroad.

A landslide a mile northwest of the Mack exit, marked by gray soil near I-70, was caused by exploratory trenches made in a coaly layer beneath the Dakota sandstone. The trenches collected surface water, saturating under-

Three fossil Camarasaurus *vertebrae, each several inches across, lie embedded in a sandstone layer in the Morrison formation near exit 2.* —Felicie Williams photo

lying shale, some of which slid out from under the sandstone. Just west of the slide the railroad passes below the highway and enters a side canyon that leads to the Colorado River.

The highway continues northwestward, following a hogback of Dakota sandstone, and then curves around the north end of the Uncompahgre Plateau among cuestas of Jurassic rock, notably some colorful green and purple shales of the Morrison formation. Mountains in view to the southwest from milepost 5 are La Sal Mountains in Utah.

Near exit 2, dinosaur skeletons were found in the Morrison formation in 1981. Displays along a 1.5-mile trail explain the fossil finds and point out geologic features and dinosaur bones.

Colorado National Monument
28-mile (45-km) loop from Grand Junction to Fruita

Deep canyons, sheer shadowed cliffs, and glowing pink sandstone characterize Colorado National Monument's scenery. This is typical Plateau Country landscape, with warm-hued sedimentary rocks in horizontal layers.

The Uncompahgre Plateau, a nearly flat-topped highland about 25 miles wide and 100 miles long, is outlined by faults and monoclines. The flat

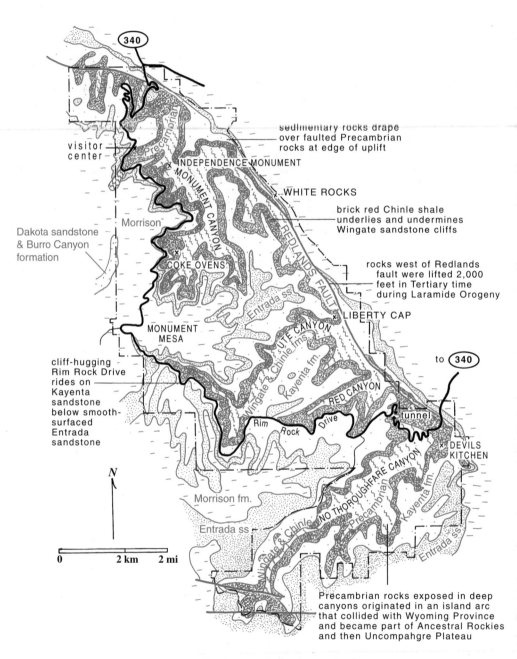

340

sedimentary rocks drape
over faulted Precambrian
rocks at edge of uplift

visitor
center

INDEPENDENCE MONUMENT

Precambrian

MONUMENT CANYON

WHITE ROCKS

brick red Chinle shale
underlies and undermines
Wingate sandstone cliffs

Morrison

REDLANDS FAULT

Dakota sandstone
& Burro Canyon
formation

COKE OVENS

rocks west of Redlands
fault were lifted 2,000
feet in Tertiary time
during Laramide Orogeny

Entrada ss.

LIBERTY CAP

MONUMENT
MESA

UTE CANYON

Wingate & Chinle fms.

cliff-hugging
Rim Rock Drive
rides on
Kayenta
sandstone
below smooth-
surfaced
Entrada
sandstone

Kayenta fm.

to 340

RED CANYON

Rim   Rock   Drive

tunnel

DEVILS
KITCHEN

N

Morrison fm.

NO THOROUGHFARE CANYON

Kayenta fm.

0   2 km   2 mi

Entrada ss

Wingate & Chinle

Precambrian

Entrada ss.

Precambrian rocks exposed in deep
canyons originated in an island arc
that collided with Wyoming Province
and became part of Ancestral Rockies
and then Uncompahgre Plateau

Geology of Colorado National Monument.

Massive Triassic-Jurassic Wingate sandstone of Colorado National Monument bends in a dramatic monocline across the fault that edges the Uncompahgre Plateau. Precambrian rocks appear in the dark gray hill at the far left. —Halka Chronic photo

surface of the dark Precambrian metamorphic rocks that form its base is visible in the national monument below the layered Mesozoic strata of its cliffs. At both the north and south ends of the monument road (Rim Rock Drive), the sedimentary rocks drape blanketlike over the fault that edges the plateau, bending rather than breaking. But between the two access routes, uplift broke the Mesozoic strata along the Redlands fault, and the 800 feet of displacement there created a sharp cliff face.

Entering the monument from either end, Rim Rock Drive crosses a Dakota sandstone cuesta, bands of rainbow-hued Morrison shale, and a little red and pink Jurassic and Triassic sandstone and shale. Then abruptly it crosses the fault that edges the uplift, where the dark reddish gray gneiss of its Precambrian base floors canyons and forms scrub-covered slopes below the high pink cliffs.

Like Precambrian rock in the Black Canyon of the Gunnison and in the cores of most of Colorado's ranges, the gneiss started out about 1.8 billion years ago as sedimentary and volcanic rock piled up along the edge of the Wyoming Province. At least three times since then it has formed the roots of mountain systems. It has been broken, folded, partly melted, recrystallized, intruded by granite, and eroded over and over again to the point where its story is hard to decipher. Walking trails within the canyons provide good close-up views of this ancient rock.

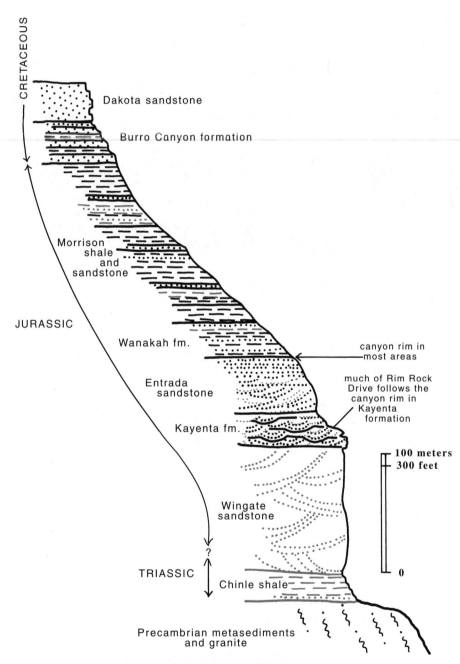

CRETACEOUS

Dakota sandstone

Burro Canyon formation

Morrison shale and sandstone

JURASSIC

Wanakah fm.

Entrada sandstone

Kayenta fm.

Wingate sandstone

?

TRIASSIC

Chinle shale

Precambrian metasediments and granite

canyon rim in most areas

much of Rim Rock Drive follows the canyon rim in Kayenta formation

100 meters
300 feet

0

Stratigraphic units in Colorado National Monument.

Horizontal beds of Wingate sandstone reveal the large-scale crossbedding characteristic of sand dune deposits. At the base of the cliff is a pile of bush- and tree-size sandstone boulders, tumbled from its face.
—Felicie Williams photo

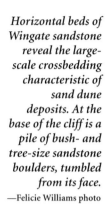

The upper surface of the Precambrian gneiss is thought to have been smoothed off during a 600-million-year erosion interval at end of the Precambrian era. Then, as part of the Ancestral Rockies Orogeny, a mountain range known as Uncompahgria was faulted up here in Pennsylvanian time. For about 100 million years, erosion scoured away at the range. Dark iron-bearing minerals from the Precambrian gneiss became the source of iron oxides that abundantly tint Mesozoic rocks of the Plateau Country.

Triassic sand and mud of a broad delta or floodplain were deposited in alternating layers on the newly smoothed surface and now form the red Triassic Chinle formation at the base of the cliffs. Later, as deserts crept across the region, a thick layer of sand accumulated and cemented into the Wingate sandstone of the pink cliffs. An old road, the Serpents Trail, now open to hikers, is a good place to see this formation. Large-scale crossbedding and rounded, pitted sand grains of uniform size identify the formation as dune sand. A dearth of fossils makes it hard to tell its exact age, but dinosaur footprints in the lower part of the Wingate appear to be Triassic, while those on its upper surface resemble Jurassic prints found elsewhere.

The ledges above the Wingate precipices are the Kayenta formation, deposited in river floodplains and hardened by the addition of silica carried in groundwater. The resistant rock forms protective caps on cliffs, buttes, and pinnacles, and it surfaces many overlooks. It contains scoured out and refilled channels and short, steep torrent-caused diagonal laminations.

The smoothly rounded salmon pink cliff above it, and above the road for much of Rim Rock Drive, is part of the Entrada sandstone, another dune sandstone that can be recognized all over southwestern Colorado and southeastern Utah. A short hike from the visitor center leads into a cool, dark box canyon in Entrada sandstone.

Set back above the rim in the southern part of the monument, marine shales and sands of the Wanakah formation separate colorful Morrison shales and sandstones from the Entrada sandstone. (Wanakah beds are included with the Morrison formation on the map.) In 1900, Elmer Riggs of the Field Museum in Chicago dug the bones of a new kind of dinosaur, the largest found at the time, *Brachiosaurus altithorax*, from the Morrison formation a mile east of the monument boundary. At that time there were no nearby bridges across the Colorado River; Riggs had to ferry his precious bones across the river by barge!

Since then, many other fossils have been discovered in the region, including *Stegosaurus, Apatosaurus, Camarasaurus, Brachiosaurus*, and a new member of the Ankylosaur family. A bone excavated from Morrison rocks 30 miles southeast of here came from a *Brachiosaurus* as tall as a six-story building. Remains of smaller dinosaurs, lizards, crustaceans, early mammals, fish, amphibians, and plants have also been found in this area. In addition to the museum at the monument visitor center, Dinosaur Journey Museum in Fruita has displays about recent discoveries—and there are new ones every year.

The Burro Canyon formation, stream-deposited coarse sediments and shale, and the Dakota sandstone, an ocean beach deposit, top the ridge west of the park.

Look eastward across the valley to the Book Cliffs. Sandstones of the Cretaceous Mesaverde group, which cap them, are at approximately the same elevation as the surface of the Entrada sandstone here. This gives a measure of the uplift of Colorado National Monument and the Uncompahgre Plateau: normally 600 feet of Wanakah and Morrison shale, 100 feet of Burro Canyon formation, 100 feet of Dakota sandstone, and 4,500 feet of Mancos shale separate the Entrada sandstone from the Mesaverde sandstone. Added together, these figures show that total uplift of the Uncompahgre Plateau relative to the Book Cliffs is in the neighborhood of 5,300 feet.

This sum is also a measure of erosion. Colorado was fairly featureless at the end of Eocene time, with the Uncompahgre uplift buried in its own debris, and Lake Uinta, between the uplift and the Rockies, filled with mud, sand, and organic matter. Regional uplift in mid-Tertiary time increased the energy of the rivers, resulting in excavation of Grand Valley. Glacial meltwater of Pleistocene time deposited outwash gravel on the valley floor,

In early 2000, a sudden rockfall in the Salt Wash member of the Morrison formation blocked Rim Rock Road. Piles of sharply angled boulders below the cliffs show that such rockfalls are fairly common. —Felicie Williams photo

now terraces that mark Ice Age floodplain levels. In this dry climate, melting winter snow and sudden summer cloudbursts continue to wash away loose sediment and cut downward into solid rock.

More subtle processes also change the rocks. Sandstone and shale, perhaps surprisingly, are rarely dry inside. When moisture between sand and shale grains freezes and expands—a nightly occurrence for about half of every year—it loosens grains and small pieces of rock. Moisture also weakens rock by dissolving the cement between grains, though it may harden rock surfaces when evaporation draws dissolved cement toward the surface.

Some rocks acquire a patina of dark iron and manganese oxides called desert varnish. Shiny black streaks on vertical cliff faces are desert varnish; duller ones are algae and mosses living on moisture that trickles down from above. Plants wear away rock by utilizing mineral nutrients between the grains, or by prying rocks apart with their roots. Sandblasting by strong desert wind knocks sand grains free as it swirls among the canyons, sometimes undercutting cliff faces and pinnacles.

Through all these processes, softer shaly layers erode more quickly, leaving harder sandstones above them unsupported. Eventually, overhanging rock breaks loose in thundering rockfalls, contributing to the shaping of the tall pinnacles and steep-sided chasms of Colorado National Monument.

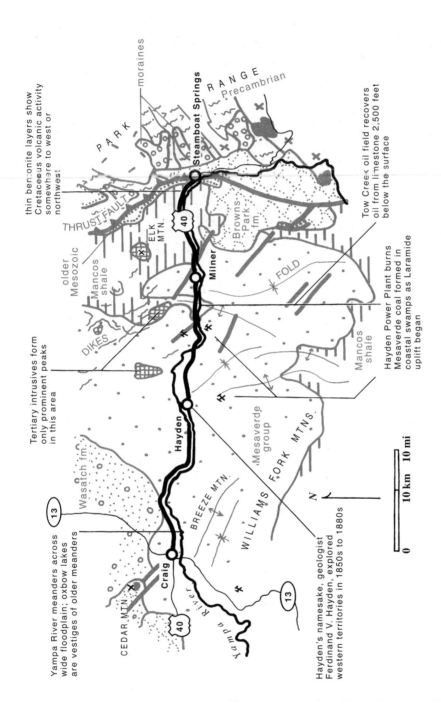

thin bentonite layers show
Cretaceous volcanic activity
somewhere to west or
northwest

Tertiary intrusives form
only prominent peaks
in this area

Yampa River meanders across
wide floodplain; oxbow lakes
are vestiges of older meanders

Hayden's namesake, geologist
Ferdinand V. Hayden, explored
western territories in 1850s to 1880s

Tow Creek oil field recovers
oil from limestone 2,500 feet
below the surface

Hayden Power Plant burns
Mesaverde coal formed in
coastal swamps as Laramide
uplift began

PARK

moraines

RANGE
Precambrian

Steamboat Springs

THRUST FAULT

older
Mesozoic

Mancos
shale

ELK
MTN.

DIKES

40

Milner

Browns-
Park fm.

FOLD

Mancos
shale

Wasatch fm.

Hayden

Mesaverde
group

BREEZE MTN.

WILLIAMS FORK MTNS.

13

CEDAR MTN.

Craig

40

13

Yampa River

N

0   10 km   10 mi

Geology along U.S. 40 between Steamboat Springs and Craig.

U.S. 40
Steamboat Springs—Craig
43 miles (69 km)

Hot springs at Steamboat Springs supply the swimming pool at the east end of town; several also bubble up in a city park west of the downtown area. Water in these springs, heated by contact with hot rocks deep below the surface, rises rapidly along the fault zone that outlines the western edge of the Park Range. Scattered patches of gray travertine deposited by hot springs are visible along the base of Quarry Mountain southwest of town.

Parts of Steamboat Springs lie on stream-deposited glacial gravel or on the Browns Park formation, deposited in Miocene and Pliocene time as a mixture of dune sand, volcanic ash, silt, and pebbly stream-deposited gravel. West of town, the valley of the Yampa River is underlain by Mancos shale dipping west off the flank of the Park Range. The fine gray shale, well exposed in river and highway cuts, contains Cretaceous marine fossils; its thin orange seams are bentonite clay layers formed by gradual chemical change in layers of volcanic ash.

The low peak north of the highway, Elk Mountain, is a small Tertiary igneous intrusion. Several similar intrusions and many dikes occur north and west of it, a few of them perhaps representing the plumbing for volcanoes that contributed volcanic ash to the Browns Park formation. However, we have no way of telling whether they ever reached through to the surface.

Near Milner, the lower part of the Mesaverde group forms bluffs north and south of the valley. Formerly called the Mesaverde sandstone, this unit has been divided into several formations, with the old name retained for the group. Widespread in western Colorado and adjacent states, the Mesaverde group was deposited near the shore of the Cretaceous sea as it retreated eastward; therefore the group as a whole becomes somewhat older

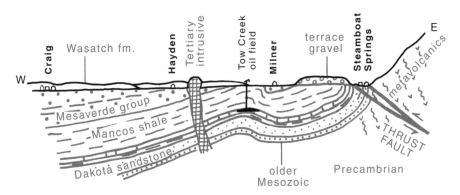

Section along U.S. 40 between Steamboat Springs and Craig.

westward. The sea backed away gradually, with millions of years passing before it completely departed from Colorado.

Mesaverde sandstones can be recognized by their light tan color and beach sand texture. They tend to develop deep, fist-size cavities as they weather; when the cavities are numerous and close together, with only thin walls between them, they are known as honeycomb weathering. Watch for these features in bluffs near the highway. In some places the massive sandstone layers break away to form large, shallow caves.

Coal layers in the Mesaverde group are rarely exposed at the surface because coal breaks down rapidly as it weathers. Some coal is visible in roadcuts, however. The coal formed, over time, from the abundant plant material that grew in swamps and marshes similar to those bordering Atlantic and Gulf Coast states today. Several underground and open-pit mines in this area recover coal from seams 3 to 20 feet thick. Some is shipped to other markets, and some is used locally to generate electricity.

A little west of Milner, U.S. 40 crosses a low but well-defined anticline. Watch the directions of dip in highway cuts to spot the crest of the anticline, where rocks that are dipping (sloping downward) toward the east level off and then dip west. Coal is mined along this anticline, and oil wells at its crest take advantage of oil's tendency to rise through porous rock, in this case fractured Cretaceous limestone about 2,500 feet below the surface. As the limestone arches in the anticline, it forms a natural trap for oil, which can't escape upward through overlying nonporous shale. Some deeper wells in this field reached the Dakota sandstone, but they hit hot water instead of oil. Here, as at Steamboat Springs, temperatures increase unusually rapidly with depth.

On the western side of the anticline the highway crosses younger and younger layers of the Mesaverde group. Slope-forming shales alternate with cliff-forming sandstone. More coal mines appear, many of them are new open-pit mines on sites of older underground workings. Mines are required by law to backfill and revegetate open-pit areas after the coal has been removed, so some of them are difficult to spot.

A double peak on the skyline far to the north, known as Bears Ears, is a late Tertiary intrusion. To the south, horizontal basalt flows of about the same age top the Flat Tops on the White River uplift, a broad dome-shaped anticline that forms a westward extension of the Rocky Mountains. The uplift is transitional between eastern faulted-anticline ranges and the western plateaus.

A big power plant near Hayden burns coal from a mine a few miles south. The town was named for Ferdinand V. Hayden, a pioneer geologist who explored and mapped western Colorado in 1869 and 1870. His *Geological*

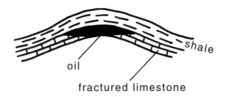

Unable to escape upward through impermeable shale, oil and gas collect in upside-down pools in a variety of geologic traps.

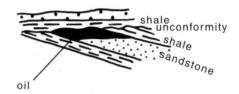

and Geographical Atlas of Colorado, published in 1877, still makes fascinating reading.

West of Hayden, flat-lying light brown shale and sandstone appear above the Mesaverde group. The lower 520 feet or so of these rocks are Cretaceous marine deposits, but upper parts contain Tertiary plant fossils and are part of the nonmarine Wasatch formation, stream-sorted sand, silt, and clay washed off the Park Range and the White River uplift. At Craig the Wasatch formation is more than a thousand feet thick. Rounded, knobby concretions in the sandstones formed when groundwater rich in minerals flowed through the sediments and deposited some of its minerals around small bits of wood, decaying leaves, or fossil shells. Once the concretions started to form, more dissolved minerals adhered to them and they grew gradually to the size you see here.

A Tertiary intrusion southeast of Craig lifted sedimentary strata, including coal-bearing beds of the Mesaverde group, into a dome known to geologists as Breeze anticline or Craig dome. Though they aren't the predominant rocks in this area, the few signs of Tertiary igneous activity here remind us that in southwestern Colorado in Tertiary time, numerous large, explosive volcanoes created vast, high-piled tiers of lava and volcanic tuff—today's West Elk and San Juan Mountains.

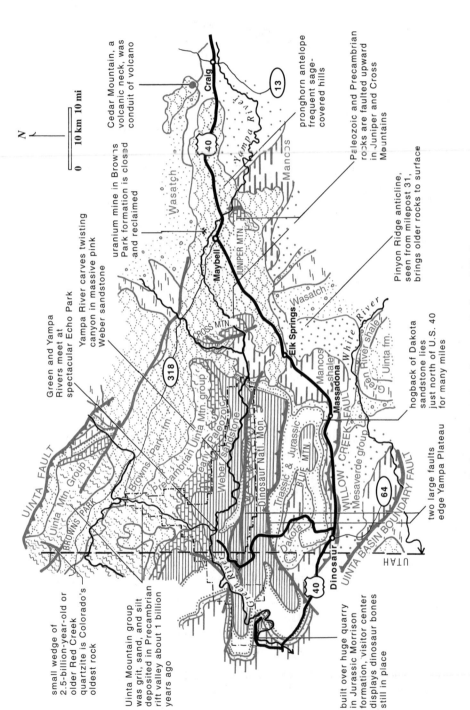

Geology along U.S. 40 between Craig and Utah.

N

0   10 km   10 mi

Cedar Mountain, a volcanic neck, was conduit of volcano

pronghorn antelope frequent sage-covered hills

Paleozoic and Precambrian rocks are faulted upward in Juniper and Cross Mountains

uranium mine in Browns Park formation is closed and reclaimed

Green and Yampa Rivers meet at spectacular Echo Park

Yampa River carves twisting canyon in massive pink Weber sandstone

Pinyon Ridge anticline, seen from milepost 31, brings older rocks to surface

hogback of Dakota sandstone lies just north of U.S. 40 for many miles

small wedge of 2.5-billion-year-old or older Red Creek quartzite is Colorado's oldest rock

Uinta Mountain group was grit, sand, and silt deposited in Precambrian rift valley about 1 billion years ago

built over huge quarry in Jurassic Morrison formation; visitor center displays dinosaur bones still in place

two large faults edge Yampa Plateau

Craig

13

40

Maybell

JUNIPER MTN.

Wasatch

Mancos

Yampa River

CROSS MTN.

Elk Springs

Massadona

White River

Mancos shale

Green River shale

Uinta fm.

318

Browns Park fm.

Precambrian Uinta Mtn. group

early Paleozoic

Weber sandstone

Dinosaur Natl. Mon.

Triassic & Jurassic

BLUE MTN.

Mesaverde group

Dakota

WILLOW CREEK FAULT

UINTA BASIN BOUNDARY FAULT

UTAH

Dinosaur

40

64

UINTA FAULT

BROWNS PARK

Yampa River

Wasatch

U.S. 40
Craig—Utah
90 miles (145 km)

Except for Cedar Mountain, a light-colored Tertiary volcanic plug north-west of town, Craig is surrounded by Cretaceous sedimentary rocks. Sand-stone and shale of the Mesaverde group underlie hills to the north and south. An open-pit mine visible southwest of Craig obtains coal from the Mesaverde group, fueling the nearby power plant.

Looking back at the Park Range you can see the horizontal surface of the Tertiary pediment quite well. North and Middle Parks, as well as all this area west of the Park Range, were filled with Miocene-Pliocene deposits nearly to the level of that pediment. Since then, the Yampa River and its tributaries have removed vast quantities of these poorly consolidated de-posits and swept them westward into basins between the ranges.

At about milepost 82, U.S. 40 passes onto some of the remaining Miocene-Pliocene sediments, white crossbedded sandstone and yellowish siltstone of the Browns Park formation. These rocks contain a high proportion of volcanic ash, suggesting that at least some of the small intrusions east and north of here may have served as the conduits of explosive volca-noes. However, the volcanic ash may have come from eruptions in the West Elk and San Juan Mountains farther south, or from other volcanic centers farther west.

In many places the Browns Park formation is coated with desert pave-ment of closely packed, rounded pebbles winnowed from the sediments as wind blows away the finer sand, silt, and clay in which they are embedded. Many of the pebbles consist of Precambrian igneous and metamorphic rocks from the Rocky Mountains to the east, carried here by Miocene and Pliocene streams. Some large-scale crossbedding in the Browns Park for-mation indicates that parts of this unit originated as dunes, with layers of clay representing interdune areas.

At milepost 66, Juniper Mountain appears to the south, jutting up through the Browns Park formation. This little range, which contains both Paleozoic and Precambrian rocks, tells us that the basic structure of the Uinta uplift farther west may extend at least this far east. The Yampa River flows right through the northern end of Juniper Mountain near milepost 64, maintaining a course it had established when the mountain was cov-ered completely with sediments of the Browns Park formation. The river's meandering course upstream from Juniper Mountain may be due to par-tial damming by the hard rocks of the range.

Cretaceous strata are faulted upward in a sharp ridge northwest of Maybell. At Maybell, U.S. 40 turns southwestward to avoid the rugged terrain of the Uinta uplift, basically a large faulted anticline that extends over a hundred miles westward, getting higher and higher as it goes. East-trending faults along the north edge of the uplift, just north of the Wyoming state line, mark the edge of the Wyoming Province, the ancient continent of 2.5 billion years ago. Faults along both the northern and southern edges of the Uinta uplift date from Proterozoic time, between 1.4 billion and 950 million years ago, when the area between them dropped downward, forming a trough that filled with sediments now known as the Uinta Mountain group. Since then, movement along the same faults has continued, alternately lifting and lowering the area between them. During the Laramide Orogeny, as the Colorado Plateau pushed northward toward the old province, the strip between the faults split lengthwise into several long, east-west blocks as it was thrust upward over adjacent rocks, forming the Uinta uplift.

In the Maybell area, groundwater in the Browns Park formation contains tiny amounts of uranium leached from volcanic ash, enough so that it is not safe to drink. Under favorable chemical conditions, groundwater deposited some of its uranium, and for a time several uranium mines were active. After they closed, mine and mill tailings were allowed to blow in the wind, eventually covering 182 acres of land with a thin veneer of radioactive waste. In 1995 and 1996, the waste was collected and buried under a cement and gravel seal.

West of Maybell, prominent crossbedding in the Browns Park formation indicates that this part of the formation was once a sand dune area. Poorly consolidated, as are most Tertiary deposits, some of the sandstone is now weathering back into sand and forming new dunes.

Gray shales and sandstones of the Wasatch and Green River formations form hills to the south. The Wasatch formation consists of fine silt and sand washed from rising highlands in early Eocene time. The Green River formation above it accumulated as sand and fine mud on the bottom of ancient Lake Uinta, which in Eocene time filled an irregular basin between uplifts in northwestern Colorado, southwestern Wyoming, and northeastern Utah. Fossil fish, crocodiles, insects, and plants occur in the Green River formation. It is also known for its rich oil shale deposits, an immense and as yet untapped energy reserve.

Cross Mountain, the forested range north of milepost 45, lies at a right angle to the east-west trend of the Uinta uplift. Stop at the wide turnout at milepost 41 for a good view of the range and its surroundings. Similar in structure, though not in size, to the faulted anticlines that make up the main ranges of the Colorado Rockies, Cross Mountain retains the Paleozoic rocks that arch across it. Notice the gray cliff of Mississippian Leadville limestone curving over its Precambrian core.

At Cross Mountain the Yampa River again follows an unexpected course, boldly cutting westward through the range. The river then tackles the Uinta uplift, carving a deep and beautiful canyon into its very heart. The notch of its canyon is visible from the milepost 41 turnout, almost lost amidst miles of bare rock. All of the low area through which the river flows west of Cross Mountain is an east-west block of the Uinta uplift that was not raised as high as its neighbors to the north and south; most of it is surfaced with massive, pale salmon-colored Weber sandstone. This is a dune-deposited sandstone that contains no fossils but is considered Pennsylvanian to Permian in age because it is sandwiched between Pennsylvanian marine shales and strata known to be Permian and Triassic. Geologists believe the course of the Yampa River was established while the whole region was deeply buried in Tertiary sediments, and that when the land rose during late Tertiary time, the invigorated river remained trapped in its original channel despite encounters with Juniper Mountain, Cross Mountain, and the Uinta uplift.

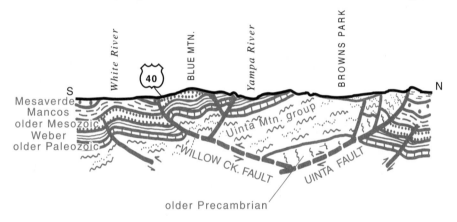

North-south section across U.S. 40 and Dinosaur National Monument.

South of Elk Springs, U.S. 40 leaves the Browns Park formation, and between here and Utah runs on Mancos shale, part of the band of Mesozoic strata that bend up toward the southern block of the Uinta uplift. The geology is much simpler than it looks: Blue Mountain, north of the highway, is half of a faulted anticline; Paleozoic and Mesozoic strata tip up against it and then bend to horizontal at its summit.

Between the highway and the mountain front, the strata become older and older. A hogback of Dakota sandstone forms the ridge just north of the highway. Up the little stream valleys that cut through the hogback are colorful green and purple shales of the Jurassic Morrison formation, then red Triassic shale and sandstone beyond them, and finally the pale pink mountain flanks of Weber sandstone. At its eastern and western ends the mountain block is surfaced with Weber sandstone; near the summit, out of sight from the highway, erosion has bared some older Paleozoic strata.

Erosion is responsible for the madly jumbled topography between the mountain and the highway, where harder, more resistant layers cap softer layers in innumerable cuestas, hogbacks, and flatirons as the strata steepen toward the mountain.

South and west of Massadona, erosion controls topography and even vegetation. Cretaceous sandstone layers are interlayered with shale and form rows of hogbacks that partly hide Blue Mountain. In this arid western country, a true desert deprived of Pacific moisture by mountains to the west, erosion is governed as much by wind as by water, and many of the minivalleys between the rows of hogbacks show no sign of stream channels. Assisted on steep slopes by gravity, wind has cleaned off the Weber sandstone surfaces of Blue Mountain and the Yampa Canyon country, as you will see if you drive north into the Colorado portion of Dinosaur National Monument. Wind is helped by occasional rain and flood, and in moist shaded areas by frost that loosens small particles of rock.

Upturned Mesozoic sedimentary rocks form rows of angular flatirons on the south flank of the Uinta uplift. The same rocks form the horizontal bands on top of the mountain.
—W. R. Hansen photo, courtesy of U.S. Geological Survey

As the highway converges with Blue Mountain, chevron-shaped flatirons of Weber sandstone lie against the flanks of the mountain. Though you can't see it, in this area the highway runs right along a large thrust fault where the Blue Mountain fault block pushed southward across Cretaceous rocks.

In the town of Dinosaur, streets are named for the ancient animals that gave the town its name. Between Dinosaur and the Utah line, rainbow hues of Jurassic and Triassic rocks appear between sculptured slopes of Dakota sandstone.

Dinosaur National Monument
For map and section see pages 316 and 319

Dinosaur National Monument was established in 1915 to preserve a famous dinosaur fossil locality in Utah. The monument extends into Colorado and includes the magnificent canyons of the Yampa and Green Rivers. It also includes Harpers Corner Scenic Drive, which climbs onto the Uinta uplift with vistas overlooking these canyons and the rivers' confluence.

Branching from U.S. 40 at milepost 5, Harpers Corner Scenic Drive passes through an eroded hogback of Dakota sandstone and crosses a narrow valley of Jurassic sedimentary rocks—the Morrison formation, tinted with green, purple, and red. The dinosaur fossil quarry is in the same Jurassic rocks about 25 miles west of here in the Utah section of the national monument. There, the bones of many species have been found, along with remains of other animals and plants. Hundreds of the great beasts apparently died together, perhaps during a drought. Later floods may have washed their bodies a short distance onto a sand bar. Many skeletons are almost complete, with skulls intact, a rarity with fossil vertebrates of any kind.

The road climbs again onto the Dakota sandstone, then at the top of the climb crosses a major east-west fault—the Yampa fault—marked by the abrupt appearance of salmon-colored Weber sandstone, the same Pennsylvanian-Permian rock visible atop Blue Mountain farther east. Look closely at the thick sandstone layers, with their broad, sweeping crossbedding, and at the red shale and siltstone beds that lie between them. These rocks were deposited as sand dunes, with reddish shales representing mudflats or interdune areas.

At either Iron Springs Bench overlook or Echo Park overlook, gaze down nearly 3,000 feet to the meeting of the Green and Yampa Rivers at Echo Park. Eastward, the Yampa fault along the north side of Blue Mountain adds a thousand feet to the depth of Yampa Canyon. Precambrian to Pennsylvanian rocks south of the fault have been thrust upward and northward

The canyon of the Yampa River cuts a ragged gash in the surface of the Weber sandstone. Its winding course dates back to a time when the river flowed across a plain of Tertiary sediments high above its present canyon. The Yampa joins the Green River behind Steamboat Rock, at the bottom of the photo. Upfaulted Blue Mountain is to the right. —W. B. Cashion photo, courtesy of U.S. Geological Survey

over Permian and Triassic rocks north of it. The steep face of Blue Mountain is mostly Weber sandstone, and the folds in this massive rock, which is at least 1,000 feet thick, are dramatic evidence of slow but powerful stresses within Earth's crust. Small side canyons cut into some of the older rocks.

North of the fault, in the low central segment of the Uinta uplift, the Weber sandstone dips gently southward. A smaller fault separates Iron Springs Bench from the lower cliffs.

The canyon of the Green River is also visible from these overlooks. It was named the Canyon of Lodore in 1869 by Major John Wesley Powell, a geologist and ethnographer who, despite the loss of an arm during the Civil War, led a daring survey trip down the uncharted waters of the Green and Colorado Rivers and the Grand Canyon. He later helped found the U.S. Geological Survey, of which he was the second director.

During his river trip, Powell and his survey party entered Colorado on the Green River, the only sizable river that flows *into* the state. (All of Colorado's other major rivers flow *out of* this highest state on the continent.) The party rowed and floated southward through Browns Park and

Hardened and firmly cemented by time, Precambrian metasediments of the Canyon of Lodore are sandstone, siltstone, and conglomerate that accumulated in a down-faulted graben that existed between 1.4 billion and 950 million years ago. Faults that edged the graben were reactivated during the Laramide Orogeny and now border the Uinta uplift.
—W. R. Hansen photo, courtesy of U.S. Geological Survey

the Canyon of Lodore, passing the confluence of the Yampa River at Echo Park and leaving the state just 20 miles south of their entry point.

Farther west are the high summits of the Uinta Mountains of Utah, the highest part of the Uinta uplift. Along the northern edge of these mountains, a small patch of Red Creek quartzite, over 2.3 billion years old, extends into Colorado. These metamorphic rocks, the state's oldest, are part of the Wyoming Province, the Precambrian nucleus of our continent.

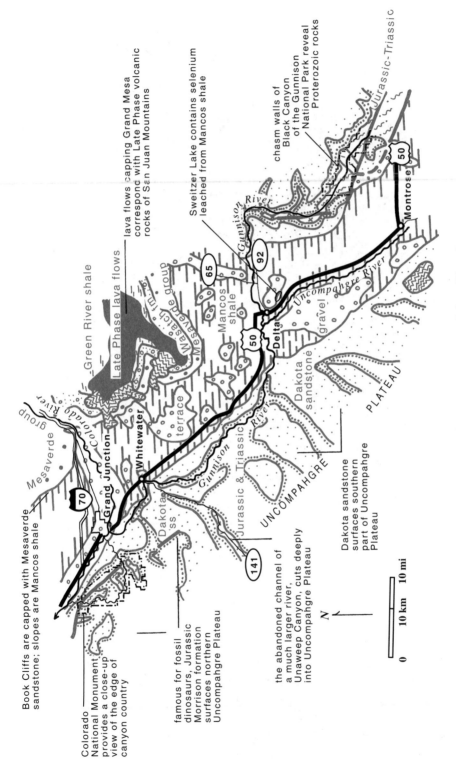

Book Cliffs are capped with Mesaverde sandstone; slopes are Mancos shale

lava flows capping Grand Mesa correspond with Late Phase volcanic rocks of San Juan Mountains

Sweitzer Lake contains selenium leached from Mancos shale

chasm walls of Black Canyon of the Gunnison National Park reveal Proterozoic rocks

Green River shale

Late Phase lava flows

Mesaverde group

Mesaverde group terrace

Mancos shale

Colorado River

Whitewater

Grand Junction

Gunnison River

Dakota ss

Jurassic & Triassic River

Dakota sandstone

Uncompahgre River

Delta

Montrose

Jurassic-Triassic

UNCOMPAHGRE

PLATEAU

Dakota sandstone surfaces southern part of Uncompahgre Plateau

the abandoned channel of a much larger river, Unaweep Canyon, cuts deeply into Uncompahgre Plateau

famous for fossil dinosaurs, Jurassic Morrison formation surfaces northern Uncompahgre Plateau

Colorado National Monument provides a close-up view of the edge of canyon country

70

141

65

92

50

50

50

N

0   10 km   10 mi

Geology along U.S. 50 between Montrose and Grand Junction.

U.S. 50
Montrose—Grand Junction
61 miles (98 km)

Montrose lies on the Uncompahgre River's floodplain, surrounded by terraces that mark the valley's level during the Ice Age, when rocks excavated by glaciers in the San Juan Mountains were carried downstream by glacial meltwater.

The Uncompahgre Plateau forms the western skyline. It is underlain by a large block of Precambrian rock faulted along both sides and pushed up nearly 7,000 feet during the Laramide Orogeny. Jurassic and Cretaceous sedimentary rocks above the Precambrian rocks either drape over the faults in monoclines, as they do near Montrose, or are broken along the faults, as they are farther north near Grand Junction. Steep dark cliffs mark the edge of another upfaulted block northeast of Montrose.

Grand Mesa, in the distance straight north, is capped by late Tertiary basalt flows that protect poorly consolidated, less resistant early Tertiary rocks on which they lie. The Green River shale just below the basalt was deposited in a large Eocene lake between Laramide uplifts. Below it, to about halfway down the slope, is the Wasatch formation, fine silt and sand deposited by streams draining the uplifts as they developed. Cretaceous sandstone and shale of the Mesaverde group underlie these poorly consolidated rocks and form the small cliff about halfway up the slope; they include sand, clay, and coal deposited along the shoreline of the receding Cretaceous sea.

Below the Mesaverde group are barren yellow and gray slopes of Mancos shale, a marine deposit also Cretaceous in age. Clay in the Mancos shale swells when it gets wet and shrinks as it dries; that and its high salt content discourage vegetation. Signs at Sweitzer Lake south of Delta warn of high selenium in the water. Mancos shale also underlies the fertile floodplain deposits of the Uncompahgre and Gunnison Rivers; it is well exposed close to U.S. 50 a few miles north of Delta.

The gently sloping surface between Grand Mesa and the highway is a pediment, an erosion surface that planes across the bedding of the sedimentary rocks. Rapid downcutting by the Uncompahgre and Gunnison Rivers, coupled with a desert climate and the soft shale of the valley floor, have made erosion the ruling force in shaping landforms here. The mountains are not surrounded by broad alluvial fans like those bordering the San Luis or Arkansas Valleys, where deposition and valley-filling are the rule.

The Uncompahgre joins the Gunnison River at Delta. The Gunnison has the larger flow, so its name is retained north to Grand Junction, where it joins the Colorado. The Colorado River was originally called the Grand

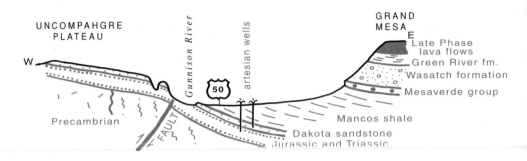

UNCOMPAHGRE
PLATEAU

Gunnison River

artesian wells

GRAND
MESA

E

Late Phase
lava flows

Green River fm.

Wasatch formation

Mesaverde group

W

50

Precambrian

FAULT

Mancos shale

Dakota sandstone

Jurassic and Triassic

Section across U.S. 50 southwest of Grand Junction.

upstream from its junction with the Green River in Utah, but the state of Colorado persuaded Congress to change its name to Colorado River (as used in Arizona and Utah) since its source is in the high ranges of this state.

North of Delta, dark boulders of basalt are strewn over yellow and gray hills of Mancos shale. These boulders either tumbled from Grand Mesa when it was larger, settled as the soft shales were eroded out from under them, or were carried down by mudflows as violent storms swept the area. The basalt is dark reddish or almost black, full of little round vesicles created by gas bubbles in the molten lava. Many vesicles are filled with white zeolite crystals. Basalt lava is particularly fluid and flows rapidly from cracks and fissures to spread into nearly horizontal layers. The many thin basalt flows that cap Grand Mesa erupted around 10 million years ago, a last gasp of the volcanism that swept western Colorado in Tertiary time. The basalt magma rose from the mantle through east-west faults seemingly associated with the rifting that created the San Luis Valley and the Rio Grande Rift farther to the southeast.

U.S. 50 gradually approaches the base of the Uncompahgre Plateau and north of milepost 54 comes close to the cuesta of Dakota sandstone that outlines the uplift. A beach sand deposited just before Cretaceous seas covered the area, the Dakota steepens along the edge of the plateau, then flattens out to form its surface. Streams that cut into the plateau expose older Jurassic shale and sandstone. Sheer cliffs in some of the deep canyons are in the bright pink Wingate sandstone, a Triassic-Jurassic formation made of windblown sands, once tall desert dunes.

A line of small brown limonite-rich mounds dot the Mancos shale east of the road between mileposts 51 and 42. Within each mound are clusters

Horizontal basalt flows that cap Grand Mesa, as well as flat terraces in the Mancos shale below, reflect the near-horizontal terrain characteristic of the Plateau Country. —Felicie Williams photo

of calcite and dolomite crystals, some as much as 2 inches long, as well as occasional fossil shellfish. A temporary change in the composition of seawater, with elevated iron and carbonate content, may have led to their development at just this one level in the shale.

Landslides mark the edge of Grand Mesa, with soft Tertiary shale contributing both slide material and slippery-when-wet skid surfaces.

Just south of Whitewater a prominent canyon shows up to the west, carved deeply into Mesozoic rocks of the Uncompahgre Plateau. This is Unaweep Canyon, occupied today by two minuscule streams flowing in opposite directions from an inconspicuous divide on the canyon floor. Slicing through 1,500 feet of sedimentary rocks and, in the heart of the plateau, 1,000 feet of hard Precambrian granite and metamorphic rocks, Unaweep Canyon is impressive even in this land of canyons. Geologists agree that a large river once flowed through it, either the Gunnison alone or the combined Gunnison and Colorado. Sometime late in Tertiary time the river altered course to flow northwest through softer sedimentary rocks. For the full story, see **Colorado 141: Whitewater—Naturita.**

Built on alluvium and soft, clayey shale that dips toward the river, this house west of Grand Junction had foundation problems even before it was completed in 1996. By 2000, an area larger than a football field had slumped. The Colorado River contributes to the problem by washing away the toes of the slump. —Felicie Williams photo

The highway climbs onto the pediment surface north of Whitewater, and two sets of cliffs come into view ahead: the lower Book Cliffs of Mancos shale capped by Mesaverde sandstone, and the upper, more distant Roan Cliffs, edging the Roan Plateau, formed of grayish pink and light gray Tertiary shale and sandstone of the Wasatch and Green River formations.

Near Grand Junction, water from the Colorado and Gunnison Rivers irrigates farms, orchards, and vineyards in Grand Valley. Most household water is piped from reservoirs on top of Grand Mesa. Artesian wells drilled in the 1930s and 1940s along the western side of the valley tap Jurassic sandstone layers that bend down along the edge of the plateau and flatten out under Grand Valley. Confined between layers of impermeable shale, the water flows down along the slanting sandstone beds, developing hydrostatic pressure as it drops below the level at which it entered the rock. These wells flow without pumping and have been doing so for more than half a century.

From the late 1960s into the 1980s Grand Junction was a major uranium center with U.S. Department of Energy offices, uranium mills, and headquarters of many uranium companies. The ore came from Permian, Triassic, and Jurassic strata in the western part of the Uncompahgre Plateau. During the uranium boom, mill tailings left after the uranium was removed were used as fill dirt around homes, college buildings, and even the police department. Sidewalks on Main Street were poured on a bed of tailings, and miles and miles of city water pipes were buried in trenches filled with them. By the time the mills shut down, people joked that Grand Junction glowed in the dark! Around 4,000 individual sites in the valley have since been cleaned up. The Department of Energy now bases a radioactive waste cleanup team here.

Near Grand Junction the Wingate formation stands as the high cliffs of Colorado National Monument.

U.S. 160
Pagosa Springs—Durango
62 miles (100 km)

Pagosa's hot springs—a large morning-glory pool and some marshy hot seeps—are just south of town across the San Juan River bridge, near which droopy gray spring deposits festoon the riverbank. The big pool discharges 200 gallons per minute, at a temperature of 120 degrees Fahrenheit. Issuing from Mancos shale, the springs are probably heated by still-hot igneous rocks of the San Juan volcanic field, the water finding quick routes upward along faults, joints, and passages dissolved in the rocks. Most springs carry minerals dissolved from rock through which they pass; in this area they are heavily charged with silica from igneous rocks. When the water cools, the silica precipitates as siliceous sinter, a form of quartz. Massive sinter deposits have deflected the course of the San Juan River so that it arcs westward around the spring area. So much water is now piped to swimming pools, fountains, and heating systems that the springs no longer overflow to deposit new sinter. The rotten-egg odor comes from hydrogen sulfide released from the water as gas.

West of Pagosa Springs, U.S. 160 rises onto a high anticline surfaced with Dakota sandstone, a Cretaceous beach deposit. From the crest of the anticline near milepost 141, the San Juan Mountains can be seen to the north. These mountains are mostly volcanic, but their volcanic outpourings overlie a broad dome of Precambrian rock—a high area near the southeast end of the Uncompahgre uplift—on whose flanks lie tilted Paleozoic and Mesozoic strata.

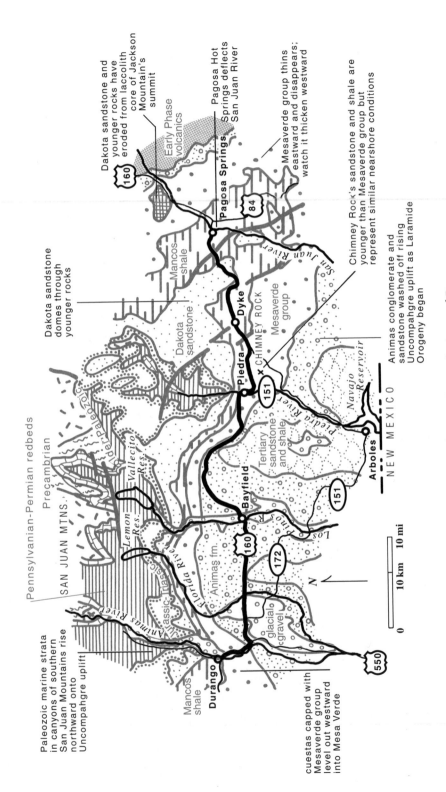

Paleozoic marine strata in canyons of southern San Juan Mountains rise northward onto Uncompahgre uplift

Dakota sandstone domes through younger rocks

Pennsylvanian-Permian redbeds

Precambrian

SAN JUAN MTNS.

older Paleozoic

Vallecito Res.

Lemon Res.

Animas River

Florida River

Triassic-Jurassic

Mancos shale

Durango

Animas fm.

glacial gravel

550

172

160

Bayfield

Los Pinos R.

151

Tertiary sandstone and shale

Piedra River

Navajo Reservoir

Arboles

151

NEW MEXICO

cuestas capped with Mesaverde group level out westward into Mesa Verde

Animas conglomerate and sandstone washed off rising Uncompahgre uplift as Laramide Orogeny began

× CHIMNEY ROCK

Piedra Dyke

151

Mesaverde group

Chimney Rock's sandstone and shale are younger than Mesaverde group but represent similar nearshore conditions

Dakota sandstone

Mancos shale

160

Pagosa Springs

84

San Juan River

Pagosa Hot Springs deflects San Juan River

Early Phase volcanics

Dakota sandstone and younger rocks have eroded from laccolith core of Jackson Mountain's summit

Mesaverde group thins eastward and disappears; watch it thicken westward

10 mi

10 km

0

N

Geology along U.S. 160 between Pagosa Springs and Durango.

Prominent cuestas to the south are capped with resistant nearshore sandstone of the Mesaverde group, which also includes layers of shale and coal. The gray slope-forming layer beneath is Mancos shale, deposited in the same widespread Cretaceous sea as the Pierre shale east of the Rockies, a sea that stretched across North America from the Arctic to the Gulf of Mexico. The two formations were given different names before geologists realized that they were deposited in the same sea before the Rockies rose.

West of the Dakota-surfaced anticline, U.S. 160 crosses the northern end of the San Juan Basin, a major oil and gas producer, and passes through Mancos shale and overlying Mesaverde sandstones. The top of the Mesaverde group is near milepost 125, at the base of Chimney Rock, a prominent landmark that rises more than 1,000 feet above the surrounding area. The hard cap of this pinnacle is the Pictured Cliffs sandstone; its slopes are dark gray Lewis shale. These two formations resemble the Mancos shale–Mesaverde group sequence in color and texture and represent a repeat run of marine and then shoreline conditions. The Lewis shale is the major source of natural gas in the San Juan Basin. Some gas wells west of milepost 98, however, produce gas from the Dakota sandstone.

At the town of Chimney Rock the Mesaverde group tilts steeply into a hogback. Pictured Cliffs sandstone farther south slopes less steeply, as do younger and younger strata still farther south, leveling out away from the San Juan Mountains. Above the Pictured Cliffs sandstone are several hundred feet more of Cretaceous lagoon and river sandstone and shale that signal the final eastward retreat of the sea as the Laramide Orogeny began.

Beds of Cretaceous coal in the San Juan Basin contain methane, which makes mining them quite dangerous. In recent years, the gas has been recovered by drilling into the coal beds and fracturing the coal. Occasionally, methane has escaped into drinking water wells, causing problems for nearby residents.

Very late Cretaceous to early Tertiary nonmarine sandstone and shale are derived from the San Juan Mountains area, which began to rise as the Laramide Orogeny began. Some of the Tertiary beds can be identified by their purple, green, and mustard yellow colors. Crossbedding of the type that signifies shifting stream channels shows up well in some roadcuts.

Approaching Durango the highway crosses a terrace of Pleistocene gravel, then follows the Animas River into town. Durango lies between a cuesta of Mesaverde sandstone and a hogback of Dakota sandstone, in a wide spot where the Animas River crosses easily eroded Mancos shale. Farther up the river, older rocks, particularly red and gray sandstone and shale of Pennsylvanian, Permian, and Triassic age, rise toward the San Juan Mountains.

Durango was once a smelter town handling large shipments of ore from Silverton and other mining centers. It is the southern terminus of a narrow-gauge railway, now a tourist attraction that carries passengers up the Animas River to the old mining town of Silverton. The railroad route passes through tilted sedimentary rocks and into the Precambrian core of the San Juan Mountains, with awe-inspiring cliffs of Precambrian gneiss and schist. It ultimately enters Tertiary volcanic rocks of the Silverton Caldera.

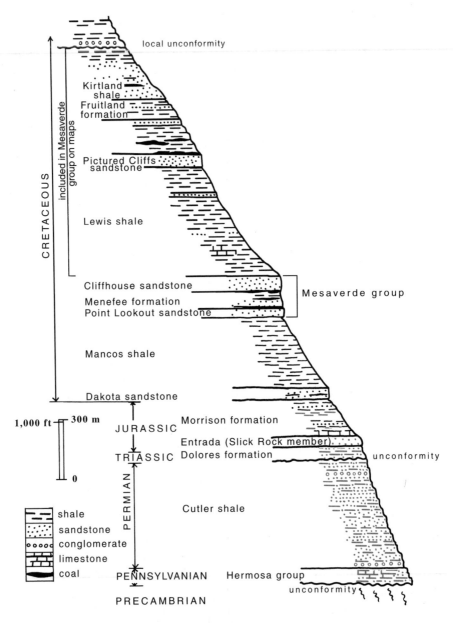

Stratigraphic column of Paleozoic and Mesozoic formations along U.S. 160 between Pagosa Springs and Four Corners.

U.S. 160
Durango—Four Corners
89 miles (143 km)

West of the Animas River at Durango, U.S. 160 passes north of Smelter Mountain, once piled with silver and uranium mill tailings. Ore from Silverton and other areas was brought to Durango for milling because coal and water were available nearby. About 2.5 million cubic yards of radioactive tailings were trucked to a safe depository in the late 1980s.

The highway follows a canyon between hills and buttes capped with the lowest of the Mesaverde group sandstones, deposited on beaches, shores, and bars of the retreating Cretaceous sea not long before the Rockies began to rise.

Coal in the Mesaverde group developed from plant material in swamps and marshy lagoons along the edge of the Cretaceous sea. In places it is thick enough to mine. North of the highway, Paleozoic and Mesozoic sedimentary rocks tilt up toward the core of the San Juan Mountains. The range is an immense pile-up of lava and ash flows erupted on the southern part of the Uncompahgre uplift in middle to late Tertiary time.

La Plata Mountains north of Hesperus contain a variety of late Cretaceous to early Tertiary intrusions, twice as old as the volcanic rocks of the San Juans. Many of these small intrusions—sills and laccoliths—squeeze between layers of red Pennsylvanian and Permian strata; dikes and stocks cut across them. Some sills are several miles long. As with many other Laramide intrusions in the Colorado Mineral Belt, metallic mineral deposits occur near these. The silver in these mountains gave them their name: *la plata* means "silver" in Spanish.

Landslides are common in the area west of Durango, where blocky, highly jointed Mesaverde sandstone lies above weak, easily eroded, slippery-when-wet Mancos shale. A marine shale, the Mancos contains fossil pelecypods, ammonites, and other marine organisms. In this area abundant tiny mica flakes give it a silvery sheen.

Sandstone units in the Mesaverde group thicken westward, so cliffs that top cuestas and mesas south of the highway get taller and taller. The uppermost third of the Mesaverde group is the Cliffhouse sandstone, the layer in which the cliff dwellings of Mesa Verde National Park are located. Gray mesa slopes are composed of Mancos shale about 800 to 1,000 feet thick. The town of Mancos is one of the oldest coal-mining towns of western Colorado; it now is an agricultural center as well.

A few miles west of Mancos, on the summit of a gentle anticline near the entrance to Mesa Verde National Park, a small gas field produces from

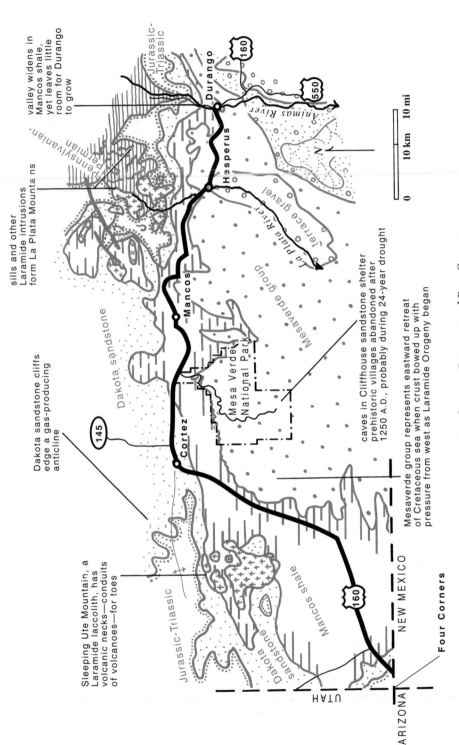

valley widens in Mancos shale, yet leaves little room for Durango to grow

sills and other Laramide intrusions form La Plata Mountains

Dakota sandstone cliffs edge a gas-producing anticline

Sleeping Ute Mountain, a Laramide laccolith, has volcanic necks—conduits of volcanoes—for toes

caves in Cliffhouse sandstone shelter prehistoric villages abandoned after 1250 A.D., probably during 24-year drought

Mesaverde group represents eastward retreat of Cretaceous sea when crust bowed up with pressure from west as Laramide Orogeny began

Durango

Hesperus

Mancos

Cortez

Mesa Verde National Park

Animas River

La Plata River

terrace gravel

Mesaverde group

Dakota sandstone

Jurassic-Triassic

Pennsylvanian-Permian

Jurassic-Triassic

Mancos shale

Dakota sandstone

160

550

145

160

NEW MEXICO

Four Corners

ARIZONA

UTAH

N

0   10 km   10 mi

Geology along U.S. 160 between Durango and Four Corners.

shallow wells drilled into the Dakota sandstone, below the Mancos shale. A number of other anticlines in the area west and northwest of Mesa Verde produce gas also, sometimes just enough for local use. Some produce carbon dioxide, which is made into dry ice or used to build up pressure in oil wells. Like oil, gas tends to rise through porous rock until it meets with impenetrable shale, so shale-topped anticlines form natural traps for gas pools. This area is on the northwest edge of the San Juan Basin, a major oil province.

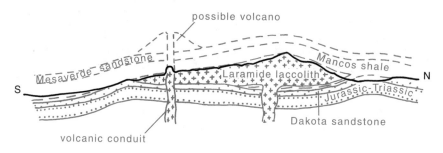

North-south section through Sleeping Ute Mountain, a laccolith intruding Mancos shale. A pair of resistant volcanic necks, once the conduits of volcanoes, form the toes of the sleeping figure.

Cortez lies on flat-lying layers of Dakota sandstone. Beyond Cortez to the southwest the isolated range of Sleeping Ute Mountain, a Laramide laccolith, is much older than volcanic rocks in the San Juans. The igneous intrusion spread out on Dakota sandstone, doming up both the Mancos shale and the Mesaverde group above it.

This area is well below the level of glaciation, but persistent Pleistocene winds spread sandy silt derived from fine glacial deposits on the gently sloping plateau northwest of Cortez. Part of a musk ox skeleton was unearthed from these sediments, a reminder of the changing climate of the last 10,000 years. Thin layers of windblown silt form ideal soil for growing pinto beans and some other crops. Windblown soil also surfaces parts of Mesa Verde, possibly increasing its attractiveness to the prehistoric people who settled there, at least until the great drought that coincides with abandonment of their villages.

From Cortez, U.S. 160 turns south between Mesa Verde and Sleeping Ute Mountain, descending gradually through the lower part of the Mancos shale. In the distance to the north, the symmetrical tip of Lone Cone, a mid-Tertiary volcanic neck, marks the western end of San Juan volcanism. Shiprock, a picturesque and famous volcanic neck in New Mexico, can be seen in the distance to the south. Mountains west of it are on the New Mexico–Arizona line.

A sandstone sentinel near the southern end of Mesa Verde was once part of the main mesa. Both wind and nighttime moisture are important erosional agents in this desert land, but running water, in the form of summer thunderstorms, carved the corrugated shale slopes. —R. W. Brown photo, courtesy of U.S. Geological Survey

Wind takes a strong part in shaping the desert scenery of the Four Corners region. As in most deserts, dust storms and sandstorms carry silt particles and sand grains, using them to abrade rock surfaces. Often the wind cleans away all the fine particles, leaving only a pebble-strewn surface known as desert pavement. Elsewhere it builds sand dunes, blows out hollows in the soil, or with the help of moisture and nighttime frost, carves fist-size holes in sandstone.

Thanks to lack of rainwater, erosion here is usually slow. But in land unprotected by vegetation, sudden downpours may wreak tremendous changes in a few short hours. Heat and cold play their part, too. Desert nights are notoriously chilly, often cold enough to fracture rock by volume changes that accompany sudden cooling. Dew that soaks into pores in the rocks freezes and expands, also fracturing rocks. Some desert rocks, particularly those exposed to the full power of the sun, develop desert varnish, a dark glossy coating of iron and manganese drawn to the surface by moisture and heat. Watch for these features as you approach Four Corners.

Southwest of milepost 4, U.S. 160 descends from the Mancos shale through older Cretaceous rocks, including a white band of limestone and some blocky Dakota sandstone in channels and gullies. The spot where four states come together—the only such spot in the United States—is on a small mesa of Dakota sandstone, surrounded by colorful but barren scenery shaped in Mesozoic rocks.

Mesa Verde National Park
15 miles (24 km) from U.S. 160 to Far View Visitor Center
5 to 12 miles (8 to 19 km) to ruins

Mesa Verde National Park was established to preserve and display unusual archeological remains—clustered dwellings on the mesa surface and cave-sheltered apartment houses of people who inhabited this area from A.D. 900 to 1250. But the dramatic mesa with its high shale slopes and nearly impregnable sandstone cliffs is as distinctive geologically as the archeological sites it preserves.

Forming the lower slopes of Mesa Verde, the Mancos shale is well exposed along the entrance road. This brownish gray shale accumulated as mud in the shallow sea that spread across the center of the continent in Cretaceous time. Many small faults offset its thin sandstone layers. It slides extremely easily, necessitating never-ending road repairs. In addition to removing landslide debris from the highway, repairs involve stabilizing slides by unloading their tops, buttressing their lower ends, and adding drainage pipes to lessen wetting of the shale. The ditch on the inner edge of the road catches small slides and rockfalls and helps drain rainwater and snowmelt from slide areas.

Above the Mancos shale are shoreline sandstones of the Mesaverde group, deposited as the Cretaceous sea retreated eastward. Originally defined as a single formation, the Mesaverde is now given group status and subdivided into three formations. The lowest of these, the Point Lookout sandstone, forms the cliffs that top the northern end of Mesa Verde. Farther south it is overlain by the shale-coal sequence of the Menefee formation, deposited in marshes and swamps close to the sea's edge. Above the Menefee formation is the Cliffhouse sandstone. The strata dip southward here; the Cliffhouse sandstone and Menefee formation have eroded off the high northern prow of Mesa Verde.

The Menefee formation erodes easily, undermining the massive, light-colored Cliffhouse sandstone. Rainwater percolating through the porous sandstone reaches the less permeable shales of the Menefee formation and flows sideways along the layers. Weakened by seepage from small springs where this flow emerges from the cliffs, the sandstone falls away or spalls off in great arcs to create the arched caves that sheltered early inhabitants. What could be more convenient than a weatherproof shelter furnished by nature, complete with a supply of running—or at least seeping—water? The cliffs of the Point Lookout sandstone below sufficed to keep out enemies and provided a ready disposal system: refuse was just tossed over the edge. Archeologists searching for clues to the daily life of early inhabitants often look along the base of the cliffs.

Slope-forming Mancos shale and cliff-forming sandstones of the Mesaverde group characterize Mesa Verde National Park. Erosion of the soft shale undermines resistant, blocky sandstone layers that form the upper cliffs. —L. C. Huff photo, courtesy of U.S. Geological Survey

Streams that drain Mesa Verde are typical of plateau areas. Upstream, each major stream branches again and again to form a treelike or dendritic pattern. Stream erosion is not severe here now, but during times of more intense rainfall, as in the rainy cycles that accompanied Ice Age glaciation, each small stream worked its way headward into the plateau, branching and rebranching, following joints in the rock to carve the narrow, steep ravines. As a result, the mesa is shaped something like a human hand, with deep canyons draining southward between its long fingers.

The first inhabitants of Mesa Verde lived on the surface of the plateau, where they built pit houses, farmed, and hunted. Farming was facilitated by a thin coating of fine, even-grained soil, wind-deposited silt dating back to Pleistocene times. Later, cave dwellings gave protection from both weather and enemies; farming still continued on the top of the mesa. Cave dwellings were occupied for less than one hundred years before their abandonment, which may have been caused by a twenty-four-year drought dated by tree-ring studies. Many other ruins in the Southwest were abandoned at about the same time.

Fires that burned large areas of Mesa Verde in the first few years of the twenty-first century have proved to be a mixed blessing. A natural part of the ecosystem, they threaten the works of man and lead to erosion by destroying plants whose roots help hold soil together. However, removal of plant cover has also led to discovery of new archeological sites.

Stream drainage produces a treelike or dendritic pattern on Mesa Verde. Because streams follow joints in the rock, many are parallel.

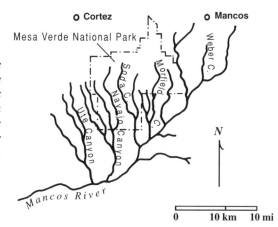

Spruce Tree House and other cave villages of Mesa Verde were built in sheltering recesses in the Cliffhouse sandstone. Springs emerging at the bottom of the porous sandstone supplied water to the villages. The springs also promoted cave formation, weakening and undermining the rock.
—Halka Chronic photo

Views of the Four Corners plateau country to the south and west are exceptionally good from a number of points along the roads between the ruins. Counterclockwise from Sleeping Ute Mountain to the west are:

- the Lukachukai and Chuska Mountains along the New Mexico–Arizona border;
- Shiprock, a large volcanic neck in northwestern New Mexico;
- the fingers of Mesa Verde to the south, pointing into the San Juan Basin;
- the Brazos Mountains of New Mexico, a fault block cored with Precambrian granite, blending northward with the San Juan Mountains;
- the San Juan Mountains, many layers of volcanic ash and lava overlying a dome of Precambrian and Paleozoic rocks;
- La Plata Mountains to the northeast and Rico Mountains to the north;
- in the distance northward, Lone Cone, marking the western end of San Juan volcanism;
- far in the distance, west of Lone Cone, La Sal Mountains of Utah, an isolated group of mid-Tertiary laccoliths;
- the Abajo Mountains of Utah, another cluster of Tertiary laccoliths in the distance north of Sleeping Ute Mountain.

Colorado 13
Craig—Rifle
90 miles (145 km)

Colorado 13 leaves U.S. 40 just west of Craig, heading south toward the White River Plateau, a broad anticlinal uplift that extends well west of the general trend of the Rockies. For some distance the highway travels on tan sandstone and gray shale of the Mesaverde group, here as elsewhere made up of three formations: a basal sandstone unit, central shale and coal layers, and an upper formation that is predominantly sandstone. These strata record the gradual, fluctuating, eastward retreat of the Cretaceous sea.

Mines along this route recover coal from the middle unit of the Mesaverde group. Some is used to fuel a power plant south of Craig. Coal seams show well in many highway cuts, as near milepost 77.

At Hamilton, the highway crosses into the Mancos shale, a marine mud that accumulated slowly in the shallow sea that covered much of North America in Cretaceous time. Sandstone of the Mesaverde group tops hills near the highway.

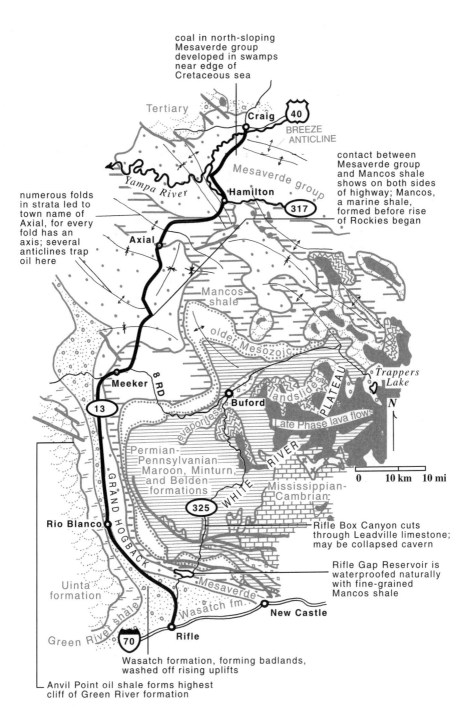

coal in north-sloping Mesaverde group developed in swamps near edge of Cretaceous sea

Tertiary

Craig 40 BREEZE ANTICLINE

contact between Mesaverde group and Mancos shale shows on both sides of highway; Mancos, a marine shale, formed before rise of Rockies began

Yampa River Mesaverde group

Hamilton 317

numerous folds in strata led to town name of Axial, for every fold has an axis; several anticlines trap oil here

Axial

Mancos shale

older Mesozoic

Trappers Lake

Meeker 8 RD landslide PLATEAU

13 **Buford** Late Phase lava flows

evaporites WHITE RIVER

N

Permian- Pennsylvanian Maroon, Minturn, and Belden formations

0   10 km   10 mi

GRAND HOGBACK

325 Mississippian- Cambrian

Rio Blanco

Rifle Box Canyon cuts through Leadville limestone; may be collapsed cavern

Rifle Gap Reservoir is waterproofed naturally with fine-grained Mancos shale

Uinta formation

Mesaverde

Wasatch fm.

New Castle

Green River shale 70 **Rifle**

Wasatch formation, forming badlands, washed off rising uplifts

Anvil Point oil shale forms highest cliff of Green River formation

Geology along Colorado 13 between Craig and Rifle.

Coal is mined south of Craig in north-south strips worked one by one from west to east. The open coal seam is in the center of the photo. East (left) of it, soil and rock have been removed preparatory to mining. To the west (right), coal has been removed and both rock and soil are being replaced, restoring the land to its original contours. This area will be reseeded with native plants. Land farther west has already been reclaimed. —Felicie Williams photo

Swinging west, Colorado 13 rounds two small dome-shaped anticlines where Mancos shale is bent up and then partly eroded away. Watch for steep dips in a resistant sandstone layer and for oil wells in valleys eroded in the centers of the domes. Shale erodes more easily than sandstone, and where the crests of these domes have been breached, ring-shaped ridges of sandstone project like broken, supersize eggshells around them. Several other anticlines and synclines are exposed along this highway.

Near Axial, the highway leads into the Mesaverde group again, with its characteristic coal beds and sandstone layers. Numerous little rounded recesses mark the sandstone cliffs and ledges. They originate when grains of sand, initially loosened by rain or frost, blow away, leaving tiny pits. Partly because they hold moisture that breaks down the cement between grains, and partly because grains of sand bombard their neighbors as they are whirled by wind eddies, the pits deepen and widen until they almost merge with one another, creating a type of surface known as honeycomb weathering. Coal seams ignited by lightning and brush fires have in some places baked layers of shale, turning them a bright brick red.

East of the road, rocks bend up in the large monocline that edges the White River Plateau. About 50 miles across, this broadly domed uplift seems to be transitional between the faulted anticline ranges farther east and the true plateaus farther west. Partly surfaced with Pennsylvanian and Permian sedimentary rocks and partly topped with nearly horizontal Tertiary basalt flows, the White River Plateau is scenically quite different from other Colorado mountains. Few roads penetrate the region, and much of it remains wilderness.

Cliffs west of Meeker are the northern end of the Grand Hogback, where Mesaverde group sandstone and shale turn up steeply along the monocline that borders the White River uplift. The hogback extends south for over 80 miles, forming a big S-shaped curve that extends to Redstone, 30 miles south of Glenwood Springs. Colorado 13 goes through the Grand Hogback west of Meeker and follows it almost to Rifle, running first in a valley eroded in a shaly part of the Mesaverde group, around 6,000 feet thick here, and then, south from about milepost 30, in soft shales at the base of the Wasatch formation.

Sand and pebbles of the Wasatch formation were washed from the rising White River and Uncompahgre uplifts in Paleocene and early Eocene

Tertiary basalt lava flows cap the Flat Tops on the White River Plateau east of Meeker, forming a scenic backdrop for remote Trappers Lake at the edge of the Flat Tops Wilderness. Hilly tree-covered glacial deposits surround the lake. —Jack Rathbone photo

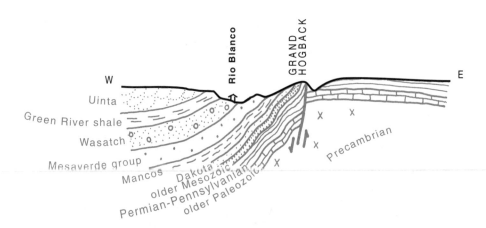

Section across Colorado 13 near Rio Blanco.

time, late in the Laramide Orogeny. Grayish pink layers of this formation show up in high cliffs west of Colorado 13. Above it lies the light gray to yellowish gray Green River shale, deposited in a large lake in late Eocene time. Near the top of the cliff, in the Green River formation, are beds of brownish oil shale that make up a huge but unused energy reserve. The most promising oil shale lies underground in untraveled parts of the Roan Plateau west of Rio Blanco.

Between mileposts 15 and 16, white sandstone at the base of the Wasatch formation is eroded into badlands and fantastic tepee-shaped mounds. Exposures of this formation and of the Green River formation above it improve southward as vegetation decreases.

From this area also, the Elk Mountains can be seen to the southeast, as can the Grand Hogback, curving southeastward beyond the Colorado River. To the south, sloping pediments surround Battlement Mesa, a basalt-capped plateau of Wasatch and Green River formations. Pliocene basalt flows resemble those on the White River Plateau.

Government Creek, parallel to Colorado 13, follows the curve of the Grand Hogback, its easiest route to the Colorado River. Its course is quite obviously controlled by geologic structure, as is the route followed by West Rifle Creek on the other side of the Grand Hogback.

For a close view of the Grand Hogback and the southern side of the White River Plateau, a 26-mile round trip on Colorado 325 leads to Rifle Gap, Rifle Falls, and Rifle Creek Box Canyon. Passing from younger to older rocks as it crosses the southwestern edge of the uplift, the route offers unsurpassed exposures of the Mesaverde group. The contact with the overlying Wasatch formation is clearly exposed, and every resistant sandstone bed stands out in sharp relief. Alternating layers of shale, sandstone, and coal tell of a seashore migrating back and forth across the land, with shale deposited in the sea, sandstone on beaches and bars, and black coaly shale in marshy areas behind the bars. Look for pinkish baked rock where coal has burned.

Colorado 325 also passes a small dam that, with the help of the Grand Hogback, holds back the waters of Rifle Gap Reservoir in a valley water-proofed by impervious Mancos shale. Crossing this valley the road goes through a hogback of Dakota sandstone, purple and green shales of the Jurassic Morrison formation, some Triassic rocks, and a full array of Paleozoic sediments.

A short walk leads to Rifle Falls, where the creek cascades over a travertine dam made of calcium carbonate precipitated when the water emerges, laden with chemicals, from the limestones of Rifle Box Canyon. Narrow, high-walled parts of this canyon, possibly once a long subterranean solution cavern, are overhung by cliffs of Mississippian Leadville limestone.

Nothing now remains at the scene of a large vanadium-uranium deposit a mile south of Rifle Falls, where radioactive minerals were mined from Jurassic and Triassic rocks from 1925 until 1977.

Colorado 62
Ridgway—Placerville
23 miles (37 km)

Ridgway lies in the valley of the Uncompahgre River, on the broad eastern slope of the northwest-southeast-trending Uncompahgre uplift. Cretaceous Mancos shale in the valley is faulted down along the Ridgway fault, which runs east-west north of town. North of the fault, soft, colorful shale of the Jurassic Morrison formation is overlain by Cretaceous Dakota sandstone. To the south, Jurassic and older strata rise toward the San Juan Mountains, where they disappear under Tertiary volcanic rocks.

Flat-topped terraces south of Ridgway are the remains of Pleistocene pediments eroded back into the mountains and covered with a veneer of gravel washed from mountain glaciers as recently as 10,000 years ago. The pediments give way to hummocky glacial deposits. About 6 miles west of Ridgway the highway crosses the moraine that marks the lowest extent of Ice Age glaciers. At the base of the steep slopes and cliffs of the mountains, landslides form an almost continuous apron where soft Cretaceous shale, weakened by abundant rain and snowmelt, slid beneath the weight of overlying volcanic rocks and glacial debris.

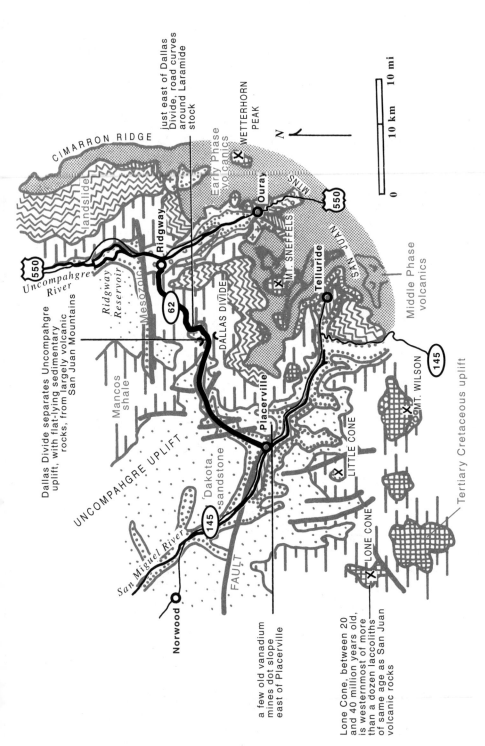

CIMARRON RIDGE

landslide

550

Uncompahgre River

Dallas Divide separates Uncompahgre uplift, with flat-lying sedimentary rocks, from largely volcanic San Juan Mountains

Ridgway Reservoir

Mesozoic

Ridgway

just east of Dallas Divide, road curves around Laramide stock

Early Phase volcanics

WETTERHORN PEAK
✕

N

Ouray

SAN JUAN MTNS

550

MT. SNEFFELS
✕

62

DALLAS DIVIDE

Middle Phase volcanics

Telluride

Mancos shale

UNCOMPAHGRE UPLIFT

Placerville

'Dakota' sandstone

LITTLE CONE
✕

MT. WILSON
✕

145

Tertiary Cretaceous uplift

145

San Miguel River

FAULT

LONE CONE
✕

Norwood

a few old vanadium mines dot slope east of Placerville

Lone Cone, between 20 and 40 million years old, is westernmost of more than a dozen laccoliths of same age as San Juan volcanic rocks

0   10 km   10 mi

Geology along Colorado 62 between Ridgway and Placerville.

Near milepost 8, the highway describes a large **S** curve as it climbs onto a small Laramide stock. Intruded as the Rockies and the Uncompahgre uplift were rising, the stock is much older than the Tertiary volcanic rocks that top mountains to the south. In roadcuts east of Dallas Divide are a few blackened wedges of shale caught up in the intrusion.

The intrusive rock is cut by many intersecting joints formed as the molten rock cooled and shrank. In places, weathering along the joints

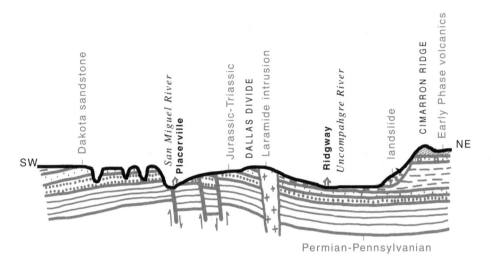

Section parallel to Colorado 62 from Ridgway to Placerville.

Mt. Sneffels, 14,150 feet in elevation, is the highest peak on the skyline south of Colorado 62. Its intrusive core is shouldered by long ridges of horizontal ash flows. Below them, Mancos shale and glacial deposits have collapsed into landslides.
—Felicie Williams photo

has produced piles of large rounded boulders. Water from rain and melting snow, soaking into joints, expands when it freezes, loosening individual crystals and small bits of rock—a process repeated night after night, year after year. And mica in the rock expands as it changes to clay, loosening other grains. The edges and corners of joint-edged blocks, attacked from two or more directions, gradually become rounded by this process of spheroidal weathering.

Vertical joints in porphyry at Dallas Divide formed as once-molten rock cooled and shrank. Long, dark gray crystals of hornblende mark this rock. The mineral chlorite imparts a greenish tinge. —Felicie Williams photo

West of Dallas Divide, the highway begins its descent toward the San Miguel River down an open, U-shaped glacial valley. Sedimentary rocks here are mostly Mesozoic continental (nonmarine) deposits. From youngest to oldest, they are:

- tan cliffs of Dakota sandstone north of the road, once Cretaceous beach and river sands;

- Morrison and Wanakah shales and shore sandstones of Jurassic age;

- Entrada sandstone, deposited as Jurassic sand dunes, easily recognized by its smooth, rounded cliffs (rocks west of a fault at mile 6.3 have dropped downward, causing a repetition of younger layers);

- Kayenta formation, Jurassic dune and water-deposited sandstone and shale;

- pinkish Wingate sandstone, dune-formed on a broad Triassic-Jurassic desert;

- Triassic Dolores shale and sandstone, fine-grained red sediments deposited by streams and rivers;
- Permian Cutler formation, a thick sequence of red sandstone and siltstone deposited in alluvial fans along the southwestern side of ancient Uncompahgria, which rose in Pennsylvanian time about where the Uncompahgre uplift is today. Its red color comes from iron oxide derived from biotite and hornblende in Precambrian rocks.

South of the highway between Dallas Divide and milepost 4, hummocky slopes are typical glacial moraine and landslide topography.

The valley narrows south of milepost 4, at the lower limit of glaciation. Lone Cone, the symmetrical peak framed by the valley walls, is cored with mid-Tertiary intrusive rock. Probably once the conduit of a San Juan volcano, it now stands out as an isolated high point surrounded by more easily eroded Mancos shale.

Layers of dark red sandstone and shale of the Cutler formation jut from both sides of the valley between milepost 3 and Placerville, a small community squeezed into the San Miguel River's narrow canyon. Placer gold was mined from river gravel here. Not much of the old mining town remains; floods long ago destroyed buildings near the river. Mudflows periodically damage the highway.

Colorado 64 and Colorado 139
Dinosaur—Loma
89 miles (143 km)

Close to the southern edge of Dinosaur, Colorado 64 crosses the Willow Creek fault, where rocks from the north are thrust southward over younger rocks. Concealed beneath soft Mancos shale of the valley floor, the fault marks the southern edge of the Uinta uplift.

The road curves south and then eastward around the edge of Coal Oil Basin. This large flat valley is structurally an anticline with its summit eroded away. Sandstones of the Mesaverde group, deposited along the shore as the Cretaceous sea receded, dip outward from the center of the anticline. Formed during the Laramide Orogeny by pressure from the southeast, the anticline is fault-edged: the Willow Creek fault cuts off its northern edge and the Uinta Basin boundary fault runs along the base of the hills south of Rangely. Exploratory wells have passed through both faults, proving their existence and defining the offset along them.

Shallow drilling near Rangely in 1902 located oil in the Mancos shale. But not until 1933, after geologists had mapped the Rangely anticline, did

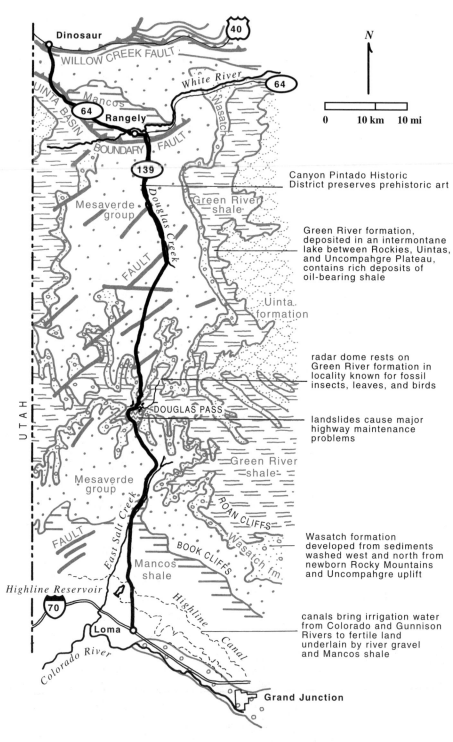

Dinosaur

WILLOW CREEK FAULT

US 40

N

0   10 km   10 mi

64

White River

Mancos

UINTA BASIN

64

Rangely

Wasatch

BOUNDARY FAULT

139

Douglas Creek

Canyon Pintado Historic
District preserves prehistoric art

*Green River
shale*

Mesaverde
group

FAULT

Green River formation,
deposited in an intermontane
lake between Rockies, Uintas,
and Uncompahgre Plateau,
contains rich deposits of
oil-bearing shale

*Uinta
formation*

radar dome rests on
Green River formation in
locality known for fossil
insects, leaves, and birds

DOUGLAS PASS

landslides cause major
highway maintenance
problems

*Green River
shale*

Mesaverde
group

East Salt Creek

ROAN CLIFFS

Wasatch fm.

FAULT

BOOK CLIFFS

*Mancos
shale*

Wasatch formation
developed from sediments
washed west and north from
newborn Rocky Mountains
and Uncompahgre uplift

U T A H

Highline Reservoir

70

Highline Canal

Loma

canals bring irrigation water
from Colorado and Gunnison
Rivers to fertile land
underlain by river gravel
and Mancos shale

Colorado River

Grand Junction

Geology along Colorado 64 and Colorado 139 between Dinosaur and Loma.

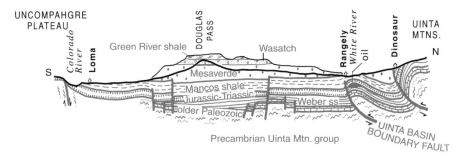

Section parallel to Colorado 139 between Dinosaur and Loma.

deep wells reveal the richness of the Rangely Oil Field, the most productive oil field in Colorado. Both oil and gas are produced from the porous Weber sandstone, formed as dunes along the northern edge of Uncompahgria in Pennsylvanian and Permian time.

Underground oil tends to travel upward through porous, permeable rocks, floating on groundwater. Gas, in turn, floats on the oil. Here, both oil and gas are trapped in pools in the arching strata of the anticline, kept from migrating farther upward by overlying layers of impermeable shale.

In 1958 oil companies began injecting water into rocks below the oil to drive more of it up into their wells. To their surprise, the procedure also caused a series of small earthquakes, seemingly by lubricating and reactivating nearby faults. Since that time, geologists have demonstrated here and elsewhere that adding water to fault surfaces frequently brings about fault movement. When we know more about this process—where, when, and how much water to inject—we may be able to prevent large earthquakes by creating small ones to relieve pressure along faults. Starting in 1986, carbon dioxide was also injected here to improve oil production. Near milepost 13 a pulloff leads to a grasshopper-shaped oil pump and an informative display about the oil field.

At Rangely, Colorado 64 crosses the White River, cloudy with clay from the Uinta, Wasatch, and Green River formations, through which it flows on its way from the White River Plateau. Rangely is a support town for oil, gas, and coal production; learn more about its history at Rangely Museum.

Colorado 139 turns south from Colorado 64 about a mile east of Rangely. There it crosses the Uinta Basin boundary fault, and rock layers become nearly horizontal. The Mesaverde group, with gentle shale slopes underlying cliffy sandstone layers, is exposed along the sides of the valley. Its sandy layers were deposited on beaches and in channels along the shifting shoreline as the Cretaceous sea retreated. Beds of coal developed in marshy or swampy areas protected from the sea's waves, while silty mud settled in bays and lagoons. Bright red shale and sandstone mark some slopes, baked red by burning coal seams; some of the baked rocks extend for miles. The coal was probably ignited by lightning or brush and forest fires.

Many small sheds dotting this valley are pump stations for gas from the Douglas Creek field, produced from faulted and gently folded Pennsylvanian through Cretaceous sedimentary rocks. This valley is the erosion-breached top of another gentle anticline, the Douglas Creek arch.

High bluffs on either side of the valley are soft, poorly cemented early Tertiary rock—the Wasatch formation, pinkish tan sandstone and siltstone composed of sediments washed off the surrounding uplifts as they rose. They are topped with paler Green River shale, deposited as mud on the bottom of Lake Uinta, which in Eocene time filled the broad basin between the Uinta and Uncompahgre uplifts and the western edge of the Rocky Mountains.

As the highway climbs toward Douglas Pass, the difficulties of building roads on steep shale slopes become obvious. In spite of nearly horizontal bedding, the shale is weak and prone to slide, especially when it is wet.

Mushroom-shaped rocks near milepost 63 on Colorado 139 are remnants of a hard sandstone layer undermined by erosion of softer siltstone below.
—Felicie Williams photo

This landslide, which dammed a stream and created a small lake, has now been smoothed and seeded. Plant roots soak up surface water and hold the soil in place. —Felicie Williams photo

Just south of Douglas Pass, before the first curve, outcrops close to the base of the Green River shale contain beds of oil shale, easy to identify by their dark brownish color and oily smell. In the form of a waxy compound called kerogen, the oil is quite difficult to remove from the shale. But east of here, where saturation is high and oil-bearing beds are thicker, it may someday be extracted, providing oil prices become high enough to balance extraction costs.

In the first few miles south of Douglas Pass the road winds around a particularly large landslide. Watch for loose rocks on the highway, especially if the ground is wet from rain or melted snow. All the classic features of landslides can be seen here.

Landslides increase in number when earthquakes shake land already prone to slides. A large earthquake in 1882, felt from Denver to Salt Lake City, may have had its epicenter between Rangely and Meeker, within 50 miles of Douglas Pass. The quake measured between VI and VII on the Mercalli scale, a scale based on damage caused and surface motion reported. Three campers, knocked off their feet, later described great rocks rolling down mountainsides and trees snapped off by the shock. The quake may have been caused by movement on a fault extending southeastward from

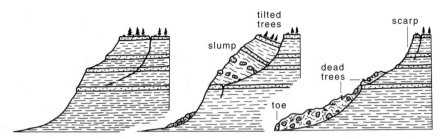

Cracks in the surface are the first sign of an impending slide. Water draining down the cracks further weakens underlying rock and creates mud that will lubricate the slide.

A slump forms when rock and soil slide as a unit. Movement may stop here, but the sloping block channels more water into the cracks, encouraging further sliding.

With enough water and enough momentum, the slump breaks up and flows, creating a hummocky jumble of rock, mud, and uprooted trees. Colorado 139 cuts across such a landslide just south of Douglas pass.

Development of a landslide

the Uinta Mountains, or by movement along the Colorado Lineament, a wide Precambrian fault zone that extends from the Grand Canyon of Arizona to the north shore of Lake Superior.

The road descends through the Wasatch formation, less than 100 feet thick here, and around the big slide into the Mesaverde group. The Mesaverde sandstones are locally honeycombed with pits scoured out as wind blows away sand grains loosened by freezing and thawing of damp rock. White crusts are calcium carbonate and salt wicked from the rock by evaporating water. Here, too, reddish shale layers in the Mesaverde group were baked by burning coal seams.

At the mouth of the canyon of East Salt Creek, the highway enters desolate badlands of Mancos shale. Beveled terraces record an earlier, higher level of the valley floor. South of the Highline Canal, where productive floodplain sediments above the Mancos shale are irrigated, the landscape abruptly changes to one of fertile farms.

The Uncompahgre uplift, now visible to the south, rose during the Laramide Orogeny, 72 to 40 million years ago. There, Triassic, Jurassic, and early Cretaceous sedimentary rocks lie on a smoothly eroded surface of dark Precambrian metamorphic rocks.

Colorado 65 and Colorado 92
Interstate 70—Delta via Grand Mesa
61 miles (98 km)

Leaving I-70 in De Beque Canyon, Colorado 65 ascends Plateau Creek eastward between cliffs of nearly horizontal sandstone layers of the Mesaverde group. Repeated sequences of marine shale, shoreline sandstone, and swamp-formed coal in the group show us that the Cretaceous sea's retreat was fluctuating and irregular—a come-and-go affair.

Near milepost 51 the highway turns south to climb a broad, gentle slope of glacial outwash that lies on top of the Wasatch formation. The pinkish tan sandstone and siltstone of the Wasatch formation, exposed in cliffs and buttes near the town of Mesa, are river and stream deposits carried in early Eocene time into a low area, the Piceance Basin, from the newly risen Rockies and the Uinta and Uncompahgre uplifts. As with most Tertiary sedimentary rocks, the Wasatch formation is poorly cemented and good exposures of the rock itself are rare.

The light gray Green River shale, above the Wasatch formation, accumulated in Lake Uinta, a broad Eocene lake that filled the Piceance Basin

Weathering in soft Wasatch sandstone highlights patterns of horizontal, wavy, or crossbedded laminations like those seen in modern-day stream deposits. —Felicie Williams photo

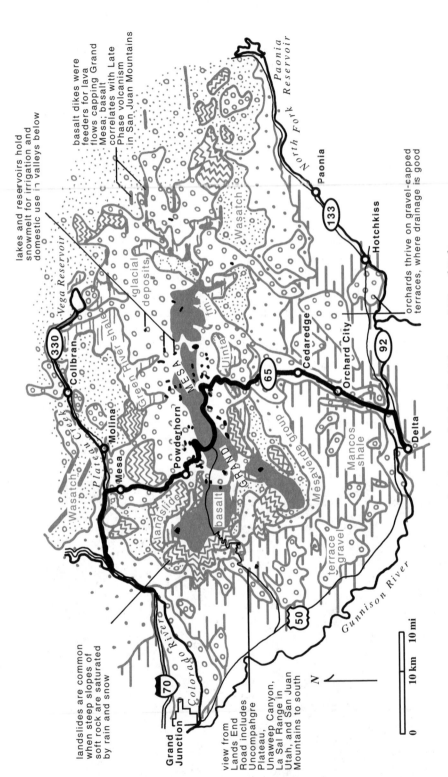

lakes and reservoirs hold
snowmelt for irrigation and
domestic use in valleys below

basalt dikes were
feeders for lava
flows capping Grand
Mesa; basalt
correlates with Late
Phase volcanism
in San Juan Mountains

orchards thrive on gravel-capped
terraces, where drainage is good

landslides are common
when steep slopes of
soft rock are saturated
by rain and snow

view from
Lands End
Road includes
Uncompahgre
Plateau,
Unaweep Canyon,
La Sal Range in
Utah, and San Juan
Mountains to south

Geology along Colorado 65 and Colorado 92 between I-70 and Delta.

after the mountains had worn down to rounded forms but before the uplift and volcanism of Miocene-Pliocene time. Outcrops of these rocks are rare, too, but they appear in roadcuts where Colorado 65 climbs steeply a few miles above Powderhorn. Both formations are well exposed in the Roan Cliffs, in the distance to the north.

A bend in Colorado 65 sliced into the soft slopes of an old earthflow, initiating a small landslide (snow-covered in this photograph). *Berms of boulders now block the slide, but for safety's sake the Colorodo Highway Department posts "no parking" signs.* —Felicie Williams photo

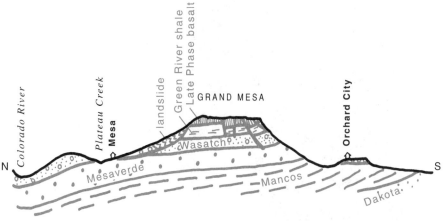

Section across Grand Mesa through Mesa and Orchard City.

Early Tertiary deposits might have been completely eroded from Grand Mesa if basalt flows had not protected them. Around 10 million years ago, as many as twenty lava flows poured from deep fissures activated by uplift and stretching of the crust. Several east-west dikes and small volcanic plugs near the eastern end of Grand Mesa may have been the conduits for these flows.

Seen from above or on a geology map, Grand Mesa's lava flows define a Y-shaped pattern. They seem to have coursed down two converging valleys, cooling and hardening there. Now, 10 million years later, softer rocks that once edged the valleys have eroded away, leaving the lava-filled valleys as the highest parts of Grand Mesa—over 10,000 feet in elevation. Lava flows have formed similar examples of topography reversal elsewhere.

Grand Mesa's basalt flows were originally more extensive. Along with the mesa's many lakes, basalt rubble on mesa slopes tells the story of erosion here. Though more durable than surrounding sedimentary rock, basalt shrinks and cracks as it cools, forming deep joints through each flow. Glacial meltwater of Ice Age times and, more recently, rain and snowmelt trickle downward through these joints, lubricating the soft shale below and turning it back into mud. Fractured basalt blocks slide or sink, leaving low spots that fill with lakes. Large blocks slide down the steep edges of the mesa, their tops tilted backward, forming depressions that also fill with water. Undermined boulders of basalt tumble down cliffs at the mesa's edge.

In a roadcut east of Mesa Lakes, two dark basalt flows are separated by a reddish soil layer several feet thick, baked by heat from the upper flow. The lower flow, its surface more highly fractured than that of the upper flow, had time to weather deeply between the two eruptions. Note figure for scale. —Felicie Williams photo

For a panoramic view of the west side of Grand Mesa and much of far western Colorado, follow Lands End Road to the edge of the mesa. Clockwise from the south are volcanic peaks of the San Juans, then the gentle rise of the Uncompahgre Plateau with the deep notch of Unaweep Canyon. Behind the plateau rise La Sal Mountains of Utah, a cluster of Tertiary laccoliths. The Gunnison River and its tributaries have shaped the wide valley between Grand Mesa and the Uncompahgre Plateau, cutting thousands of feet into Tertiary and Cretaceous sedimentary rocks and carrying hundreds of cubic miles of debris north into the Colorado River. To the northwest the Book Cliffs, with Mesaverde sandstones capping soft Mancos shale, curve around the northern end of Grand Valley. Above the Book Cliffs to the north, the Wasatch and Green River formations appear in the Roan Cliffs.

Descending the south side of Grand Mesa, Colorado 65 follows the slumped north shore of Island Lake, then curves around a gentle valley lined with landslides and glacial deposits, with big basalt boulders mixed in with shaly soil. To the east are the West Elk Mountains, a range of mid-Tertiary laccoliths and volcanic lava and ash flows, with their own collection of glacial deposits, rockfalls, and landslides.

The highway follows a Pleistocene pediment into Cedaredge. Erosion scoured several levels of pediments here, each beveling the Mancos shale during a stable period in the erosional history of the area. Their gently sloping surfaces make them easy to irrigate, and the loose glacial outwash with which they are coated improves their agricultural value. The clayey, salty Mancos shale, which forms badlands between Cedaredge and Delta, doesn't support much vegetation.

West of Cedaredge, red sandstones in the Mesaverde group were baked the color of fired clay by burning coal seams. Underground coal mines to the east obtain coal from the Mesaverde group, which forms the lower slopes on this side of Grand Mesa.

Colorado 141
Whitewater—Naturita
97 miles (156 km)

The Ancestral Rocky Mountain range of Uncompahgria, a fault-block range that rose in Pennsylvanian time, extended farther east and north than the present Uncompahgre uplift, which rose during the Laramide Orogeny. Cored with Precambrian igneous and metamorphic rocks and covered with early Paleozoic sedimentary rocks, Uncompahgria was almost completely eroded away by Triassic time. Mesozoic sediments were thus deposited

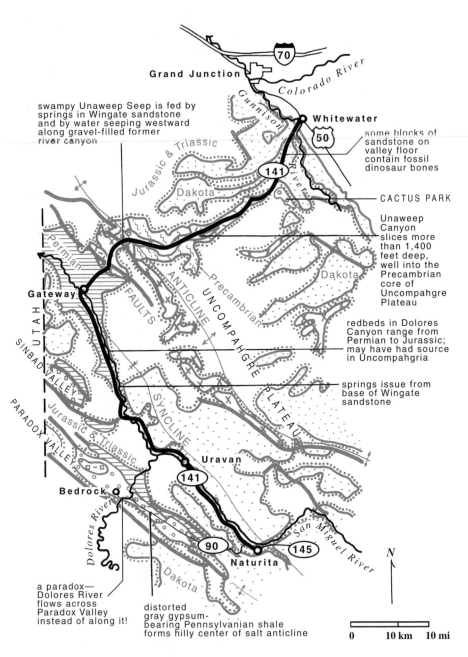

swampy Unaweep Seep is fed by springs in Wingate sandstone and by water seeping westward along gravel-filled former river canyon

some blocks of sandstone on valley floor contain fossil dinosaur bones

CACTUS PARK

Unaweep Canyon slices more than 1,400 feet deep, well into the Precambrian core of Uncompahgre Plateau

redbeds in Dolores Canyon range from Permian to Jurassic; may have had source in Uncompahgria

springs issue from base of Wingate sandstone

a paradox— Dolores River flows across Paradox Valley instead of along it!

distorted gray gypsum-bearing Pennsylvanian shale forms hilly center of salt anticline

N

0   10 km   10 mi

Geology along Colorado 141 between Whitewater and Naturita.

directly on its Precambrian core. After crossing the Gunnison River west of Whitewater, Colorado 141 encounters these Mesozoic strata, tilted by the present uplift, from youngest to oldest:

- about 200 feet of interbedded light green to brown channel sandstone and floodplain shale—the Dakota sandstone and Burro Canyon formations, both Cretaceous—at roadside level for several miles;

- colorful green, red, and purple shale, limestone, and sandstone of the Jurassic Morrison formation, formed on continental flats and river floodplains; its lower, sandier half is known as the Salt Wash member;

- the Wanakah formation, also Jurassic, similar to the Morrison formation but deposited near the eastern shore of a shallow sea;

- the Entrada sandstone, crossbedded, wind-deposited pink and tan Jurassic sandstone about 50 feet thick; its lower part a distinctive smooth-surfaced cliff-forming rock known as the Slick Rock member;

- 100 feet of tan to pink river-deposited sandstone of the Jurassic Kayenta formation, forming a series of ledges;

- prominent 350-foot red cliffs of wind-deposited Wingate sandstone (milepost 143), ranging in age from Triassic to Jurassic;

- the Chinle formation, up to 200 feet of dark red Triassic siltstone and sandstone, a continental deposit lying directly on dark Precambrian rocks.

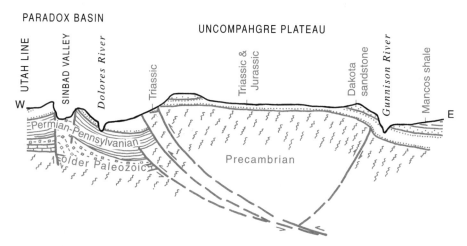

Section across the Uncompahgre Plateau just south of Unaweep Canyon. Movement along the western faults lifted the west side of the plateau more than the east side, creating its long eastern slope. Thrusting along the deep faults has foreshortened the crust and pushed Paleozoic strata downward on the western side. Similar thrusting along many of the same faults in Pennsylvanian time formed the range of Uncompahgria and the Paradox Basin west of it.

Mysterious, empty Unaweep Canyon, walled with hard Precambrian gneiss, schist, and granite, must have been excavated by a major river far more powerful than the two tiny streams that drain it today. —S. W. Lohman photo, courtesy of U.S. Geological Survey

Near milepost 143 the highway enters the steep-walled, flat-floored heart of Unaweep Canyon. One of the most unusual and interesting canyons in Colorado, the gorge cuts crosswise through the core of the Uncompahgre uplift, slicing deeply into hard Precambrian metamorphic and intrusive rocks.

Unaweep means "canyon with two mouths" in the Ute language. On either side of an almost imperceptible divide, East Creek and West Creek flow in opposite directions. Each is too small to wash away rock that falls from the canyon walls, so it just piles up below the cliffs.

The sheer size of this gorge proves a powerful river established a westward course across the uplift, probably during Tertiary time when the uplift was covered by thick debris fans from the eroding Rocky Mountains. River gravels in the canyon and in Cactus Park, another abandoned gorge, contain volcanic and sedimentary boulders like those carried by the modern Gunnison River, so the Gunnison River alone may have carved Unaweep Canyon. But the canyon lines up with the course of the Colorado River above Palisade, so the Colorado and Gunnison Rivers may have joined at the mouth of Cactus Park, cutting the canyon with their combined flow.

In any case, a mighty river quickly trenched through the broad apron of soft Tertiary deposits. Then, imprisoned in a self-carved valley, it deepened its trench through underlying Mesozoic sedimentary layers, and finally, attacked the hard Precambrian rocks that core the uplift.

Another river, channeling easily through softer Tertiary and Mesozoic sedimentary rocks north of the Uncompahgre uplift, eventually "captured" Unaweep Canyon's flow. The Gunnison River now joins the Colorado River at Grand Junction, whence their combined waters are excavating a new canyon through the north end of the uplift.

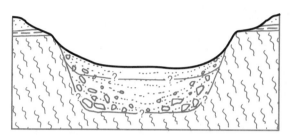

Attempts have been made to measure the depth to bedrock in Unaweep Canyon (left) *using reflected sound waves. One study suggests the canyon was once twice as deep as it is now, more than rivaling the Black Canyon of the Gunnison* (right) *for depth. Other studies suggest that fill fallen from the sides is only 100 to 300 feet deep. None are conclusive.*

Precambrian granite, gneiss, and schist wall Unaweep Canyon between mileposts 143 and 115. Strongly veined with light-colored pegmatites, the ancient rocks are highly jointed as well. In most places the original textures and minerals of the rock were so altered by metamorphism that we can't easily unravel their history. Near milepost 130, however, banding in the gneiss suggests water-deposited shale and sandstone layers, relic textures perhaps of original nearshore marine sediments.

About 1.7 billion years ago these rocks were tightly folded and metamorphosed and intruded by granite as they collided with the edge of the protocontinent, the Wyoming Province. About 1.4 billion years ago a similar collision southeast of Colorado caused intrusion of some younger granite.

The Precambrian rocks rise westward into higher and higher cliffs until they are cut off abruptly near milepost 116 by one of the large faults along the southwestern edge of the Uncompahgre uplift. Near the highway the fault is hidden by rock debris and vegetation. In the western distance the clustered laccoliths of La Sal Mountains in Utah rise above Unaweep Canyon's cliffs.

In Thimble Rock, Precambrian schist is cut by nearly horizontal veins of light pegmatite and dikes of dark diorite. The stone ruins were built of Wingate sandstone from the other side of the canyon. —Felicie Williams photo

On the palisade north of Gateway, the sedimentary strata include Permian redbeds at the base and reach to the Morrison formation at the top. The highest cliffs are Wingate sandstone. —F. W. Cater photo, courtesy of U.S. Geological Survey

West of the fault, sedimentary rocks appear again near the highway, in reverse order from the sequence east of Unaweep Canyon, with two additional units at their base:

- the Permian Cutler, in places thousands of feet thick, consisting of fine alluvial deposits washed westward off Uncompahgria;

- the Triassic Moenkopi formation, in this area siltstone and fine sandstone deposited on floodplains and coastal mudflats as Uncompahgria became little more than a region of low hills.

Both of these formations are intensely colored by fine particles of red iron oxide—the mineral hematite—derived from dark iron-bearing minerals such as biotite and hornblende in the Precambrian rocks of the uplift.

Between Gateway and Naturita, Colorado 141 threads its way up the Dolores River through another impressive canyon, rising gradually into successively younger formations. Above the sharp ledges of the Kayenta

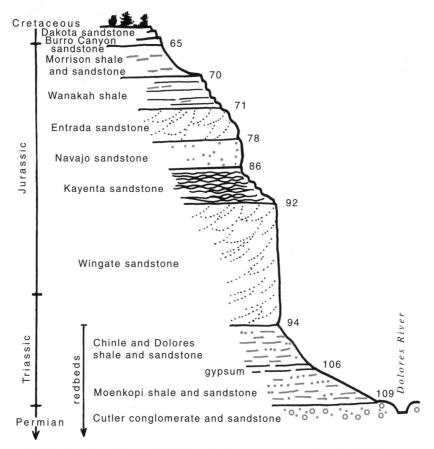

Stratigraphic diagram for Dolores River canyon. Numbers give highway mileposts where contacts between formations appear at roadside level.

Near Uravan, the rounded Slick Rock member of the Entrada sandstone lies on flat-bedded, blocky Navajo sandstone. Both are Jurassic in age. The Navajo sandstone extends west to Zion National Park in Utah and south into Arizona and New Mexico. —F. W. Cater photo, courtesy of U.S. Geological Survey

formation and below the slickrock cliffs of the Entrada sandstone is a 50-foot layer of light-colored, crossbedded, wind-deposited Navajo sandstone. This rock thickens westward and southward, becoming more and more massive and often tinted bright pink, reaching fame as a major scenery-former in southern Utah, northeastern Arizona, and northwestern New Mexico.

Half a mile south of milepost 88, Roc Creek empties into the Dolores River. Soft yellow carnotite, an ore of uranium, radium, and vanadium, was discovered in 1881 at the head of this creek. In this region, carnotite and other uranium minerals sometimes fill spaces between sand grains and adhere to and replace bits of wood that were deposited in stream channels in sedimentary rocks, particularly in the Salt Wash member of the Morrison formation. Between about 1900 and the early 1920s, the long, narrow ore

Gravel deposits near Mesa Creek contain fine placer gold carried by Pleistocene rivers from the San Juans. In 1889 and 1890 miners built a hanging flume, a wooden ditch suspended from the canyon wall, to carry water for washing gold from the gravel. The remains of the flume can be seen below the highway between mileposts 82 and 81. Coke from the oven at mile 83.5 is said to have been used to make spikes for the flume. —Felicie Williams photo

bodies yielded radium for medical use—and for the Paris laboratory of Marie Curie.

The former town of Uravan takes its name from two mineral products that brought it into existence: uranium and vanadium. In the 1930s and 1940s the mill extracted vanadium for hardening steel; in the 1940s the tailings were reprocessed for uranium. The mill is now closed, most of the buildings have been torn down, and wastes are being removed. Plastic-lined evaporation ponds near the highway concentrate wastes taken from un-lined ponds.

From Vancorum, Colorado 90 runs west to Paradox Valley, a collapsed salt anticline that's worth a side trip. In Pennsylvanian and Permian time, the area west of Uncompahgria bowed down, forming the Paradox Basin, which filled with more than 20,000 feet of sediments, including 5,000 feet of salt, gypsum, and potash formed by evaporation of seawater in the nearly landlocked basin. These evaporites are capable of slowly flowing and are more buoyant than most rock. Upward-flowing salt created several long salt anticlines, including those that eventually formed Paradox Valley, Sinbad Valley, and Big Gypsum Valley, each with a northwest trend determined by the trend of Precambrian faults in rocks below them.

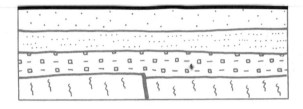

Salt and other evaporites were deposited in a landlocked basin, then overlain by thick layers of younger rocks.

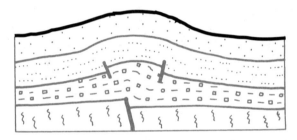

The weight of overlying rocks caused the salt to flow upward, especially where it lay on an uneven surface above a Precambrian fault.

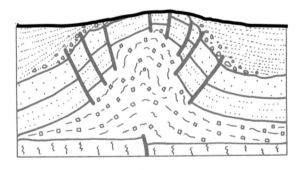

Upward-flowing salt arched overlying rocks, which later eroded away.

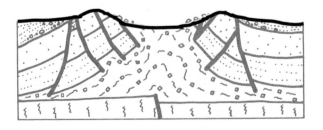

Exposed salt was dissolved and carried away by rainwater and groundwater, causing collapse of unsupported rocks along the edges of the valley.

The development of Paradox Valley.

Colorado 145
Naturita—Telluride
54 miles (86 km)

Naturita lies at the southeastern end of the Paradox Valley salt anticline. Look back along that valley from the top of the hill a mile east of Naturita to see Mesozoic rocks tilted by upward flow of Pennsylvanian salt.

The general trend of both topographic and geologic features in this area is northwest-southeast. Deep below the sedimentary rocks, faults that developed in Precambrian time 1.6 billion years ago have been active on and off ever since, controlling features like the position and trend of this and neighboring salt anticlines.

Between Naturita and Norwood, Colorado 145 stays on the surface of the Dakota sandstone, shortcutting a bend in the San Miguel River. To the southeast rise Mt. Sneffels, Wilson Peak, and Dolores Peak in the San Juan Mountains, as well as Lone Cone farther south. All the highest peaks of this part of the San Juans are composed of mid-Tertiary intrusive rock, some of them probably conduits that once led upward to erupting volcanoes.

A few miles east of Norwood the highway drops into the canyon of the San Miguel River, carved in rocks of Jurassic and Triassic age. Sedimentary formations exposed along this road are almost identical with those described in the preceding road guide, but they are more often cloaked in vegetation as you approach the San Juan Mountains, which form a storm center where rainfall is more plentiful. Near milepost 87 the redbeds at the base of the sedimentary sequence show up well across the river, where they include a purple conglomerate packed with pebbles of almost uniform size. Steep faults cut the sedimentary rocks here, one of them running along the north wall of the canyon. Rocks north of the fault dropped as much as 500 feet relative to those south of the fault, accounting for the lopsided shape of the canyon. A fault near milepost 90 cuts across the canyon with a vertical displacement of about 200 feet, bringing the Morrison formation against much older redbeds.

At milepost 89, watch for an old boulder-filled channel across and above the river, occupied by the San Miguel River in Pleistocene time. Boulder heaps below it result from placer mining to recover gold washed, along with the gravel, from the San Juan Mountains. The modern San Miguel River periodically floods its narrow valley; in 1909 it carried away all but two buildings from Placerville.

Between mileposts 78 and 77, a well-exposed dike near the road cuts obliquely across the canyon; its flat face shows up in an adjacent gully. Dikes become increasingly common as we approach the San Juans, reminders of the volcanic violence responsible for these mountains. Some dikes resist

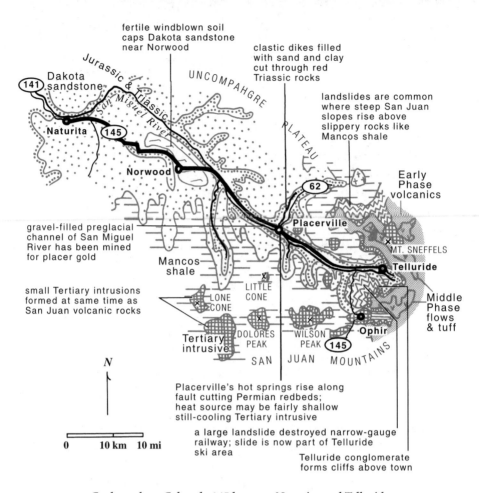

fertile windblown soil
caps Dakota sandstone
near Norwood

clastic dikes filled
with sand and clay
cut through red
Triassic rocks

landslides are common
where steep San Juan
slopes rise above
slippery rocks like
Mancos shale

Jurassic & Triassic

Dakota
sandstone

UNCOMPAHGRE

141

Naturita 145

San Miguel River

Norwood

62

PLATEAU

Placerville

Early
Phase
volcanics

MT. SNEFFELS

Telluride

Middle
Phase
flows
& tuff

gravel-filled preglacial
channel of San Miguel
River has been mined
for placer gold

Mancos
shale

small Tertiary intrusions
formed at same time as
San Juan volcanic rocks

LITTLE
CONE

LONE
CONE

Tertiary
intrusive

DOLORES
PEAK

WILSON
PEAK 145

Ophir

SAN JUAN MOUNTAINS

N

0   10 km   10 mi

Placerville's hot springs rise along
fault cutting Permian redbeds;
heat source may be fairly shallow
still-cooling Tertiary intrusive

a large landslide destroyed narrow-gauge
railway; slide is now part of Telluride
ski area

Telluride conglomerate
forms cliffs above town

Geology along Colorado 145 between Naturita and Telluride.

erosion and stand out as ridges. Softer, more easily eroded dikes weather into slots between walls of baked sedimentary rocks.

Small mines in Entrada sandstone east of Placerville are vanadium mines, some dating from as long ago as 1910, when vanadium was used primarily to color red and orange glass and glazes. Although small amounts of vanadium are widely distributed in various minerals, as well as in coal and petroleum, vanadium ores of commercial value are rare. The element is now used in high-strength, high-performance steel alloys used in power tools and construction.

East of Placerville watch for moraines of the San Miguel glacier and of tributary glaciers coming in from the south and north. Born in cirques high above Telluride, the San Miguel glacier widened and straightened this valley. Its lowest moraine, an irregular hill of unsorted rocks, sand, and silt, cuts across the valley floor at milepost 74 at an elevation of about 8,500 feet, generally the lowest elevation reached by glaciers in the San Juans.

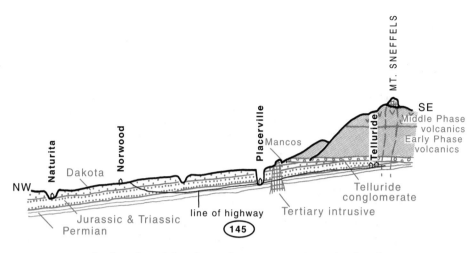

Section along Colorado 145 between Naturita and Telluride.

The valley becomes distinctly U-shaped a few miles above the moraine—a sure sign of glacial erosion.

Landslides are common in this region, with four factors operating in their favor:

- unstable volcanic rocks that often contain layers of poorly consolidated volcanic ash;
- even less stable Mancos shale underlying the heavy volcanic rocks;
- valley walls oversteepened by glaciers;
- mountain storms and heavy snows that saturate both soil and rocks, adding weight and slipperiness.

With highway and railroad cuts and clear-cutting of trees to provide timbers for mines and construction, modern man also contributed to landslide formation. The highway crosses a slide near Society Turn, the junction to which Telluride high society used to ride in their carriages on Sunday afternoons. Mountain Village and most of the Telluride ski area lie on another large landslide. And slides visible across the valley terminated the career of the *Galloping Goose,* a rail-riding bus that plied a narrow-gauge railway to Ridgway.

The Telluride spur of Colorado 145 continues up the glaciated valley of the San Miguel, its floor filled with more than 500 feet of Pleistocene lake deposits formed at a time when the river was dammed by glacial moraines. U-shaped hanging valleys and well-developed glacial cirques frame Telluride. Hanging valleys are typically created by small glaciers unable to keep up, in terms of erosive power, with the bigger glacier that occupies the valley they flow into. Bridal Veil Falls cascades from one of them, with a free fall of 350 feet.

Mesozoic rocks rise eastward toward the San Juan dome, where the Uncompahgre uplift curves and rises around a batholith of hard Precambrian granite. Pushed up early in the Laramide Orogeny, the domed area was soon exposed to erosion, so that rocks above the Dakota sandstone, present above Society Turn, do not appear at the eastern end of the valley. The valley walls above Telluride clearly display the angular unconformity where the Telluride conglomerate overlies a beveled surface of tilted Mesozoic rocks. The conglomerate is about 250 feet thick and forms a reddish vertical cliff.

Above the Telluride conglomerate are Early Phase welded tuff layers of the Tertiary volcanic episode. The long ridge on the skyline north of Telluride, and Greenback Mountain around the head of the valley to the south, are capped by Middle Phase lava flows and welded tuff. As much as 5,000

Viewed from the gondola station high above town, Telluride nestles in a glacial valley. Hanging valleys formed by tributary glaciers are visible higher up. Catching the sun, the lowest band of cliffs, which are below horizontal layers of lava and welded tuff, is the Telluride conglomerate. During the mining heyday some miners lived year-round in Savage Basin, below Imogene Pass (center skyline). The road down to town was considered too terrifying for frequent use. —Felicie Williams photo

feet of regional uplift followed Middle Phase volcanism, giving rivers and glaciers the power to cut the deep valleys we see here today.

At the height of the mining boom, these mountains were honeycombed with hundreds of miles of interconnected mine tunnels, with hundreds of portals and dumps. One tunnel extended through the divide to Red Mountain, about 7 miles away. Ore was dug and blasted from veins associated with the many mid-Tertiary igneous intrusions in this region. The surrounding forest was completely destroyed, cut for mine timbers, buildings, and fuel. The Idarado Mine, one of the largest mines in the area, for a time produced 1,650 tons of gold, silver, lead, copper, and zinc ore per day!

Most of the mines are now sealed off and camouflaged by second-growth spruce and aspen forest. Mill tailings that once disfigured the beautiful glacier-carved valley above Telluride have been removed or smoothed over and planted with suitable vegetation. As a ski area and resort town, Telluride is finding gold in another form in "them thar hills."

For More Information

Most recent geology textbooks contain good up-to-date discussions of general geology. For Colorado geology, visit local libraries and bookstores, including those in state and national parks. Geologic libraries at universities and colleges or at the U.S. Geological Survey or the Colorado Geological Survey in Denver have out-of-print publications. Geologic maps are available for perusal at many of these libraries.

Geologists employed by museums, state and federal surveys, colleges, and universities are often willing to answer specific questions. Members of rock, mineral, and fossil clubs are particularly knowledgeable about their immediate areas.

Geologic maps that cover 1° by 2° areas of Colorado at a scale of 1:250,000 (1 inch equals approximately 4 miles) are listed below, with their U.S. Geological Survey numbers. These maps are far more detailed than the ones in this volume, with excellent descriptions of geologic and geographic features. For even greater detail, the U.S. Geological Survey produces 7.5-minute geological quadrangle maps of much of the state. Their map distribution office is at the Federal Center, Sixth Avenue and Kipling Street, in Denver.

WYOMING			NEBRASKA
Rowley, P. D., 1985 Vernal Quadrangle USGS Map I-1526	Tweto, O., 1976 Craig Quadrangle USGS Map I-972	Scott, G.R., et al., 1986 Greeley Quadrangle USGS Map I-1626	Scott, G.R., 1980 Sterling Quadrangle USGS Map I-1092
Cashion, W.B., 1973 Grand Junction Quad. USGS Map I-736	Tweto, O., et al., 1978 Leadville Quadrangle USGS Map I-999	Bryant, B., et al., 1981 Denver Quadrangle USGS Map I-1163	Sharps, J.A., 1980 Limon Quadrangle USGS Map I-1250
Williams, P.L., 1964 Moab Quadrangle USGS Map I-360	Tweto, O., et al., 1976 Montrose Quadrangle USGS Map MF-761	Scott, G.R., et al., 1978 Pueblo Quadrangle USGS Map I-1022	Sharps, J.A., 1976 Lamar Quadrangle USGS Map I-944
Haynes, D.D., et al., 1972 Cortez Quadrangle USGS Map I-629	Steven, T.A., et al., 1974 Durango Quadrangle USGS Map I-764	Johnson, R.B., 1969 Trinidad Quadrangle USGS Map I-558	Scott, G.R., 1968 La Junta Quadrangle USGS Map I-560

NEW MEXICO

1° x 2° geological maps of Colorado.

Though many others exist, we've selected the following books and maps because they are not too technical and lead beyond this volume, either covering aspects of Colorado's geology in more detail or describing the less-traveled roads.

Arnold, C., and R. Hewett. 1989. *Dinosaur Mountain: Graveyard of the Past.* New York: Clarion Books, 48 pages, color photographs.

Baars, D. L. 1992. *The American Alps: The San Juan Mountains of Southwest Colorado.* Albuquerque: University of New Mexico Press, 194 pages, illustrations, color.

Braddock, W. A. 1988. *Geologic Cross Sections of Rocky Mountain National Park and Adjacent Terrain.* Estes Park, Colo.: Rocky Mountain Nature Association, color.

Braddock, W. A., and J. C. Cole. 1990. *Geologic Map of Rocky Mountain National Park and Vicinity, Colorado.* U.S. Geological Survey MI Map I-1973, map scale 1:50,000, color.

Bryant, B., and P. L. Martin. 1988. *The Geologic Story of the Aspen Region: Mines, Glaciers, and Rocks.* U.S. Geological Survey Bulletin 1603, 53 pages, illustrations, some color.

Chronic, H. 1984. *Pages of Stone, Geology of Western National Parks and Monuments, Vol. 1: Rocky Mountains and Western Great Plains.* Seattle: The Mountaineers, 168 pages, illustrations, photos, some color.

————. 1984. *Time, Rocks, and the Rockies: A Geologic Guide to Roads and Trails of Rocky Mountain National Park.* Missoula, Mont.: Mountain Press, 120 pages, illustrations, photos.

Chronic, J. and H. 1972. *Prairie, Peak, and Plateau: A Guide to the Geology of Colorado.* Colorado Geological Survey Bulletin 32, 126 pages, color photographs, drawings, maps.

Collins, D. B. 1985. *Scenic Trips into Colorado Geology: Uncompahgre Plateau—Montrose, Ridgway, Norwood, Naturita, Uravan, Gateway, Delta.* Colorado Geological Survey Special Publication 27, color; map scale 1:250,000, color.

Davis, M. W., and R. K. Streufert. 1990. *Gold Occurrences of Colorado.* Colorado Geological Survey Resource Series 28, 101 pages, illustrated.

Dino Productions. 1987. *Pathway to the Dinosaurs.* Colorado Geological Survey Miscellaneous Information Series 28, map scale 1:2,000,000.

Eckel, E. B., and others. 1997. *Minerals of Colorado.* Golden, Colo.: Fulcrum Publishing, color photographs, maps.

Griffitts, M. O. 1990. *Guide to the Geology of Mesa Verde National Park.* Mesa Verde Museum Association, Inc., 88 pages.

Hansen, W. R. 1971. *Geologic Map of the Black Canyon of the Gunnison River and Vicinity, Western Colorado.* U.S. Geological Survey MI Map I-584, map scale 1:31,680, color.

Hansen, W. R., P. D. Rowley, and P. E. Carrara. 1983. *Geologic Map of Dinosaur National Monument and Vicinity, Utah and Colorado.* U.S. Geological Survey MI Map I-1407, map scale 1:50,000, color.

Hopkins, R. L., and L. B. Hopkins. 2000. *Hiking Colorado's Geology.* Seattle: The Mountaineers, 239 pages, illustrations.

Jenkins, J. T., Jr., and J. L Jenkins. 1993. *Colorado's Dinosaurs.* Colorado Geological Survey Special Publication 35, 74 pages, illustrated, color.

Kious, W. J., and R. I. Tilling. 1996. *This Dynamic Earth: The Story of Plate Tectonics.* U.S. Geological Survey, available as Colorado Geological Survey Miscellaneous Information Series 62, 77 pages, color.

Laing, D., and N. Lampiris. 1980. *Aspen High Country, the Geology, a Pictorial Guide to Roads and Trails.* Aspen, Colo.: Thunder River Press, 132 pages, illustrations, color photos.

Lockley, M. G. 1990. *A Field Guide to Dinosaur Ridge.* Friends of Dinosaur Ridge and the University of Colorado at Denver Dinosaur Trackers Research Group, 32 pages, photos, illustrations.

———. 1995. *Dinosaur Tracks and Other Fossil Footprints of the Western United States.* New York: Columbia University Press, 338 pages, illustrations.

Lockley, M. G., B. J. Fillmore, and L. Marquardt. 1997. *Dinosaur Lake—The Story of the Purgatoire Valley Dinosaur Tracksite Area.* Colorado Geological Survey Special Publication 40, 64 pages, color.

Murphy, J. A. 1995. *Geology Tour of Denver's Buildings and Monuments.* Historic Denver Association, available as Colorado Geological Survey Miscellaneous Information Series 53, 96 pages, illustrations.

National Park Service. 1982. *Great Sand Dunes.* Washington D.C.: National Park Service, map scale 1:70,000.

Parker, B. H., Jr. 1992. *Gold Panning and Placering in Colorado: How and Where.* Colorado Geological Survey Information Series 33, 83 pages, maps.

Pearl, R. M. 1972. *Colorado Gem Trails and Mineral Guide, Third Edition.* Athens, Ohio: Swallow Press, Ohio University Press, 222 pages, photos, maps.

————. 1969. *Exploring Rocks, Minerals, Fossils in Colorado*. Athens, Ohio: Swallow Press.

Prather, T. 1999. *Geology of the Gunnison Country*. Gunnison, Colo.: B & B Printers, 149 pages, illustrations (some color), maps, photos.

Rathbone, Jack, D. Spearing, L. I. Goldman, and M. W. Longman. 1995. *Mountains and Canyons: A Photographic Description of the Rocky Mountain Region*. Denver: Rocky Mountain Association of Geologists, 100 pages, maps.

Raup, O. B. 1996. *Geology along Trail Ridge Road, Rocky Mountain National Park, Colorado*. Helena, Mont.: Falcon Press Publishing Co. and Rocky Mountain Nature Association, 73 pages, color photos, fold-out map.

Scott, R. B., and others. 2001. *Geologic Map of Colorado National Monument and Adjacent Areas, Mesa County, Colorado*. U.S. Geological Survey Geologic Investigations Series I-2740, map scale 1:24,000, color.

Silbernagel, B. 1996. *Dinosaur Stalkers: Tracking Dinosaur Discoveries of Western Colorado and Eastern Utah*. Grand Junction, Colo.: Pyramid Printing, 64 pages, photos, maps, figures.

Stewart, K. C., and R. C. Severson. 1994. *Guidebook on the Geology, History, and Surface-Water Contamination and Remediation in the Area from Denver to Idaho Springs, Colorado*. U.S. Geological Survey Circular 1097, 55 pages, illustrations, photos.

Trimble, D. E. 1980. *The Geologic Story of the Great Plains*. U.S. Geological Survey Bulletin 1493, reprinted as Colorado Geological Survey Miscellaneous Information Series 48, 54 pages, 30 figures, color.

Tweto, O. 1983. *Geologic Sections across Colorado*. U.S. Geological Survey MI Map I-1416, map scale 1:500,000, color.

————. 1979. *Geologic Map of Colorado*. U.S. Geological Survey and Colorado Geological Survey, map scale 1:500,000, color.

Voynick, S. M. 1994. *Colorado Rockhounding: A Guide to Minerals, Gemstones, and Fossils*. Missoula, Mont.: Mountain Press Publishing Company, 371 pages, some color photos.

Glossary

alluvial fan. A fan-shaped wedge of gravel and sand deposited by a mountain stream where it runs out onto a level or nearly level plain.

alluvium. Sediments deposited by rivers and streams.

amazonite. Feldspar colored bluish green by minute amounts of lead.

ammonites. An extinct group of shell-forming mollusks related to the modern chambered nautilus. Both coiled and straight-shelled forms are known.

andesite. A medium- to dark-colored volcanic rock containing a high proportion of feldspar.

anticline. A fold that is convex upward. When eroded, an anticline has the oldest rocks in the center.

aquamarine. Light bluish green gem variety of the mineral beryl, found in pegmatites.

aquifer. A porous rock layer from which water may be obtained.

artesian well. A well in which water level rises above the top of the water-bearing layer.

ashfall. Volcanic ash falling from clouds rising from an erupting volcano.

ashflow. A ground-hugging cloud of hot volcanic gases and ash ejected explosively from a volcano, usually flowing rapidly downhill.

basalt. A dark gray to black, finely crystalline volcanic rock with a high proportion of iron and magnesium minerals, often with small vesicles or gas bubbles.

basement. A general term for igneous and metamorphic rocks, usually Precambrian, lying below the sedimentary rock sequence.

batholith. A large mass of light-colored, coarse-grained igneous rock with more than 40 square miles of surface exposure, intruded as molten magma, often formed at least in part by melting and recrystallization of older rocks.

bedrock. The solid rock that lies below loose surface material.

bentonite. Clay formed from decomposition of certain types of volcanic ash.

beryl. A light green mineral occurring in pegmatites; emerald and aquamarine are gem varieties.

biotite. Black (iron- and magnesium-rich) mica.

blowout. A hollow in sand or silt formed by wind erosion, often many feet in diameter.

brachiopod. A shell-bearing marine invertebrate that has two bilaterally symmetrical shells.

breccia. Rock consisting of angular fragments in a finer matrix.

butte. An isolated hill or small mountain, usually with a horizontal top and steep sides.

calcite. A common mineral, calcium carbonate ($CaCO_3$), the principal ingredient of limestone.

caldera. A large more or less circular basin-shaped volcanic depression formed by collapse over a depleted magma chamber.

caprock. A comparatively resistant rock layer, either sedimentary or volcanic, forming the top of a mesa, butte, or cuesta.

carbonaceous. Containing carbon or coal derived from organic material.

carbonate. Rocks containing carbon and oxygen in combination with sodium, calcium, or other elements, particularly as in limestone or dolomite.

chert. A hard, compact, dull or shiny variety of quartz in which individual silica grains are microscopically small.

chlorite. A group of green micalike metamorphic minerals that forms at fairly low temperatures and pressures.

cinder cone. A small cone-shaped volcano formed as volcanic cinders are ejected from a volcanic vent.

cirque. A deep, steep-walled, usually semicircular scoop excavated from a mountainside by the head of a glacier.

columnar jointing. Jointing that results in long polygonal columns of rock, caused by shrinking as volcanic flows cool. Columns are usually perpendicular to cooling surfaces.

concretion. A round or nodular lump, a concentration of minerals deposited around a central nucleus, usually harder than surrounding rock.

conglomerate. Rock composed of rounded or subangular rock fragments in a fine-grained matrix of sand or silt.

cordierite. A faintly bluish metamorphic mineral that often forms a knobby texture in schist.

crossbedding. Laminae that slant obliquely between the main horizontal layers of a sedimentary rock (generally sandstone), caused by movement of water or wind during deposition.

cuesta. A ridge with a long gentle slope capped by a resistant layer, and a short steep slope of eroded underlying weaker rock.

dacite. A medium- to light-colored volcanic rock with a high proportion of quartz and feldspar.

dendritic drainage. A treelike pattern of irregularly branching streams.

desert pavement. A surface concentration of pebbles and other rock fragments resulting when wind blows away finer dust and sand.

desert varnish. A dark shiny surface of iron and manganese oxides that forms on long-exposed rock surfaces in deserts.

diabase. A fine-grained intrusive rock of the same composition as basalt.

dike. A thin, tabular body of igneous rock that cuts across layering in surrounding rock, or across massive rock.

dip. The angle at which a rock layer is inclined, measured below horizontal.

dolomite. A limestone-like rock containing magnesium carbonate as well as calcium carbonate.

earthflow. A slow flow of soil and weathered rock lubricated with water.

epidote. A green metamorphic mineral common near metamorphosed limestone.

escarpment. A cliff or steep slope edging a region of higher land.

evaporite. A mineral deposited from highly mineralized or salty water as a result of evaporation.

exfoliation. A weathering process in which curved sheets crack off an exposed rock mass, usually caused by a combination of frost wedging and chemical weathering.

extrusive rocks. Igneous rocks that cool on Earth's surface; volcanic rocks.

fault. A break in the rock along which rocks on either side have moved relative to each other.

fault scarp. A cliff formed by a fault, usually modified by erosion.

feldspar. A group of abundant rock-forming minerals, usually light colored, common in igneous rocks.

flatiron. Triangular remnant of a resistant layer steeply tilted against the flank of a mountain.

foliation. A platy texture caused by flat minerals oriented the same way throughout the rock.

formation. A named, recognizable, mappable unit of rock.

frost wedging. Breakup of rocks by water freezing in joints.

geomorphology. A branch of geology that deals with Earth's surface features or landforms.

gneiss. A coarse-grained metamorphic rock with bands of granular crystalline minerals such as quartz and feldspar, and fine or platy dark minerals such as biotite.

gouge. Fine, puttylike clayey material between the walls of a fault or between a vein and the rock surrounding it.

graben. A long valley downdropped between two more or less parallel faults.

granite. Coarse-grained intrusive igneous rock with feldspar and quartz as principal minerals.

ground moraine. Material deposited on the ground surface by a melting glacier.

groundwater. Subsurface water filling pore spaces, cracks, or solution channels in rock or soil.

group. A stratigraphic unit consisting of several formations, usually originally a single formation subdivided after subsequent research.

grus. Loose coarse sand formed by in-place weathering of coarse-grained rock such as granite.

gypsum. A common evaporite mineral ($CaSO_4$) used in manufacturing plaster.

halite. The mineral name for common salt ($NaCl$), formed from evaporation of salt water.

hanging valley. A valley whose floor is substantially higher than the floor of the valley into which it drains.

hard rock. A mining term for igneous and metamorphic rocks, which are usually harder than sedimentary rock.

hematite. An ore of iron (Fe_2O_3), often red and powdery (red rust).

hogback. A sharp ridge produced by erosion of steeply tilted rock layers, one of which is more resistant than the others.

honeycomb weathering. Weathering of rock (usually sandstone) by wind and water, producing deep, close-together, fist-size holes in the rock surface.

hornblende. A dark mineral common in igneous and metamorphic rock, often forming needle-shaped crystals.

hornblende gneiss. Dark gneiss containing hornblende as the most abundant mineral.

hydrostatic pressure. Pressure caused by the weight of water in pipes or water-bearing rock layers.

hydrothermal. Caused by or related to hot water.

ice cap. Glacial ice that spreads in all directions over a high, relatively flat surface.

igneous rock. Rock formed by solidification of molten rock material (magma).

injection gneiss. Gneiss containing sheets of granite injected under great pressure deep in Earth's crust.

intrusive rock. Igneous rock that has hardened from molten rock material (magma) without reaching the surface.

island arc. A belt of volcanic islands resulting from collision between tectonic plates.

joint. A fracture in rock along which no appreciable movement has occurred.

kaolinite. A type of clay usually formed by decomposition of feldspar minerals.

karst. A distinctive type of landscape where water has dissolved limestone layers, causing abundant caves, sink holes, and solution valleys, often with red soil residue.

kerogen. A solid hydrocarbon found in oil shale.

kettle lakes. Small lakes filling low spots left in moraines by melting ice.

kimberlite. Rock containing minerals from Earth's mantle, intruded very quickly through the crust.

laccolith. A body of intrusive rock that squeezed between rock layers, doming those above.

lateral moraine. A ridgelike moraine deposited along the side of a valley glacier.

latite. Volcanic rock with large feldspar crystals in a fine or glassy matrix, with little or no quartz.

Law of Superposition. The rule that, for undisturbed sedimentary or volcanic rocks, younger rocks rest on top of older rocks.

leached. Depleted of elements and minerals by slowly moving water.

limonite. A brown to yellow powdery iron oxide mineral (yellow rust).

lithosphere. The uppermost solid part of Earth's surface including the crust and the uppermost mantle.

lode. A deposit of valuable minerals in solid rock, in contrast to placer deposits, which are in gravel.

magma. Molten rock from which igneous rocks eventually solidify.

magnetite. A black or dark gray, strongly magnetic iron mineral (Fe_3O_4).

marble. A metamorphic rock created by heating and recrystallization of limestone.

massif. An entire mountain mass or group of mountains that behaves geologically as a single unit.

mesa. A tableland or flat-topped mountain, usually capped by a resistant rock layer and edged on more than one side with cliffs or steep slopes.

metamorphic rock. Rock formed from older rock that has been changed by great heat, pressure, chemical fluids, or a combination of them.

mica. A group of minerals characterized by readily separating into thin platy flakes with shiny surfaces.

mid-ocean ridge. A submarine mountain range with a central rift valley, where volcanic activity creates new oceanic crust.

migmatite. Rock made of layers of metamorphic and igneous minerals, formed by injection or partial melting during intense metamorphism.

monocline. A simple fold or flexure in stratified rock in which horizontal strata bend upward or downward and then level out again.

monzonite. A light-colored intrusive rock containing abundant feldspar minerals but very little quartz and few dark minerals.

moraine. An accumulation of unsorted gravel, boulders, and dirt deposited by a glacier.

muscovite. White or light brown mica.

mylonite. A very fine-grained metamorphic rock formed by grinding and compression of rock within a fault zone.

nahcolite. A white mineral ($NaHCO_3$) mined for soda ash and sodium bicarbonate for glass making and food preparation.

nunatak. An isolated hill or peak that projects through a glacier and is surrounded by glacial ice.

oil shale. Shale containing kerogen, a waxy substance from which petroleum may be obtained by heating.

ore. A rock containing enough valuable minerals to be economically mineable.

orogeny. Formation of mountains by faulting, folding, uplift, and sometimes intrusive igneous activity.

outcrop. Bedrock exposed at Earth's surface.

outwash. Gravel and sand carried from glaciers by meltwater and deposited below the actual glaciated area.

oxbow lake. A crescent-shaped lake formed in an abandoned river bend.

paleontology. A branch of geology that deals with plant and animal remains and the life of the past.

palisade. A high scenic cliff or line of cliffs at the edge of a river or lake.

paternoster lakes. A series of small lakes in a glacially eroded valley.

pediment. A gently inclined erosion surface carved in bedrock at the base of a mountain or mountain range, usually in a desert or near-desert environment.

pegmatite. Exceptionally coarse-grained igneous rock usually occurring in veins and dikes near the margins of batholiths, generally like granite in mineral composition but sometimes with additional rare minerals.

peneplain. Land surface worn down by erosion to a nearly flat plain.

phenacite. A very hard white beryllium mineral, found rarely in pegmatites.

phonolite. One of a variety of fine-grained volcanic rocks that emit a ringing sound when struck with a hammer.

placer. A gravel or sand deposit containing particles of gold or other valuable minerals.

plateau. A large, relatively high area bounded on one or more sides by cliffs or steep slopes.

pluton. An igneous intrusion that cools deep in the crust or a body of rock developed by melting of older rocks deep in the crust.

plutonic. Pertaining to igneous rocks formed at great depth.

porphyry. An igneous rock with large mineral grains (often feldspar, quartz, or hornblende) surrounded by much smaller or even microscopic grains.

primary texture. An original sedimentary or igneous texture that formed as the rock formed, before metamorphism.

pyrite. A metallic, brass-colored iron mineral (FeS_2), often called "fool's gold."

pyroxene. A group of dark minerals common in igneous rock.

quartz. A hard, glassy mineral, silicon dioxide (SiO_2), one of the most common rock-forming minerals.

quartzite. A metamorphic rock formed from sandstone cemented by silica.

radiometric date. Age of a rock, usually in years before present, determined by measuring the amount of decay of radioactive elements in its minerals.

recessional moraine. A glacial moraine formed during a temporary pause in the retreat of a glacier.

redbeds. Red, pink, and purple sedimentary rocks, usually sandstone, siltstone, and shale.

rhodochrosite. A pink magnesium mineral ($MnCO_3$) that occurs in veins.

rhyolite. Light-colored, fine-grained volcanic rock, often with scattered large quartz and feldspar crystals.

rift. A deeply faulted valley where two parts of Earth's crust are separating.

rift fault. One or a pair of deep faults that reach down through the crust, forming the boundaries of a rift.

rock glacier. A glacierlike body of angular broken rock lubricated by interstitial ice and moving slowly like a true glacier.

sandstone. Sedimentary rock composed of grains $\frac{1}{16}$ to 2 millimeters in diameter.

scarp. Cliff or steep slope formed either by erosion or faulting.

schist. Metamorphic rock whose parallel orientation of abundant mica flakes or hornblende needles causes it to break easily along parallel planes.

shale. Platy sedimentary rock formed from silt or clay, breaking easily parallel to bedding.

shear zone. A zone or group of roughly parallel faults; rock within the zone is often deformed or broken up.

silica. Silicon dioxide (SiO_2), occurring as quartz and as a major part of many other minerals.

siliceous sinter. Hot spring deposits composed largely of silica.

silicified. Impregnated with silica.

sill. A sheet of igneous rock intruded between sedimentary rock layers.

siltstone. Sedimentary rock composed of grains $\frac{1}{156}$ to $\frac{1}{16}$ millimeter in diameter.

sinkhole. A depression caused by collapse of the ground into an underlying limestone cavern.

slickenside. A polished surface resulting from movement of rock against rock along a fault.

soft rock. A mining term referring to sedimentary rock, which is generally much softer than igneous or metamorphic rock.

stalactite. A calcium carbonate "icicle" hanging from the roof of a limestone cave.

stalagmite. A conical or columnar deposit of calcium carbonate building upward from the floor of a limestone cave.

stock. An igneous intrusion with an exposure smaller than 40 square miles (100 square kilometers) at the surface.

strata. Layers or beds of sedimentary rock. Singular is stratum.

stratified. Formed in layers, as sedimentary rock.

striation. A scratch or furrow on a rock surface, caused by a glacier.

subduction. The downward movement of a plate of Earth's crust as it goes beneath the edge of another plate.

syncline. A downward fold. When eroded, a syncline has the youngest rocks in its center.

tailings. Waste debris from ore-processing mills.

talus. A mass of fallen angular rocks piled below a steep outcrop that was their source.

tantalite. A heavy black pegmatite mineral containing tantalum.

telluride. A compound of the element tellurium with a metal, often containing gold.

tepee butte. A small cone-shaped butte formed by erosion around local harder material in shale.

terminal moraine. Unsorted bouldery glacial debris dumped at the lower end of a glacier at the time of its greatest extent.

thrust fault. A low-angle (less than 45 degrees) fault in which one side is pushed up and over the other side by horizontal compression.

travertine. Chemically deposited calcite occurring in caves or around springs and hot springs.

trench. A long deep depression in the ocean floor formed where an oceanic plate is subducted beneath a continental plate.

trilobite. A small, three-lobed Paleozoic marine arthropod.

tuff. A rock formed of compacted volcanic ash and cinder.

type section. The place at which a sedimentary unit is best or most typically displayed, and for which it is usually named.

unconformity. A substantial break in the geologic record, separating younger strata from older rocks.

vein. A tabular mass of minerals filling a fracture or fault in older rock, often containing ore.

volcanic ash. Fine material ejected into the air from a volcano.

water table. The upper surface of groundwater, below which soil and rock are saturated.

welded tuff. Volcanic ash deposited from ashflows, fused by its own original heat, by hot gases, and by the weight of overlying layers.

xenolith. A piece of other rock enveloped in igneous rock.

xenotime. A brown, yellow, or red pegmatite mineral containing rare earth elements.

zeolites. A group of minerals that often form masses of radiating needles in basalt vesicles.

zoning. Grouping of minerals in bands, for example in a pegmatite or an ore deposit, determined by variations in the chemical environment when the minerals crystallized.

Index

HALKA CHRONIC

FELICIE WILLIAMS

About the Authors

Halka Chronic received a doctorate in geology from Columbia University in 1949. Her career has taken her all over the globe—from teaching at Haile Sellassie University in Ethiopia to identifying Geologic Landmarks for the United States Park Service in the southern Rockies. She spent more than thirty years in Boulder, Colorado, working and raising her daughters.

She now lives in Sedona, Arizona, and considers herself semiretired. She volunteers at a public school taking children on field trips and is on the Alumni Committee of the Museum of Northern Arizona. She has written three other *Roadside* guides for Mountain Press—Utah, Arizona, and New Mexico.

Felicie Williams grew up with Colorado's rocks. She and her three sisters spent their summers in the field with their geologist parents, so pursuing a career in geology was an easy decision for her. She earned a bachelor's degree in geology from the University of Colorado in Boulder and a master of science in geology from the University of British Columbia.

Felicie lives in Grand Junction, Colorado, with Mike Williams—another geologist—and their two children. She enjoyed the adventure she and her mother, Halka Chronic, shared while studying Colorado's inspiring geology and distilling it into words, drawings, and photos for this new edition.

We encourage you to patronize your local bookstore. Most stores will order any title they do not stock. You may also order directly from Mountain Press, using the order form provided below or by calling our toll-free, 24-hour number and using your VISA, MasterCard, Discover or American Express.

Some geology titles of interest:

_____ROADSIDE GEOLOGY OF ALASKA	18.00
_____ROADSIDE GEOLOGY OF ARIZONA	18.00
_____ROADSIDE GEOLOGY OF SOUTHERN BRITISH COLUMBIA CAN: $25.00 US: 20.00	
_____ROADSIDE GEOLOGY OF NORTHERN and CENTRAL CALIFORNIA	20.00
_____ROADSIDE GEOLOGY OF COLORADO, 2nd Edition	20.00
_____ROADSIDE GEOLOGY OF CONNECTICUT and RHODE ISLAND	26.00
_____ROADSIDE GEOLOGY OF FLORIDA	26.00
_____ROADSIDE GEOLOGY OF HAWAII	20.00
_____ROADSIDE GEOLOGY OF IDAHO	20.00
_____ROADSIDE GEOLOGY OF INDIANA	18.00
_____ROADSIDE GEOLOGY OF MAINE	18.00
_____ROADSIDE GEOLOGY OF MASSACHUSETTS	20.00
_____ROADSIDE GEOLOGY OF MONTANA	20.00
_____ROADSIDE GEOLOGY OF NEBRASKA	18.00
_____ROADSIDE GEOLOGY OF NEW MEXICO	18.00
_____ROADSIDE GEOLOGY OF NEW YORK	20.00
_____ROADSIDE GEOLOGY OF OHIO	24.00
_____ROADSIDE GEOLOGY OF OREGON	16.00
_____ROADSIDE GEOLOGY OF PENNSYLVANIA	20.00
_____ROADSIDE GEOLOGY OF SOUTH DAKOTA	20.00
_____ROADSIDE GEOLOGY OF TEXAS	20.00
_____ROADSIDE GEOLOGY OF UTAH	20.00
_____ROADSIDE GEOLOGY OF VERMONT & NEW HAMPSHIRE	14.00
_____ROADSIDE GEOLOGY OF VIRGINIA	16.00
_____ROADSIDE GEOLOGY OF WASHINGTON	18.00
_____ROADSIDE GEOLOGY OF WISCONSIN	20.00
_____ROADSIDE GEOLOGY OF WYOMING	18.00
_____ROADSIDE GEOLOGY OF THE YELLOWSTONE COUNTRY	12.00
_____GEOLOGY UNDERFOOT IN NORTHERN ARIZONA	18.00
_____GEOLOGY UNDERFOOT IN SOUTHERN CALIFORNIA	14.00
_____GEOLOGY UNDERFOOT IN DEATH VALLEY AND OWENS VALLEY	16.00
_____GEOLOGY UNDERFOOT IN ILLINOIS	18.00
_____GEOLOGY UNDERFOOT IN CENTRAL NEVADA	16.00
_____GEOLOGY UNDERFOOT IN SOUTHERN UTAH	18.00

Please include $3.50 for 1-4 books, $5.00 for 5 or more books to cover shipping and handling.

Send the books marked above. I enclose $_____

Name_____

Address _____

City/State/Zip _____

☐ Payment enclosed (check or money order in U.S. funds)

Bill my: ☐VISA   ☐MasterCard ☐Discover ☐American Express

Card No. _____ Expiration Date:_____

Security No._____Signature _____

MOUNTAIN PRESS PUBLISHING COMPANY
P.O. Box 2399 • Missoula, MT 59806 • Order Toll-Free 1-800-234-5308
E-mail: info@mtnpress.com • Web: www.mountain-press.com